Species and Speciation
in the Fossil Record

Species and Speciation
in the Fossil Record

EDITED BY WARREN D. ALLMON
AND MARGARET M. YACOBUCCI

THE UNIVERSITY OF CHICAGO PRESS   CHICAGO AND LONDON

Warren D. Allmon is director of the Paleontological Research Institution in Ithaca, New York, and professor in the Department of Earth and Atmospheric Sciences at Cornell University.
Margaret M. Yacobucci is professor of geology at Bowling Green State University.

The University of Chicago Press, Chicago 60637
The University of Chicago Press, Ltd., London
© 2016 by The University of Chicago
All rights reserved. Published 2016.
Printed in the United States of America
25 24 23 22 21 20 19 18 17 16   1 2 3 4 5

ISBN-13: 978-0-226-37744-5 (cloth)
ISBN-13: 978-0-226-37758-2 (e-book)
DOI: 10.7208/chicago/9780226377582.001.0001

Library of Congress Cataloging-in-Publication Data
Names: Allmon, Warren D., editor. | Yacobucci, Margaret M. (Margaret Mary), editor.
Title: Species and speciation in the fossil record / edited by Warren D. Allmon and Margaret M. Yacobucci.
Description: Chicago : The University of Chicago Press, 2016. | Includes bibliographical references and index.
Identifiers: LCCN 2015048963| ISBN 9780226377445 (cloth : alk. paper) | ISBN 9780226377582 (e-book)
Subjects: LCSH: Evolutionary paleobiology. | Fossils. | Evolution (Biology)
Classification: LCC QE721.2.E85 S64 2016 | DDC 576.8/6—dc23 LC record available at http://lccn.loc.gov/2015048963

Contents

Taking Fossil Species Seriously

Warren D. Allmon and Margaret M. Yacobucci

In their landmark book, Coyne and Orr (2004, 1) note that "one of the most striking developments in evolutionary biology during the last 20 years has been a resurgence of interest in the origin of species." This resurgence is well-evidenced by the enormous volume of literature, including not just hundreds of individual papers but also at least two dozen books and journal special issues in the past two decades. The corner of taxonomy that addresses the "species problem" now even has its own name: *eidonomy* (Dubois 2011).

What exactly are all these people writing about? A lot of it is new information, both from ecology and from the continuing tsunami of molecular sequence data and what it reveals about phylogeny and genetic differences associated with speciation. Some is focused on new or newly sharpened theoretical paradigms, such as "adaptive" or "ecological" speciation, or on new mathematical models. A lot of it is controversial—the continuation of the protean and seemingly indestructible "species problem," or arguments over which "species concept" (among a list now approaching 30; Wilkins 2006) is best.

Yet largely missing from this flood of new research and rethinking is paleontology, so much so that amidst the multitude of species concepts, one might be forgiven for proposing yet one more. It might be called the "Janus concept" (Allmon 2011), and it would describe how most paleontologists recognize and study species in the fossil record. According to the Janus concept, species are both real and imaginary, central and marginal, the focal point of macroevolution and incomparable to anything studied by neontologists. Formal adoption of this concept would acknowledge the reality of modern paleobiology, which appears to be able to maintain these mutually incompatible views of species without much difficulty. Paleontologists talk about and use species but don't uniformly respect them; we constantly discuss the evolution of species, but are simultaneously a bit queasy about taking them seriously.

Although you won't see it stated explicitly very often, something close to the Janus concept has long been, and remains, implicit in much of paleobiology. (This might in some ways be similar to the 19th- and early 20th-century views on the atom [Knight 1967]. It could be inferred from indirect evidence that matter was comprised of atoms, even though the atoms themselves could not be directly "seen." This analogy demonstrates that a lot of good science can be done even if direct observation is not possible.)

Consider the following. "Species" and "speciation" feature prominently in a large and growing body of paleontological studies of macroevolutionary pattern and process, from studies of individual lineages or clades (see almost any issue of *Journal of Paleontology*), to tempo and mode (e.g., Eldredge and Gould 1972; Gould and Eldredge 1977; Eldredge and Cracraft 1980; Cracraft 1981; Barnosky 1987; Prothero 1992; Erwin and Anstey 1995; Jackson and Cheetham 1999; Geary 2009; Brett 2012; Hunt 2012), to wider studies of origination and extinction (e.g., Jablonski 1986a,b, 1987, 1997, 2008; Allmon et al. 1993, 1996; Sepkoski 1998; Lieberman 1999, 2001a,b; Jablonski and Roy 2003; Etienne and Apol 2008; Benton 2010). Even more broadly, much of the "paleobiological revolution" of the 1970s–1980s (Sepkoski 2012) was based on thinking about what species look like in the fossil record and what we can learn about them from studying fossils (e.g., Eldredge and Gould 1972; Stanley 1979, 1982, 1986; Eldredge 1979, 1985; Gould 2002).

Yet amidst all this discussion of "species," the paleobiological literature is also filled with qualifiers and cautions about using species in the fossil record, or equating such species with those recognized among living organisms. Much of this skepticism has been expressed by neontologists. Coyne and Orr, for example, state that recognition of fossil species is necessarily based on no more than "*reasonable guesses* about the likelihood of reproductive isolation from discontinuities between phenotypes" (2004: 45–46). Paleontologists themselves, however, have frequently seemed scarcely more optimistic (e.g., Schopf 1979; Gingerich 1985; Jablonski et al. 1986; Raup and Boyajian 1988; Hoffman 1989; Smith 1994; Levinton 2001; Pearson and Harcourt-Brown 2001; Forey et al. 2004; Benton 2010,

1479). For example, it is common to defend the use of higher taxa (families or genera) in large-scale paleodiversity studies in part because species are seen as challenging to recognize or differentiate in the fossil record (e.g., Sepkoski 1998; see discussion in Hendricks et al. 2014).

An exception to this pattern of paleontological pessimism about species is cladistics/phylogenetic systematics. Several paleontologists were involved in the early years of the "phylogenetic revolution" (e.g., Eldredge and Cracraft 1980; Cracraft 1981), and some of these authors did indeed discuss the nature of species. More recently, there has been some paleontological debate on the perennial topic of phylogeny vs. classification, with paleontologists arguing for (e.g., Brochu and Sumrall 2008) or against (e.g., Benton 2000) the PhyloCode, which has touched on the issue of species. Yet, other than the substantial consequence that cladistic techniques are now standard operating procedure for many paleontologists (e.g., Smith 1994; Foote and Miller 2006), the impact of these discussions on paleontological thinking about species seems to have been limited.

The Janus concept raises at least two important issues for evolutionary paleontology. The first is simple intellectual consistency and coherence. Do paleontologists think that the things they call "species" are in some real sense equivalent to what neontologists call "species"? If not, what are they saying when they continue to use the term? The second and more substantive concern is that much of macroevolutionary theory—at least in its current form—is based on and can only be tested with concepts of species as presumably real biological units, and that when paleontologists say "species" they mean something equivalent enough to what neontologists mean to allow for common discussion.

What, however, does it mean for such discussions and fossil-based macroevolutionary theories that paleontologists are not entirely sure what we mean when we say "species"? Perhaps it doesn't matter. Perhaps the morphological patterns that are paleontologists' stock-in-trade are valid, interesting, and important no matter what hierarchical level they represent. If so, perhaps paleontologists should be clearer about that, so that the indisputable importance of the fossil record for understanding the history and evolution of life can be more precisely understood.

This book seeks to address the Janus concept of species in paleontology. The chapters focus on fossil animals, because as we were planning the volume we quickly found that to include meaningful considerations of plants and protists would make the project too large. Within animals, we have tried with mixed success to include chapters from across the phyla.

Trilobites, however, are unfortunately not covered, and we are regrettably light on vertebrates.

Exploration of the "species problem" in paleontology requires that we first understand its origins within the development of the Modern Synthesis and of paleobiology as a discipline. Both Sepkoski (chapter 1) and Miller (chapter 2) provide such reviews herein, while highlighting paleontology's unique contributions. Sepkoski argues that the development of quantitative ways of documenting and analyzing variation with fossil assemblages allowed paleontologists to integrate paleobiology into the Modern Synthesis. Miller points to a growing modern consensus around the lineage species concept, noting that only paleontologists can provide the temporal perspective that this view of species requires. Allmon (chapter 3) agrees with Miller, and attempts to push this line of thought to its logical conclusion, making specific recommendations for how paleontologists should talk about species.

Paleontologists who work with different metazoan clades face a variety of challenges when attempting to recognize and define species from fossil specimens. Several of the chapters in this volume explore these taxon-specific challenges from their authors' own perspectives (e.g., Budd and Pandolfi (chapter 7), Schweitzer and Feldmann (chapter 9), Ausich (chapter 10), Bemis (chapter 11)). While identifying a variety of difficulties, from poor preservation, homoplasy, and cryptic speciation to collecting biases and overenthusiastic splitting of taxa for biostratigraphic purposes, these authors also conclude that, with careful interpretation and a clear species concept, fossil species may be sufficiently robust for meaningful paleobiological analyses.

The fossil record provides evidence of evolutionary change on longer timescales, which most paleontologists think is essential for understanding both the process and the implications of speciation. The tempo and mode of speciation have long been of interest to paleontologists, an emphasis reflected in many of this volume's chapters. Hageman (chapter 5) presents an integrated model of quantitative morphological and molecular change set in a temporal context in order to determine rates of evolution both within and between species. Liow and Ergon (chapter 6) explore the question of whether speciation is more likely to occur early or late in a species' stratigraphic lifespan, while Hopkins and Lidgard (chapter 13) assess whether the choice of traits used for species discrimination affect our ability to distinguish stasis from directional or random change. Yacobucci (chapter 8) tackles the long-standing question of why certain clades, in this case ammonoid cephalopods, show much higher rates of speciation than others. Allmon and Sampson (chapter 4) formalize a proposal for how paleontologists can be more explicit in their hypotheses of causes for speciation, by breaking the process into four discrete stages. These theoretical efforts are complemented by the data-driven approaches of Prothero et al. (chapter 14) and Stigall (chapter 12), who investigate potential controls on the tempo and mode of speciation in terrestrial mammals and marine invertebrates.

We hope that reading this book will prompt paleontologists to reflect critically on how they view fossil species. What species concept do they envision when they work, and how do they operationalize that concept in order to recognize and define species within their own fossil group of interest? Do they collect the data necessary to document evolutionary rates and to assess the tempo and mode of fossil lineages? What can they conclude about the biotic and abiotic processes that may affect speciation potential over geologic time?

Paleontologists can contribute tremendously to contemporary evolutionary biology, but only (we think) if we can agree on a biologically meaningful species concept for fossil clades and employ it in our "deep time" perspective to explore the speciation process. We also hope that neontologists will read this book, and take away not only some sense that paleontologists *do* understand the complexity and challenge of recognizing "species" in the fossil record, but also the very real opportunities to study at least some phenomena in that record that approximate species as recognized in living animals. While agreement cannot be expected on all of the points discussed in these chapters, we feel strongly that neontologists and paleontologists can agree that the subject of species is central to evolutionary biology, and that to understand the origin and significance of species in evolution requires insights from both living and fossil organisms.

Acknowledgments

As editors, we are very grateful to the authors and reviewers of the chapters included here, for their interest, time, effort, and especially patience when things did not move as quickly as we all would have liked. We also profusely thank Christie Henry at the University of Chicago Press for believing in the project from the start and encouraging us through to the end. As authors, we are also very grateful to Michael Foote, Dana Geary, Christie Henry, and Chuck Mitchell for discussion and/or comments on earlier versions of this introduction.

References

Allmon, W. D. 2011. "The 'Janus concept' of paleontological species: Are fossil species critically important or are we just kidding ourselves?" *Geological Society of America Annual National Meeting, Abstracts w/Program.* 43(5): 207 (no. 78-1)

Allmon, W. D., G. Rosenberg, R. W. Portell, and K. S. Schindler. 1993. "Diversity of Atlantic coastal plain mollusks since the Pliocene." *Science.* 260:1626–1628.

Allmon, W. D., G. Rosenberg, R. Portell, and K. Schindler. 1996. "Diversity of Pliocene-Recent mollusks in the Western Atlantic: extinction, origination and environmental change." In *Evolution and environment in tropical America.* J. B. C. Jackson, A. G. Coates and A. F. Budd, eds. University of Chicago Press, Chicago, pp. 271–302.

Barnosky, A. D. 1987. "Punctuated equilibrium and phyletic gradualism: Some facts from the Quaternary mammalian record." *Current Mammalogy.* 11: 109–147.

Benton, M. J. 2000. "Stems, nodes, crown clades, and rank-free lists: Is Linnaeus dead?" *Biological Reviews.* 75: 633–648.

Benton, M. J. 2010. "Naming dinosaur species: the performance of prolific authors." *Journal of Vertebrate Paleontology.* 30(5): 1478–1485.

Brett, C. E. 2012. "Coordinated stasis reconsidered: A perspective at fifteen years." In *Earth and life. Global biodiversity, extinction intervals, and biogeographic perturbations through time*, ed. J. A. Talent, pp. 23–36. Springer, Dordrecht.

Brochu, C. A., and C. D. Sumrall. 2008. "Phylogenetics and the integration of paleontology within the life sciences." In *From evolution to geobiology: Research questions driving paleontology at the start of a new century*, ed. P. H. Kelley and R. K. Bambach, 187–204. Paleontological Society Papers, volume 14.

Coyne, J. A., and H. A. Orr. 2004. *Speciation.* Sinauer Associates, Sunderland, MA, 545 p.

Cracraft, J. 1981. "Pattern and process in paleobiology: The role of cladistics analysis in systematic paleontology." *Paleobiology.* 7(4): 456–468.

Dubois, A. 2011. "Species and 'strange species' in zoology: Do we need a 'unified concept of species'?" *Comptes rendus Palevol.* 10(2): 77–94.

Eldredge, N. 1979. "Alternative approaches to evolutionary theory." *Bulletin of the Carnegie Museum of Natural History.* 13:7–19.

Eldredge, N. 1985. *Unfinished synthesis: Biological hierarchies and modern evolutionary thought.* Oxford University Press, Oxford, UK, 256 p.

Eldredge, N., and J. Cracraft. 1980. *Phylogenetic patterns and the evolutionary process: Method and theory in comparative biology.* Columbia University Press, New York, 349 p.

Eldredge, N., and S. J. Gould. 1972. "Punctuated equilibria: An alternative to phyletic gradualism." In *Models in paleobiology*, ed. T. J. M. Schopf, 82–115. Freeman, Cooper & Co., San Francisco.

Erwin, D. H., and R. L. Anstey. 1995. "Speciation in the fossil record." In *New approaches to speciation in the fossil record*, ed. D. Erwin and R. Anstey, 11–38. Columbia University Press, NY.

Etienne, R. S., and M. E. F. Apol. 2008. "Estimating speciation and extinction rates from diversity data and the fossil record." *Evolution*. 63(1): 244–255.

Foote, M., and A. I. Miller. 2006. *Principles of paleontology*. 3rd ed. W. H. Freeman, New York, 480 p.

Forey, P. L., R. A. Fortey, P. Kenrick, and A. B. Smith. 2004. "Taxonomy and fossils: A critical appraisal." *Philosophical Transactions of the Royal Society of London B*. 359: 639–653.

Geary, D. H. 2009. "The legacy of punctuated equilibrium." In *Stephen Jay Gould: Reflections on his view of life*, ed. W. D. Allmon, P. H. Kelley, and R. M. Ross, 127–145. Oxford University Press, New York.

Gingerich, P. D. 1985. "Species in the fossil record: Concepts, trends, and transitions." *Paleobiology*. 11(1): 27–41.

Gould, S. J. 2002. *The structure of evolutionary theory*. Harvard University Press, Cambridge, MA, 1,433 p.

Gould, S. J., and N. Eldredge. 1977. "Punctuated equilibria: The tempo and mode of evolution reconsidered." *Paleobiology*. 3(2): ll5-l5l.

Hendricks, J. R., E. E. Saupe, C. E. Myers, E. J. Hermsen, and W. D. Allmon. 2014. "The generification of the fossil record." *Paleobiology*. 40(4): 511–528.

Hoffman, A. 1989. *Arguments on evolution: A paleontologist's perspective*. Oxford University Press, New York, 274 p.

Hunt, G. 2012. "Measuring rates of phenotypic evolution and the inseparability of tempo and mode." *Paleobiology*. 38(3): 351–373.

Jablonski, D. 1986a. "Larval ecology and macroevolution in marine invertebrates." *Bulletin of Marine Science*. 39: 565–587.

Jablonski, D. 1986b. "Background and mass extinctions: The alternation of macroevolutionary regimes." *Science*. 231: 129–133.

Jablonski, D. 1987. "Heritability at the species level: Analysis of geographic ranges of Cretaceous mollusks." *Science*. 238: 360–363.

Jablonski, D. 1997. "Body-size evolution in Cretaceous molluscs and the status of Cope's rule." *Nature*. 385: 250–252.

Jablonski, D. 2008. "Species selection: Theory and data." *Annual Review of Ecology, Evolution, and Systematics*. 39: 501–524.

Jablonski, D., and K. Roy. 2003. "Geographical range and speciation in fossil and living molluscs." *Proceedings of the Royal Society London B*. 270: 401–406.

Jablonski, D., S. J. Gould, and D. M. Raup. 1986. "The nature of the fossil record: A biological perspective." In *Patterns and processes in the history of life*, ed. D. M. Raup and D. Jablonski, 7–22. Springer-Verlag, Berlin..

Jackson, J. B. C., and A. H. Cheetham. 1999. "Tempo and mode of speciation in the sea." *Trends in Ecology and Evolution*. 14(2): 72–77.

Knight, D. M. 1967. *Atoms and elements: A study of theories of matter in England in the nineteenth century*. Hutchinson, London, 167 p.

Levinton, J. S. 2001. *Genetics, paleontology, and macroevolution*. 2nd ed. Cambridge University Press, New York, 617 p.

Lieberman, B. S. 1999. "Turnover pulse in trilobites during the Acadian Orogeny." Proceedings of the Appalachian Biogeography Symposium. *Virginia Museum of Natural History Special Publication* 7: 99–108.

Lieberman, B. S. 2001a. "A test of whether rates of speciation were unusually high during the Cambrian radiation." *Proceedings of the Royal Society London B.* 268: 1707–1714.

Lieberman, B. S. 2001b. "Analyzing speciation rates in macroevolutionary studies." In *Fossils, phylogeny, and form: An analytical approach*, ed. J. M. Adrain, G. D. Edgecombe, and B. S. Lieberman, 340–58. Plenum Press, New York.

Pearson, P. N., and K. G. Harcourt-Brown. 2001. "Speciation and the fossil record." In *Encyclopedia of Life Sciences*, 1–5. John Wiley & Sons, New York.

Prothero, D. R. 1992. "Punctuated equilibrium at twenty: A paleontological perspective." *Skeptic.* 1:38–47.

Raup, D. M., and G. E. Boyajian. 1988. "Patterns of generic extinction in the fossil record." *Paleobiology.* 14:109–125.

Schopf, T. J. M. 1979. "Evolving paleontological views on deterministic and stochastic approaches." *Paleobiology.* 5: 337–352.

Sepkoski, D. 2012. *Rereading the fossil record: The growth of paleobiology as an evolutionary discipline.* University of Chicago Press, Chicago.

Sepkoski, J. J., Jr. 1998. "Rates of speciation in the fossil record." *Philosophical Transactions of the Royal Society, London, B.* 353: 315–326.

Smith, A. B. 1994. *Systematics and the fossil record: Documenting evolutionary patterns.* Blackwell, London, 223 p.

Stanley, S. M. 1979. *Macroevolution: Pattern and process.* W. H. Freeman, San Francisco, 332 p.

Stanley, S. M. 1982. "Speciation and the fossil record." In *Mechanisms of speciation*, C. Barrigozzi, 41–49. Alan R. Liss, Inc., New York.

Stanley, S. M. 1986. "Population size, extinction, and speciation: The fission effect in Neogene Bivalvia." *Paleobiology.* 12:89–110.

Wilkins, J. S. 2006. "A list of 26 species concepts." http://scienceblogs.com/evolving thoughts/2006/10/01/a-list-of-26-species-concepts/

The "Species Concept" and the Beginnings of Paleobiology

David Sepkoski

In the late 1940s, the discipline of paleontology took major steps towards becoming more fully integrated into the community of evolutionary biology. A primary reason for this was the involvement of George Gaylord Simpson in the development of the Modern Synthesis. Along with works by Theodosius Dobzhansky (*Genetics and the Origin of Species*), Ernst Mayr (*Systematics and the Origin of Species*), Sewall Wright, and others, Simpson's contributions to the emerging Synthesis between genetics, population biology, and paleontology were extremely important in reorienting the priorities of evolutionary biologists and paleontologists. Simpson's *Tempo and Mode in Evolution* (1944) and the revised follow-up, *Major Features of Evolution* (1953), were read by generations of paleontologists, and in large part defined the scope of paleontological contributions to evolutionary theory between 1950 and the 1970s. (On the growth of evolutionary paleontology—or "paleobiology"—see Sepkoski 2009, 2012).

The resulting Synthesis defined evolutionary biology as a study of the movement (via inheritance and mutation) of *genes* within *populations*. One of the most important aspects of the synthetic approach was the development of a quantitative understanding of gene flow in populations, which allowed biologists to confirm that Darwin's qualitative assessment of the sufficiency of natural selection to produce evolution agreed with the modern understanding of genetics; some historians, including Provine (1971), view this as *the* major accomplishment of the Synthesis. Herein, however, were the seeds of a potentially intractable problem for paleontologists:

fossils leave no record of genetic information, and it is notoriously hard to reconstruct assemblages of fossils that correspond to living populations. So how could paleontologists accommodate the idiosyncratic evidence of the fossil record to the increasingly precise resolution expected in the growing field of population genetics?

This chapter will address the 'species question' in paleontology, or the problem paleontologists faced during the 1940s and 1050s in accommodating fossil data to the populational understanding of species promoted by geneticists in the Modern Synthesis. One strategy adopted by paleontologists was to redefine the very notion of the "paleontological" species as a "historical" concept that did not grant species independent ontological existence. Debates carried out in the journals *Evolution* and *Journal of Paleontology* in the early 1950s, for example, show that considerable divergence of opinion existed over whether the term "species" meant the same thing in both paleontology and neontology. I will argue, however, that the solution to the paleontological species problem ultimately came not from arguments about definitions but from a new methodological approach: a select group of paleontologists, led by the American Museum of Natural History (AMNH) invertebrate specialist Norman Newell, began to develop statistical, quantitative methods for evaluating and interpreting fossil populations. This approach was important for three reasons: (1) it posed a solution to the problem of how to incorporate "populational thinking" into paleontology; (2) it introduced greater analytical rigor (of the type increasingly expected in mathematical population biology) into paleontology; and (3), it introduced new theoretical possibilities for interpreting the evolutionary significance of the fossil record, which contributed to the further growth of evolutionary paleobiology.

Paleontology in the Modern Synthesis

Historians have emphasized that the synthetic project was, in large part, an institutional project. The architects of the Modern Synthesis called for dialogue between a variety of evolutionary disciplines, including genetics, ecology, systematics, zoology, anatomy, and paleontology, which necessitated the orchestration of what Joe Cain has called "institutionalized cooperation." According to Cain, this allowed the architects to cross "disciplinary boundaries in pursuit of common problems . . . to ensure inclusion and to elevate the status of fields and practices otherwise deemed marginal within biology" (Cain 1993, 2). This is part of the process of "unification" that Betty Smocovitis describes as involving several discrete steps, beginning with Dobzhansky's successful translation of mathematical population genetics into terms comprehensible to the average naturalist, and concluding with the establishment of the Society for the Study of Evolution and its journal, *Evolution* (Smocovitis 1996, 99–127).

Dobzhansky's *Genetics and the Origin of Species* (1937) provided the wake-up call to biologists, but it was Julian Huxley and Mayr who took the lead in advancing the institutional agenda of the Synthesis. Shortly after the publication of Dobzhansky's monograph in 1937, Huxley began organizing a movement in Britain to redefine the field of systematics in light of advances in genetics. This activity culminated in the book *The New Systematics* (1940), which Huxley edited for the British Systematics Association, which intended "to integrate the various studies of divergence and isolation and relate them to taxonomic groups and evolutionary mechanisms" (Cain 1993, 4–5). Meanwhile, Huxley pursued a similar reform project in America, and was able to generate interest among a number of important biologists, including Alfred Emerson, Dobzhansky, and Mayr, sufficient to launch a working group called the Society for the Study of Speciation in 1940 (Cain 1993, 7; Smocovitis 1994, 1–2). This group was short-lived, as the intervention of the war and the arrival of Dobzhansky to the faculty at Columbia shifted the center of the Synthesis to New York, where the AMNH also played a prominent role. There the Columbia biologist L. C. Dunn oversaw publication of the Columbia Biological Series of monographs, whose titles included Dobzhansky's *Genetics and the Origin of Species*, Mayr's *Systematics and the Origin of Species* (1942), Simpson's *Tempo and Mode in Evolution* (1944), Bernhard Rensch's *Evolution above the Species Level* (1959), and other seminal works of evolutionary biology. This series was an enormously effective tool for promoting the agenda of the Synthesis, and was centered around Dobzhansky's influential interpretation of population genetics. It also provided a vehicle for members of disciplines outside genetics to promote their own theoretical legitimacy: as Smocovitis argues, these monographs "were written by individuals who, engaging in dialogue with Dobzhansky, in turn legitimated as they grounded *their* disciplines with Dobzhansky's evolutionary genetics" (Smocovitis 1996, 134).

Simpson and other paleontologists firmly believed that paleontologists could not become part of the community of evolutionary biology if their work was only presented to, and read by, other paleontologists. For this reason, paleontologists who were committed to the project of the Modern Synthesis, like Simpson, Norman Newell, Carl Dunbar, Benjamin Burma, and others, actively sought to present their ideas in the journal *Evolution*, the newly established organ of the Society for the Study of Evolution, rather than in traditional paleontological outlets like *Journal of Paleontology*. During the early 1950s, this led to the somewhat awkward circumstance in which paleontologists actively carried out parallel public debates over the species problem in two separate forums—*Evolution* and *Journal of Paleontology*—that only partially overlapped. However, because I am most interested in understanding how paleontologists self-consciously attempted to integrate themselves into the broader evolutionary community, I will focus my discussion on a series of arguments that were presented in *Evolution* between 1947 and 1951 and their aftermath.

From one perspective, it was a legitimate triumph for paleontologists that their discipline was recognized so prominently in the institutionalization of the synthetic theory. Without question, this was largely due to Simpson's efforts, which were undeniably heroic. It would be a mistake, however, to conclude that paleontology was, whether in 1940, 1944, 1946, or afterwards, a fully equal and respected partner in the community of neo-Darwinian evolutionary biology. In fact, considerable pressure was exerted by biologists to ensure friendly paleontologists' "cooperation" in adhering to the synthetic party line. In 1944, the Princeton biologist Glenn Jepsen stressed to Mayr that "paleontology presents good evidence, as you know, that evolution proceeds by microgenetic rather than macrogenetic alterations and that this evidence is in harmony with experimental genetics," and he appealed to Mayr's colleague, the invertebrate paleontologist Kenneth Caster, "I hope you will be willing to make a statement on this subject" (quoted in Cain 1993, 12). The concern here was that many paleontologists had in the past been "seduced" by macromutations and saltations as explanatory mechanisms for major evolutionary change. Indeed, Mayr recalled many years later that "most paleontologists were either saltationists or orthogenesists, while those we believe to have been neo-Darwinists failed to write general papers or books" (Mayr 1980, 28). As a response, Cain concludes, biologists "effectively controlled the identity of biology through the synthesis period," and "'synthesis,' from this perspective, meant the expansion of laboratory work together with the subsumption of descriptive studies by field and museum workers who were 'brought into line'" (Cain 1993, 17–18).

The trend in paleontological statements about evolution, then (as evidenced for example in published annual presidential addresses of the Paleontological Society from that period), reflects paleontology's move closer to the mainstream of biological evolutionary theory (and away from saltationism and orthogenesis), but also indicates the extent to which the "disciplining" efforts of synthetic biologists were successful in engineering agreement with neo-Darwinian principles. Paleontologists would certainly benefit from greater participation in the evolutionary biology community— more secure institutional positions, greater respect for their data, better access to mainstream publications and conferences, and a larger stake in theoretical discussions all followed over the next few decades. But there was a cost as well: as Patricia Princehouse argues, "in large part the Modern Synthesis served to sideline major research traditions in paleontology" (Princehouse 2003, 21). One of those traditions involved approaching macroevolutionary analysis of the fossil record with confidence that paleontology had unique access to patterns and processes of evolution undetectable by genetics or systematics.

The Species Question

As early as 1940, Huxley situated the species problem at the heart of the emerging evolutionary Synthesis. In his introduction to *The New Systematics*, Huxley defined the project of the "new systematics" as "detecting evolution at work," and in particular answering the question of "how discontinuity in groups is introduced into the biological continuum," or in other words answering whether the divisions imposed by taxonomy correspond in some natural sense to real populations of organisms (Huxley 1940, 2). Huxley's answer was fairly unequivocal: in most cases "species can be readily delimited, and appear as natural entities, not merely convenient fictions of the human intellect," and he concluded that "species are in some sense valid natural groups" (Huxley 1940, 11, 16). The same year, Mayr offered his now-famous definition of the "biological species concept": species are groups of actually or potentially interbreeding natural populations, which are reproductively isolated from other such groups (Mayr 1940). This definition was reprinted two years later in *Systematics and the Origin of Species* (Mayr 1942), where it had wide influence in the emerging Synthesis.

While Mayr's discussion of the species concept sidestepped questions about the ontological reality of biological species, it is important to note that this definition effectively excluded paleontologists from the discussion. Paleontologists have no way of grouping fossils into interbreeding populations or of assessing reproductive isolation, and so are limited to estimating degrees of morphological similarity. Unlike Huxley, who had rather dismissively claimed that paleontology "can only give us information concerning the course of evolution, and not concerning its mechanism," Mayr did acknowledge the challenge of addressing what he calls "allochronic" (temporal) species in an imperfect fossil record. Ultimately, however, the best he could offer is that paleontologists and biologists mean different things when they discuss species: as he put it, "the 'species' of the paleontologist is not necessarily always the same as the 'species' of the student of living faunae," since paleontologists have no choice but to impose arbitrary divisions based on breaks in stratigraphic sequences (Mayr 1942, 154).

One of the first responses to this problem from the paleontologists came from Norman Newell, who published a paper titled "Infraspecific Categories in Invertebrate Paleontology" in the third issue of the journal *Evolution*. Newell, in 1947, was recently arrived as curator of invertebrate paleontology at the American Museum of Natural History, working under G. G. Simpson. It can fairly be said that nobody did more to promote the agenda of evolutionary paleontology in the 1950s and '60s than Newell, and his influence, measured directly through his work and indirectly through his mentoring of students and younger paleontologists, was profound. Newell's hand touched nearly every major aspect of paleobiology during his career, and he can be said to have been directly responsible for, in no particular order: the investigation of broad patterns in the fossil record, the development of quantitative approaches to fossil databases, the study of the evolutionary significance of mass extinctions, and the creation of the subdiscipline of paleoecology. Throughout his career, Newell also tirelessly promoted the institutional agenda of paleobiology, and he trained many of the leaders of the movement's next generations.

Newell's 1947 paper was clearly a call to action for both paleontologists and biologists. The opening line proclaimed that "evolution as a modern philosophy requires the synthesis of paleontology, genetics, and neontology," and Newell proceeded to diagnose just what he thought paleontology could and should contribute to that Synthesis (Newell 1947, 163). It is noteworthy that Newell favored the term "neontology" here and in later publications; he regarded paleontology and biology as sister disciplines, and in part his championing this notion influenced important members of both communities. Newell's target in this essay was the gap between the paleontologic and neontologic understandings of "species," the closure of which, he argued, was a crucial step in facilitating greater synthesis. The problem involved the common paleontological practice of basing taxa on single type specimens, which, in Newell's mind, failed consider "the variability of organisms in taxonomy." In other words, whereas biologists understood "species" as populations of organisms exhibiting graded variability, paleontologists tended to assign specimens that differed only slightly from one another to separate taxa.

The solution, according to Newell, was for paleontologists to adopt the biological concept of "subspecies" and to develop greater sophistication in establishing methods for discerning the true relationships between related organisms. His definition of subspecies as "entire populations, or races, which have become differentiated through some degree of isolation" was drawn from Mayr's 1942 *Systematics and the Origin of Species*, and explicitly conceived of these populations as geographic and populational units (Newell 1947, 164). Of course, paleontologists are unable to establish genetic relationships between fossil organisms, but Newell argued that paleontology would benefit simply from paying closer attention to the populational and biogeographical vocabulary of biology. For instance, he urged paleontologists to abandon the imprecise concept "variety," which does not necessarily connote a distinct population, in favor of the biological "subspecies," which usually does.

Newell's argument was more than simply a semantic one: if paleontologists were indeed to participate fully in the Modern Synthesis, they must accept the principle that "in all probability, evolution invariably has been accompanied by gradual morphological change" (Newell 1947, 167). Since the time of Darwin paleontologists had been aware of the likelihood of missing transitional sequences in the fossil record, but before the Synthesis, paleontologists did not feel particularly constrained by a theoretical necessity to extrapolate a continuous gradation of forms between taxa. Theories like orthogenesis and saltationism allowed alternatives to perfectly graded sequences. However, accepting the synthetic definition of evolution meant accepting Mayr's and Dobzhansky's populational understanding of taxa, in which species are inherently unstable, variable entities and divisions between taxonomic groups are often very subtle. So Newell's argument was important not only because it drew attention to the asymmetry between paleontological and biological taxonomic definitions, but also because it issued a challenge to paleontologists: if they wanted to sit at the table, they would have to find a way to make their data workable within the conceptual vocabulary of biology and genetics.

This was no easy task, and paleontologists would spend the next several decades attacking and arguing about this problem. Newell's initial suggestions involved quantitative strategies that looked back towards Simpson's earlier work and forward to the approaching quantitative revolution in numerical taxonomy (on the growth of the quantitative taxonomy movement, see Hagen 2003): one strategy required establishing empirical criteria for determining a population's inherent variability, which involved adapting the "normal curve" of variation for particular kinds of organisms (fig. 1.1). Another approach used stratigraphy to estimate the effects of time and geography on speciation—in effect treating evolution not just as a *vertical* sequence of forms, but as one that involves significant *horizontal* branching due to geography (barriers and migrations) and infraspecific variation (Newell 1947, 167–169).

Newell's discussion of the paleontological species concept played into a larger discussion and debate that began to unfold on the pages of *Evolution* and carried on throughout the 1950s. It must be remembered that the synthetic conceptual approach Newell was promoting was still quite new and did not enjoy universal support even among biologists. In taxonomic circles, there was far from complete agreement about even the basic philosophical integrity of the concept of the "species" itself. In 1949 paleontologist Benjamin Burma published a contentious opinion piece in *Evolution* in which he argued, from a logical perspective, that given the reality of indefinitely grading breeding populations, "when we try to deal with larger aggregates of individuals, our categories become more and more abstract and empty of any real meaning." Therefore, he concluded, "'species' have only a subjective existence" (Burma 1949, 370). This position drew immediate criticism from Mayr, who pointed out the reality of "sharp" discontinuities between populations in every environment on earth, arguing that "the arrangement of organic life into well-defined units is universal," hence "there can be no argument as to the objective reality of the gaps between local species in sexually reproducing organisms" (Mayr 1949, 371). This attempt at resolution did not end the debate. The next year paleontologist Maxim K. Elias published a paper in *Evolution* that drew further attention to the temporal component in species definition, and concluded that "for the paleontologist both genera and species are mere temporal cross-sections of an endless continuity of changes, no segment of which . . . can possibly constitute an objective reality" (Elias 1950, 177).

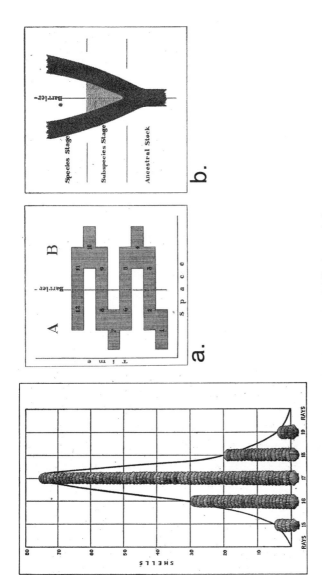

Figure 1.1 Newell's quantitative approaches to estimating normal variability in populations. Left: the normal curve of morphological variation in scallops from an Atlantic beach. a: Geographic fragmentation of successive populations (the numbered rectangles) accompanying vertical differentiation of a phyletic line. b: A barrier divides a population, resulting in isolation and limiting gene flow. From Newell 1947, 166, 169.

When Simpson stepped into the conversation in 1951 with a longer meditation on the problem, he advocated a practical solution. He began by noting that the "species question" is a central and perennial discussion in evolutionary theory, and that while "its endless discussion is sometimes boring and seemingly fruitless," it is nonetheless "not wholly futile" (Simpson 1951, 285). In addition to the debate that had been running in *Evolution* for two years, a parallel discussion, less concerned with philosophy than with taxonomic practice, had been taking place in the pages of *Journal of Paleontology* at the very same time; the *JP* debate consisted of Weller 1949, Jeletzky 1950, Bell 1950, and Wright 1950. Simpson offered to try to navigate these two disputes, though he observed the extreme difficulty involved in "combin[ing] some of their apparent but not really conflicting views into one consistent statement," especially since he "agree[d] with most of what all the authors" had written. A large portion of the disagreement could be attributed to semantics, which he essentially waved away by avoiding "such terms as 'real,' 'natural,' or 'objective,'" and substituting the terms "arbitrary" or "nonarbitrary" to refer to classification procedures that either do or do not group organisms "on the basis of pertinent, essential continuity" (Simpson 1951, 286). Then, following Newell and others, Simpson endorsed the genetic, populational account used by neontologists as an example of the most important, nonarbitrary definition of species (though he admitted that other definitions are valid), provided they were modified to properly take notice of the temporal factor.

When Simpson arrived at the practical issues involved in paleontological taxonomy, however, he acknowledged significant difficulties: paleontology and neontology have different kinds of data, and paleontology is more often confronted by "discontinuities" of both "observation" and "record" in reconstructing ancient populations (Simpson 1951, 290–291). Granting that paleontologists must deal with "special questions" involving "succession or sequence," he elaborated a procedure for paleontological classification: (1) compare multiple sample populations and estimate "morphological variation in those populations"; (2) if no significant variation is present, assume "a single population and hence taxonomic group"; (3) if differences do apply but overlap in mean variation is present, assume multiple subspecies; (4) if differences apply without overlap, assume distinct species (Simpson 1951, 291). This procedure set the quantification of mean variability for a population as a central task for paleontological taxonomy, and the technique he proposed to model the phyletic relationships between populations involved graphically representing samples as

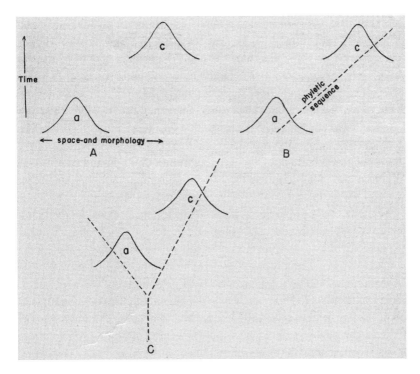

Figure 1.2 Simpson's variation curves.
Possible interpretations of data from two related fossil species separated in space and/or time. A. The data without interpretation. B. Interpretation as a single phyletic sequence (*a* and *c* are the same species by genetic definition). C. Interpretation as a branching sequence (*a* and *c* are different species by genetic or evolutionary definition). From Simpson 1951, 297.

variation curves in temporal and morphological sequence (fig. 1.2). This proposal would become the standard general method for presenting infra- and interspecific relationships in paleontology.

While Simpson had perhaps the final word among paleontologists in the initial debate in *Evolution*, the flurry of interest in the paleontological species concept continued over the next several years. In 1954, a special meeting of the British Systematics Association was convened to consider "The Species Concept in Palaeontology." This symposium, and the eventual published proceedings, were billed as a kind of follow-up to Julian Huxley's 1940 volume *The New Systematics*, which had also been sponsored and published by the Systematics Association. Huxley's volume had, at its publication, been a manifesto of sorts for the emerging Modern

Synthesis that had attempted to redefine the project of taxonomy as "that of detecting evolution at work" (Huxley 1940, 2). "Specifically," Huxley had written in his introduction, "its chief question is how discontinuity in groups is introduced into the biological continuum," or in other words whether the divisions imposed by taxonomy correspond in some natural sense to real populations of organisms. The general conclusion of the new systematics, according to Huxley, was that "species can be readily delimited, and appear as natural entities, not merely convenient fictions of the human intellect"; Huxley credited geographic isolation and the "Sewall Wright effect" for producing the discontinuity that allowed systematists to make sharp distinctions between taxa (Huxley 1940, 11).

While Huxley's grand vision for systematics earned considerable notoriety and is remembered as one of the major texts of the Modern Synthesis, David Hull has concluded that "in retrospect, very little that could be counted as 'new' actually appeared" in the volume (Hull 1988, 102). The collection also had very little to say about paleontology, which as a discipline was represented by only a single paper (out of 21 in the volume), and about which Huxley dismissively remarked that "owing to the nature of its data, can only give us information concerning the course of evolution, and not concerning its mechanism" (Huxley 1940, 3n3). The 1954 meeting then was a chance to revisit the issue in light of the more recent debates about the species concept between and among paleontologists and biologists, or as Errol White, who chaired the symposium, reflected in the published proceedings, "for the purpose of promoting an exchange of views between zoologists and palaeontologists on the application of the more recently developed taxonomic methods to palaeontological problems" (White 1956, iii). The meeting, like its predecessor, drew a mostly British panel of speakers, including both paleontologists and neontologists, with the single American contribution coming from Newell, who spoke on fossil populations. This fact seems rather odd given that the majority of paleontological discussion about the species problem was being conducted by American scientists, but perhaps the organizers reasoned, as Huxley had in 1940, that relying mostly on British contributors would "facilitate the co-ordination of the different articles" (Huxley 1940, v).

British paleontologist P. C. Sylvester-Bradley began the volume with an introductory essay about "the new palaeontology," which he defined as "more than anything else . . . the comparison of vertical with horizontal variation," or the introduction of the "three-dimensional" species concept (Sylvester-Bradley 1956, 4). This theme reappears in many of the essays, variously referred to as "the time factor," or the "chronospecies," and is enthusiastically applauded by the paleontologists who contributed to the volume (Thomas 1956, 17; Rhodes 1956, 38). Many of these papers simply revisited the debate carried out in *Evolution*, although F. H. T. Rhodes's summary offered the constructive proposal that "a combination of the concepts of Dobzhansky and Simpson may provide an acceptable solution" (Rhodes 1956, 49). Interestingly, a number of the authors explicitly rejected Huxley's confidence in the reality of biological species, including J. B. S. Haldane's rather scathing two-page contribution, which concluded that "in a complete palaeontology all taxonomic distinctions would be as arbitrary as the division of a road by milestones" (Haldane 1956, 96).

Indeed, White's summation of the meeting leaves the distinct impression that not all participants were convinced that major strides had been taken. Noting the "curious lack of emphasis" on problems with the fossil record and the "tacit assumption of the gradualness of evolution," White mused,

> It is only just to the older palaeontologists to point out that some of the claims made for the new taxonomic approach by its more enthusiastic devotees go far beyond the facts. Awareness of the difficulties presented by the three-dimensional palaeontological species is not a new development engendered by the 'New Systematics'; neither has 'the synthesis of taxonomic and evolutionary ideas' nor its 'assimilation into palaeontological thought' resulted from the work of the last fifteen years.

White's comments were directed squarely at Sylvester-Bradley's enthusiastic essay (from which the quotations were taken), and his conclusion made no bones about his implied rebuke: "Fresh breezes are doubtless blowing through the musty halls of orthodox palaeontology, but some of us may be forgiven for thinking that in places the amount of wind is excessive" (White 1956, iv).

Despite the inconclusive outcome of the symposium, Newell's contribution made a noteworthy effort at extending some of Simpson's proposals towards practical fruition. Newell wisely sidestepped the difficulties involved in identifying species from fossil data by declaring that "the genus, which, in practice, is the smallest consistently recognizable unit, has become the working unit of palaeontology" (Newell 1956, 63). The main target of Newell's analysis was the unique set of problems paleontology faced in applying taxonomic divisions to fossil populations. Here his major

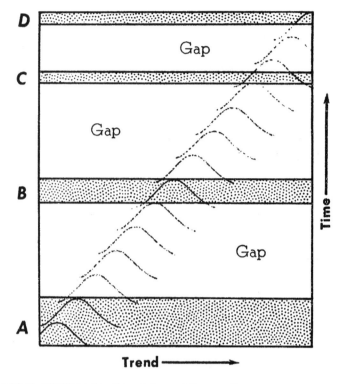

Figure 1.3 Newell's illustration of population variation over time.
From Newell, 1956, 69.

concern was preservational bias: while "the fossil record is in fact astonishingly rich and meaningful," the "time dimension" in paleontology com-
plicates matters, since "the selection of species limits in a vertical series
might be arbitrary" (Newell 1956, 67). In other words, the added dimension of time is both a boon and a hindrance to paleontology: within a given
"horizontal" sample (i.e., a group of organisms taken from the exact same
stratum or "moment" in geological time) it might certainly be possible to
distinguish taxa, including species and perhaps even subspecies or varieties. But paleontology also has a vertical dimension, and as the taxa identified from horizontal samples continue forward in time, it is extremely
difficult to discern where taxonomic limits or divisions should be placed
(fig. 1.3). This situation is further complicated by the fact that vertical
sequences are almost always interrupted, and the paleontologist is not
guaranteed to fill in these gaps by further collection. Finally, as Newell

noted, horizontal and vertical perspectives must be combined to get an accurate picture of the influence of geography on phyletic evolution: "It may be doubted . . . that appreciable portions of the evolution of a lineage are completed at one place without extensive migration accompanied by repeated segregation and reunion of local populations (geographic speciation)" (Newell 1956, 70). Nonetheless, in the face of such apparently insoluble difficulty, Newell remained confident that "properly conceived and diagnosed, palaeontological species and subspecies can be consistently recognized and studied by the same methods as those employed in neontology" (Newell 1956, 70–71).

The answer to this problem, Newell determined, was to apply quantitative analysis to the confusing array of fossil data—to let statistics do what the paleontologist is unable to accomplish using traditional descriptive techniques. In the past, paleontologists had relied on a typological basis for identifying species and higher taxa, but ecological and evolutionary study required paleontology to reorient itself to the biological population understanding; according to Newell the "crude procedure" of typology "does not measure up to modern requirements in studies of stratigraphic and evolutionary palaeontology" (Newell 1956, 71). This is mainly because the typological species concept ignored population variability, which should in each instance follow a normal population curve (à la Simpson's 1951 discussion). A type specimen is normally chosen (i.e., sampled) arbitrarily, and the paleontologist had no guarantee that it "represent[s] the most frequent condition of populations" (i.e., that it would fall in the middle of a normal variability curve). Instead, the procedure should be to select, ideally as randomly as possible, a group of examples from a population and to estimate, using "biometrical analysis," the range of variation for that population. The trick, according to Newell, "is to summarise in a reasonably accurate way the characteristics of a vast assemblage of individuals, perhaps numbering billions, by means of data provided by a few specimens" (Newell 1956, 74).

The only way such a drastic extrapolation would be justified would be if paleontologists had confidence that the few specimens chosen gave a reasonable indication of the limits of variability in their parent population. Surprisingly, Newell argued, most populations *can* be estimated in such a way, and individual samples are in fact reliable indicators of average variability provided that they are sampled *randomly*. The mistaken belief that only large and well-documented collections could be analyzed this way had meant "very little headway has been made toward the establishment of uniform practice in quantitative palaeontology." What we are seeing in Newell's proposal is the solidification of a major argument that statistical analysis can correct for the inadequacies of fossil preservation. This would be perhaps the single most important future direction in paleobiology, but it ultimately depended on a serendipitous convergence of paleontological thinking and technology. As Newell noted a few years later, "the recent application of electronic IBM computers in the solution of paleontologic problems" is "more than just another statistical technique"; rather, as he went on to predict, the advent of inexpensive, readily available digital computing meant that "in the near future, we may have at our disposal the means for more or less routine quantitative solutions of all sorts of paleontological problems involving complex interrelationships of many variables" (Newell 1959, 490). In other words, evolutionary paleontology was about to become a quantitative discipline.

Conclusion

Overall, what the debates over the paleontological species problem show us is that paleontologists, during the period following the Modern Synthesis, took the task of bringing their concepts and methods into line with biologists' very seriously. This does not mean, necessarily, that paleontologists were accepted as equal partners at the evolutionary "high table" (they weren't), nor that they were able to decisively "solve" the species problem. Indeed, despite philosophers' attempts to step in to mediate the dispute, arguments about the proper way to define and identify species continue today. One recent discussion in paleontological literature observes, for example, that "there is a strong flavor of scientific parochialism in this debate: workers in different biological disciplines typically favor their own brand of species concept" (Lieberman, Miller, and Eldredge 2007, 34).

What is particularly interesting about the responses of paleontologists during the short period of time I have discussed, however, is the approach that was taken to resolve a significant dilemma. On the one hand, paleontologists such as Simpson and Newell were invested in supporting the Modern Synthesis and in advocating an expanded role for paleontology within evolutionary biology. On the other, they were faced with a difficult choice: either to reject a central premise of that Synthesis (the biological species definition) or to recuse themselves from a vital discussion in evolutionary systematics. Rather than chose either extreme alternative,

Simpson and especially Newell redefined the discussion so that it would (a) allow meaningful participation by paleontologists, (b) accommodate paleontological data, and (c) adopt at least the *language* of the increasingly quantitative approach to evolutionary theory used by biologists of the Synthesis.

This tactic offers a microcosm of the strategy of the broader paleobiological movement, of which Newell was a great champion, that developed over the next several decades. A central goal of that movement—which culminated in the analytical, theoretical work of Stephen Jay Gould, Steven Stanley, David Raup, Jack Sepkoski, and others in the 1970s and '80s—was to appeal for greater disciplinary autonomy for paleontology while simultaneously borrowing explicitly from the methodologies of biology, ecology, and other disciplines (see Sepkoski 2012). In this way, Simpson and Newell's "solution" to the species problem in the early 1950s mirrored the "adaptive strategy" of evolutionary paleontology more broadly: to develop a meaningful way to integrate the new genetic understanding of evolution into paleontological discussions while maintaining a distinctly paleontological approach to studying evolution. This tactic may have been only partially successful in convincing biologists to take paleontology more seriously, but it had a transformative effect on paleontology itself.

Acknowledgments

Portions of this essay are modified from Sepkoski 2012. The author is grateful for permission from the University of Chicago Press to reprint the material.

References

Bell, W. C. 1950. "Stratigraphy: A Factor in Paleontologic Taxonomy." *Journal of Paleontology* 24 (4):492–496.

Burma, Benjamin H. 1949. "The Species Concept: A Discussion." *Evolution* 3 (4):369–373.

Cain, Joseph A. 1993. "Common Problems and Cooperative Solutions: Organizational Activity in Evolutionary Studies, 1936–1947." *Isis* 84:1–25.

Elias, Maxim K. 1950. "Paleontologic versus Neontologic Species and Genera." *Evolution* 4 (2):176–177.

Hagen, Joel. 2003. "The Statistical Frame of Mind in Systematic Biology from Quantitative Zoology to Biometry." *Journal of the History of Biology* 36 (2): 353–384.

Haldane, J. B. S. 1956. "Can a Species Concept Be Justified?" In *The species concept in palaeontology:A symposium*, edited by P. C. Sylvester-Bradley. London: Systematics Association.

Hull, David L. 1988. *Science as a process : An evolutionary account of the social and conceptual development of science*. Chicago: University of Chicago Press.

Huxley, Julian. 1940. *The new systematics*. Oxford: Clarendon Press.

Jeletzky, Jurij A. 1950. "Some Nomenclatorial and Taxonomic Problems in Paleozoology: With a Discussion of the Correlation of Some Uppermost Jurassic and Cretaceous Faunas on Both Sides of the Atlantic." *Journal of Paleontology* 24 (1):19–38.

Mayr, Ernst. 1940. "Speciation Phenomena in Birds." *American Naturalist* 74: 249–278.

Mayr, Ernst. 1942. *Systematics and the Origin of Species*. New York: Columbia University Press.

Mayr, Ernst. 1949. "The Species Concept—Semantics versus Semantics." *Evolution* 3 (4):371–373.

Mayr, Ernst. 1980. "Some Thoughts on the History of the Evolutionary Synthesis." In *The evolutionary synthesis: Perspectives on the unification of biology*, edited by Ernst Mayr and William B. Provine. Cambridge, MA: Harvard University Press.

Newell, Norman D. 1947. "Infraspecific Categories in Invertebrate Paleontology." *Evolution* 1 (3):163–171.

Newell, Norman D. 1956. "Fossil Populations." In *The species concept in palaeontology: A symposium*, edited by P. C. Sylvester-Bradley. London: Systematics Association.

Newell, Norman D. 1959. "Adequacy of the Fossil Record." *Journal of Paleontology* 33 (3):488–499.

Princehouse, Patricia M. 2003. *Mutant phoenix: Macroevolution in twentieth-century debates over synthesis and punctuated evolution*. Doctoral dissertation, History of Science, Harvard University, Cambridge, MA.

Provine, William B. 1971. *The origins of theoretical population genetics*. Chicago: University of Chicago Press.

Rhodes, F. H. T. 1956. "The Time Factor in Taxonomy." In *The species concept in palaeontology: a symposium*, edited by P. C. Sylvester-Bradley. London: Systematics Association.

Sepkoski, David. 2009. "The Emergence of Paleobiology." In *The Paleobiological revolution: Essays on the growth of modern paleontology*, edited by David Sepkoski and Michael Ruse. Chicago: University of Chicago Press.

Sepkoski, David. 2012. *Rereading the fossil record: The growth of paleobiology as an evolutionary discipline*. Chicago: University of Chicago Press.

Simpson, George Gaylord. 1951. "The species concept." *Evolution* 5 (4):285–298.

Smocovitis, Vassiliki Betty. 1994. "Disciplining Evolutionary Biology—Ernst

Mayr and the Founding of the Society for the Study of Evolution and Evolution (1939–1950)." *Evolution* 48 (1):1–8.

Smocovitis, Vassiliki Betty. 1996. *Unifying biology: The evolutionary synthesis and evolutionary biology*. Princeton: Princeton University Press.

Sylvester-Bradley, P. C. 1956. "The New Palaeontology." In *The species concept in palaeontology: a symposium*, edited by P. C. Sylvester-Bradley. London: Systematics Association.

Thomas, Gwyn. 1956. "The Species Conflict." In *The species concept in palaeontology: A symposium*, edited by P. C. Sylvester-Bradley. London: Systematics Association.

Weller, J. Marvin. 1949. "Paleontologic Classification." *Journal of Paleontology* 23 (6):680–690.

White, Errol. 1956. "Introduction." In *The species concept in palaeontology: A symposium*, edited by P. C. Sylvester-Bradley. London: Systematics Association.

Wright, Claud William. 1950. "Paleontologic Classification." *Journal of Paleontology* 24 (6):746–748.

The Species Problem: Concepts, Conflicts, and Patterns Preserved in the Fossil Record

William Miller III

It was the best of times, it was the worst of times, it was the age of wisdom, it was the age of foolishness, it was the epoch of belief, it was the epoch of incredulity . . . —Charles Dickens, *A Tale of Two Cities*

Introduction

In recent years, the "species problem" has mainly stood for the seemingly endless debate over *operational* species concepts: which conceptualizations or practical approaches work best in delimitation of species of various kinds. But there is more to it than that (Allmon, chapter 3, this volume). The species problem also involves making the case that species correspond to real evolutionary products or entities and visualizing what those entities could be—the *ontology* of species. And these issues are connected to approaches used in the practical business of *identifying* taxa in samples. When it comes to the fossil record, there is the additional complication associated with trying to discover or delimit species using mostly skeletal residues of organisms.

There is a strong parochial flavor to the debate over species concepts. Biologists interested in asexual or uniparental organisms attempt to delimit or visualize species differently than mammal or bird biologists; and the botanists often have another point of view (Wheeler and Meier, 2000). And using varied methods to delimit taxa and match them to real evolutionary entities—gene markers and sequencing, morphometry, methods of phylogenetic systematics, ecologic and behavioral analysis, etc.—has resulted in a proliferation of concepts (Mayden, 1997; Harrison, 1998; Hey, 2001; Coyne and Orr, 2004; Wilkins, 2009a; Hausdorf, 2011). It might seem that the new methods have actually made answering the question "What is a species?" more elusive than it has ever been (Allmon, chapter 3, this volume). To make the situation more complicated, a split has developed between taxonomists who think a resolution of the species problem is in sight (by recognizing the difference between operational criteria and an *ultimate* concept, and by realizing that many operational concepts are actually pointing to some version of a lineage concept of species: de Queiroz, 1998, 1999, 2007; Miller, 2001, 2006) and those who say that species are not really discrete entities in nature after all (a reprise of Darwin's stance on the relationship between varieties and species, and the expectation that diversity should unfold gradually and continuously: Bachman, 1998; Mallet, 1995, 2001; Wu, 2001).

Add to this the issues and limitations faced by paleobiologists, who try to keep pace with movements in neobiology, but have to face the daunting task of discovering species using, in most cases, a few skeletons of once-living organisms. This has led to our own internal division: some workers consider delimitation of species in the fossil record an impossibility (because only parts of phenotypes of organisms are preserved), while others continue to emphasize the central role of species in macroevolutionary theory (the hierarchical level in evolution where innovation and diversity are "saved" and positioned in the structure of clades [Gould, 2001]). (The chapters by Allmon [chapter 3] and Allmon and Sampson [chapter 4] in this volume evaluate this division in more detail.) The former stance can be seen in efforts to detect trends in the history of Phanerozoic diversity (resolution of patterns involving higher taxa seem more tractable and believable than reconstructing species diversity patterns). The latter position is taken in discussions of punctuated equilibria, species sorting/selection, diversity loss and recovery associated with pulses of extinction, connections of macroevolutionary patterns to development of large-scale ecologic systems, and in documentation of eruption, expansion and demise of clades (e.g., Eldredge, 1985, 1989, 2003; Vrba, 1985a, b, 1993, 2004; contributions in Erwin and Anstey, 1995; Lieberman et al., 1995; Miller, 2002, 2004; contributions in Vrba and Eldredge, 2005; Lieberman et al. 2007.)

The purpose of this essay is to explore the "species problem" from a paleobiologic point of view. I will do this by reviewing briefly the problem as it stands in neobiology; by arguing that a general concept applicable to all kinds of organisms is a possibility; and by showing how we, as paleobiologists, now have a central role in the discussion—we are no longer camp followers in this controversy, but are positioned to be central players in its resolution. When the crucial element of time (provided by the fossil record) is added to the discussion (although speciation can happen in many different ways and outcomes can vary) the nature of species-as-lineages becomes much clearer. I have been thinking about this issue for some time (Miller, 2001, 2006), and this essay presents a more personal point of view than I have previously offered.

Practical Problems, Working Concepts, and On to Theory: A Personal View

Every paleobiologist who actually puts their hands on samples and specimens has, at some point in their career, faced the practical side of the question and/or the more theoretical version: What is a species? Sorting through a large sample of richly fossiliferous Pliocene sand, disaggregating a huge block of Jurassic siltstone loaded with bones, or scanning a big slab of Ordovician limestone filled with brachiopods and bryozoans—the exact circumstances do not really matter—if we are at all introspective, we have to face the same questions. Is it possible to identify these specimens "to species" (i.e., comparable to species delineated by neobiologists), to recognize the differences between closely related forms, and do the species taxa identified or delimited correspond to real evolutionary entities? Are skeletons alone adequate for doing these things? Is every little difference in skeletal morphology indicative of a different species, or simply the intraspecific variation we have learned to expect within a species?

My own experience came early. When I was about 13 years old, I started collecting and trying to identify late Cenozoic fossil molluscs from the outer Coastal Plain of North Carolina. Looking back, my early attempts to identify those bivalves and gastropods were pretty ambitious. Among the first sources I tried to use were Julia Gardner's U.S. Geological Survey Professional Papers 199A and B (1944 and 1948, respectively)—covering an array of assiduously described and beautifully illustrated Pliocene and Miocene shells. My reaction was a mixture of shock (in the sense that formations could contain so many different species, delimited in some cases on very subtle shell features) and awe (the publications were beautifully

produced; but moreover how could anyone ever figure all of this out?).
Later I would think that Gardner had overdone it for some of the genera
(fig. 2.1), but more recently I am not so sure. I had the same kind of experience in my undergraduate paleontology class when Ken McKinney
introduced the concept of morphologically variable species, and then confronted me with a series of Cenozoic *Ecphora* gastropods. Were they all
different species, based on minor variations in shape and ornamentation,
or were most of them varieties of only a few species? And later in gradu-
ate school, when I encountered the literature on pyramidellid gastropods,
it was the same experience. Could each coastal embayment or depth zone
on the shelf really harbor its own set of unique species, or had the experts
indulged in some mighty immoderate splitting? From attempting to identify the species in a single rock sample all the way up to tracing changes in
diversity patterns through the Phanerozoic, one eventually confronts some
aspect of the "species problem."

My position, until recently, was that the *biologic species concept* was
probably the correct picture of species as evolutionary products (at least
for biparental metazoans); but because I was involved in community paleoecology, and had never tried to make connections between ecologic
patterns and processes and evolutionary theory, I really did not pay close
attention to the debate over operational species concepts heating up in the
1970s and '80s. Like the majority of my colleagues, assuming that *morphologic species* of fossils corresponded in most cases to biologic species (and
that biologic species were the real things) was adequate for me. Recently,
however, I have started to mistrust species delimitation based solely on
morphologic criteria, and I have been persuaded that the biologic concept
has some serious problems (for similar thoughts see, e.g., Lambert and
Spencer, 1995; Wheeler and Meier, 2000). As far as an "ultimate" concept
is concerned (one applicable to most kinds of organisms, and representing
the "true" nature of species), I now think G. G. Simpson probably had it
essentially right half century ago; but when it comes to practical problems
of telling closely related species apart, any combination of approaches that
delimits taxa coming close to representing real species-lineages might be
the best we can do (Claridge et al., 1997; Wheeler and Meier, 2000). Now
I realize that I *must* pay attention to the various aspects of the "species
problem," because I am convinced that one of the next frontiers in evolutionary theory involves discovering connections between macroevolutionary patterns and the origin, development, and collapse of large regional
ecosystems (Vrba, 1985a, b, 1993, 2005; Ivany and Schopf, 1996; Miller,

PELECYPODS.

Figure 2.1 Late Cenozoic species and subspecies of the bivalve *Astarte.*
From various localities in southeastern Virginia and eastern North Carolina, as recognized by
Julia Gardner (1944, pl. 12—see p. 51–61 for evaluation of taxa and locality data). Gardner's
identifications in this plate are: 1–4, *A. symmetrica*; 5–8, *A. exaltata*; 9–10, *A. roanokensis*;
11–12, *A. hertfordensis*; 13–14, *A. arata*; 15, *A. coheni*; 16–17, *A. stephensoni*; 18, *A. hertfordensis meherrinensis*; 19–20, *A.* (*Astarotha*) *rappahannockensis*; 21, *A. hertfordensis*; 22,
A. (*A.*) *griftonensis*; 23–24, *A. berryi*; 25, *A.* (*A.*) *undulata*; 26–27, *A.* (*A.*) *undulata vaginulata*;
28, *A.* (*A.*) *griftonensis*; 29–30, *A.* (*A.*) *undulata deltoidea*; 31, *A.* (*A.*) *undulata*; 32–34, *A.* (*A.*)
concentrica; 35–36, *A.* (*A.*) *undulata deltoidea*; 37, *A.* (*A.*) *concentrica conradi*; 38–39, *A.* (*A.*)
concentrica bella; 40, *A.* (*A.*) *concentrica*; and 41, *A.* (*A.*) *concentrica conradi*. Should these
forms be regarded as many separate species based on subtle differences in morphology, or
as only a few species displaying intraspecific geographic variation—a perennial problem in
paleontology?

2002, 2004; Eldredge, 2003; Eldredge et al., 2005; Erwin and Anstey, 1995; Lieberman et al., 2007), and it is impossible to get very far with that line of thinking without knowing what species are and how to discover them in collections of fossils.

Three Centuries of Species Concepts

A comprehensive history of the species concept is beyond anything I can accomplish here (see Stamos, 2003; Allmon, 2013), so I will use *On the Origin of Species* as a manageable—and familiar—starting place. Darwin (1859) certainly was not the first naturalist to discuss the properties of species, but most texts and evolution courses do not push the story back much further than him. Mayr (1982) provided a concise summary of pre-Darwinian evolutionary theory and species concepts; and Wilkins (2009a, b) has provided a more extensive history of early concepts. But starting with Darwin, we should remember that his ideas grew out of a context of late 18th-century thinking about species, and that the historians of biology still have a lot of territory to explore before publication of the *Origin* (Wilkins 2009a, b; Allmon, 2013; Eldredge, 2015). In general, many of the major themes and tensions that we recognize today are in play by the time Linnaeus is assembling and revising his *Systema Naturae*: essentialism vs. nominalism, the idea of unity owing to common descent as opposed to shared features, discontinuities separating true species vs. intergrading varieties, kinds of living things fixed since creation vs. varieties and species produced through "secondary causes." Thus, the seeds of the "species problem" were sown long ago (Stamos, 2003; Wilkins, 2009b; Allmon, 2013; Eldredge, 2015).

Darwin's attitudes toward species—at least the convictions he held after recognizing natural selection as the primary mechanism of transmutation—are well known from a few passages in the *Origin*. He claims that the true nature of species remains undecided ("No one definition has yet satisfied all naturalists," 1859, p. 44); that delimitation is more a matter of seasoned expertise than anything else ("In determining whether a form should be ranked as a species or a variety, the opinion of naturalists having sound judgment and wide experience seems the only guide to follow," p. 47); and that species are simply varieties that have very gradually matured over time ("Varieties have the same general characters as species, for they cannot be distinguished from species," p. 58). There is the possibility that

Darwin was keeping an earlier personal view of species in his pocket (the notion that species are like individual organisms, with a birth, life history, and eventual death), or avoiding the problem in order to get the word out about natural selection to as many naturalists—of all persuasions—as possible. (The latter interpretation or something similar has been suggested before; but see Mallet [1995] for another opinion.)

In recent essays, Eldredge (2009a, b; Dominici and Eldredge, 2010) has discussed the extent to which Darwin was both directly and indirectly exposed to the works of earlier naturalists (e. g., Brocchi's *Conchiologia Fossile Subapennina* published in 1814), some of which visualized species as separate creations or genealogic entities with births, separate histories, and deaths—a picture rather like the modern evolutionary or lineage concept. It may be the case that by the time of the *Origin*, he did not promote or favor this or any particular view both because he thought natural selection requires varieties to gradually become separate species and possibly because he did not want to detract from his main agenda; but he does acknowledge what Stamos (1996) calls *horizontal species* (entities that become distinct after sufficient prolonged maturation, but with ancestors grading imperceptibly into descendants through time). Or in Darwin's words, "Varieties, when rendered very distinct from each other, take the rank of species" (p. 114).

Some biologists and philosophers of science think the "species problem"—as we usually speak of it—really did not become a crucial aspect of evolutionary theory until the Modern Synthesis (Dobzhansky, 1937; Mayr, 1942, 1993; Huxley, 1943; Simpson, 1944; Eldredge, 1985; Hey, 2006; de Queiroz, 2005), with its emphasis on gene pools, population structure, geographic context, variation and polytypic species, and the possibility of isolated reproductive communities (starting with Fisher [1930], Dobzhansky [1935, 1937], and Mayr [1940, 1942]). At this stage, we can divide most taxonomists into two general camps: those who lean toward a phenetic or morphologic species concept, and those attempting to apply the biologic species concept ("Species are groups of actually or potentially interbreeding natural populations which are reproductively isolated from other such groups" [Mayr, 1942, 120]). The latter becomes widely accepted by zoologists, although taxonomy in practice—until the introduction of molecular genetic techniques of species delimitation, augmented by morphometry and ecologic and behavioral studies—is largely carried on using the most obvious anatomic features. (This was especially true for paleontologists and the museum taxonomists, who concerned themselves with practical problems of delineation using collections of fossil or preserved specimens. As Mayr [1982, 276] put it, "Where only preserved material is studied, as is true for many groups of insects and other invertebrates, the prevailing species concept is rather typological even today.") In other words, although many neobiologists and some paleontologists claimed to adhere to the biologic species concept with the introduction of the synthesis, especially when writing about evolutionary processes (e.g., Imbrie, 1957), in practice species were still classified or identified based on a limited number of phenotypic properties, a practice going right back to Linneaus (King, 1993).

By the last half of the 20th century, the debate over delimitation and reality of species began to heat up. This wasn't just about biology; there was also some social context. This was a time of departmental splits ("organismal" vs. molecular biology), finer division of specialties, proliferation of workers, diversification of methods, and the multiplication of specialist journals. In paleontology, the divide between traditional specimen-based research and new analytic approaches opened (Sepkoski, 2012). The best the paleontologists seem to have been able to do when it came to evolutionary theory and ecology was to follow the lead of the neobiologists. By the end of the century, about two dozen species concepts were in circulation, with the debate mostly taking place on the neobiology side of the aisle (Mayden, 1997; Wheeler and Meier, 2000; Hey, 2001; Wilkins, 2009b; Allmon, chapter 3, this volume). Although operational concepts are still proliferating, some neo- and paleobiologists are beginning to think that many of these apparently different concepts can be seen as amounting to *different approaches to the same unified concept* (emphasizing different stages in the speciation process or certain contingent properties of species [de Queiroz, 1998; Allmon and Sampson, chapter 4, this volume])—a picture of the ultimate nature of species not unlike Simpson's *evolutionary species concept* (1951, 1961; Miller, 2001, 2006).

A recent development involves a return to ideas Darwin would have found appealing, mostly held by molecular geneticists (Mallet, 1995, 2001; Wu, 2001). Instead of distinct entities having different phenotypes, effective isolating mechanisms, or at least separate histories as lineages consisting of metapopulations extended through time, some writers think that a continuous gradation of distinctiveness or adaptations (reminiscent of Darwin's views about varieties, species, and the reliability of delimitation based on fertility) is a more accurate picture of the organization of life above the level of local populations. Parts of genomes may diverge, possibly producing phenotypic differentiation but not necessarily involving total reproductive isolation. In Mallet's (2001, 888) words, "Speciation is a process of emerging genealogical distinctiveness, rather than a discontinuity affecting all genes simultaneously." This is obviously a genome-centered view of species and speciation, but may not be incompatible with a general lineage concept (de Queiroz, 2007)—it can be viewed as another practical approach, or a "magnified" view of certain stages in the speciation process.

Thus, at the beginning of the 21st century, different operational concepts are still used by different practitioners working on different kinds of organisms, but attitudes related to an ultimate concept have settled into three modes. (1) The biologic species concept is still used, in some cases in more sophisticated ways (Lambert and Spencer, 1995), but many neo- and paleobiologists (especially those interested in macroevolutionary theory) are thinking of species more in terms of independent lineages, consisting of populations and comprising clades, having unique properties and histories, that may or may not be strongly differentiated and reproductively isolated—essentially the evolutionary species concept (Simpson, 1951, 1961; Miller, 2001, 2006; Allmon, chapter 3, this volume). (2) The genome-centered practitioners are resurrecting Darwin's views and eschewing hierarchical thinking (i.e., not much discussion of different properties of populations/species/clades, construed as nested historical entities in nature). (3) But some ecologists, biogeographers, and paleontologists continue to think that when it comes to practical problems of delimitation, whatever works, works; and that determining the ultimate nature or ontology of species is not within our grasp, or that one particular concept cannot be used to discover or delimit species of all kinds (see the discussion in Hull, 1997)—hence the epigraph at the beginning of my essay.

The Paleontologic Perspective

Through all of this, paleontologists probably have been the most pragmatic practitioners of all (Sylvester-Bradley, 1956; Fox, 1986; and especially the review by Allmon, chapter 3, this volume). Identification and classification of species based on fossils, in some cases the remains of organisms having no close living relatives, mostly involves exclusive focus on skeletal anatomy (traditionally qualitative description of size, form, and kinds

of preserved body parts). Other phenotypic properties (physiology, behavior, certain aspects of development) are considered irretrievable in ordinary fossil deposits in many cases (fig. 2.2). (But consider the problem of
detecting cryptic species in the fossil record. If subtle phenotypic features,
developmental variations detected in growth series of specimens, paleo-
environmental context, and paleoecologic patterns are taken into account,
cryptics may in fact be identifiable [fig. 2.3].)

As with tracing the history of modern evolutionary theory, in the historical summaries of species concepts in paleontology, we usually do not
start any earlier than Lyell (1832) and Darwin (1859) (but see Allmon,
2013, for a new review). (And there really is not a sharp professional distinction between neo- and paleobiologists in the late 1700s–early 1800s,
anyway.) Again, these 19th-century naturalists or natural historians were
not the first to try to define species. The tension between Cuvier's views
and those of Lamarck—a world of fixed species periodically revamped vs.
species undergoing an uncoordinated, gradual, continuous transformation—is certainly at play when Lyell and Darwin were on the stage and is
still being debated today (Rudwick, 1985; Gould, 1977; Wilkins, 2000a, b;
Eldredge, 2015). Lyell's (1832) thinking about the nature of species, at
least in *Principles of Geology* vol. 2, is close to Cuvier's: transmutation
is not supported by evidence, species are separate entities in nature,
and characteristic variation is a specific property (not evidence of transitions) (see Wilkins, 2009a). In the earliest days of biostratigraphic practice, species were mostly viewed as real things in nature (being distinctive
and having diagnostic properties), and to be anatomically stable and persistent enough over time to be used in the practical business of ordering
and correlating sedimentary rock formations. Eldredge (2009b, 2015; Dominici and Eldredge, 2010) has shown that Brocchi (1814) was one of the
first paleontologists to state this clearly, comparing the life history of species (having some form of natural births, distinctive roles in nature, and
eventual deaths) to human lifespans. After Darwin, paleontologists antic-
ipating gradual, continuous evolutionary transitions became preoccupied
with how best to divide continuous lineages into biostratigraphically useful
segments (e. g., Trueman, 1924; Simpson, 1943, 1961; Sylvester-Bradley,
1951, 1956; Imbrie, 1957; Allmon, this volume).

My impression is that paleontologists working before the synthesis of
the mid-20th century either were hyper-pragmatists unconcerned with a
grand theory of species and speciation (a tradition that continues today),
or developed evolutionary theory that did not feature natural selection as

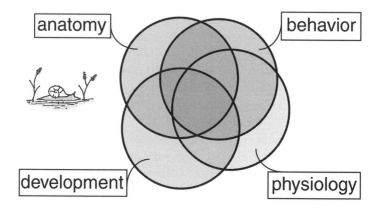

A. PHENOTYPES OF ORGANISMS

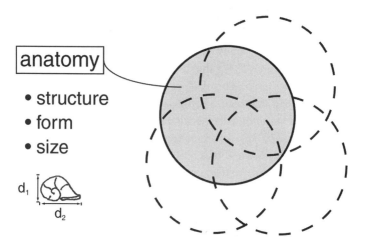

B. RECOVERED PHENOTYPES (FOSSILS)

Figure 2.2 Phenotypes of living organisms and fossils.
Phenotypes of living organisms (A), including the integrated general properties of anatomy (body parts), behavior (reaction to environmental stimuli including other organisms), physiology (chemical reactions and pathways within living organisms), and development (patterns and processes of growth and development). "Recovered" phenotypes of fossils (B) are mostly patterns of skeletal anatomy (structural parts, forms of skeletons and skeletal units, and sizes of preserved elements). The other general properties, however, could be inferred from, for example, trace fossils (behavior), growth series (development), and environmental distributions (physiology) (see fig. 2.3).

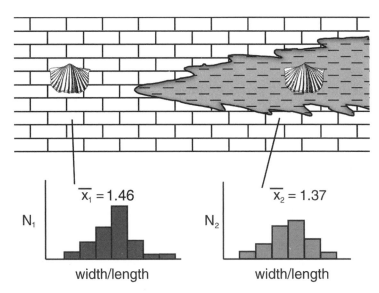

Figure 2.3 Subtle differences in skeletal morphology, paleoenvironmental context, and paleoecologic patterns could be used to delimit cryptic species in the fossil record.
In this hypothetical example, a cryptic species of the articulate brachiopod *Platystrophia* might be detected based on slight differences in morphology (N_1 vs. N_2), supported by close association with facies representing deposition in a low-diversity, muddy-bottom, oxygen-deficient environment (right portion of diagram).

the primary mechanism (Gould, 1977; Allmon, 2013). We got on with the tasks of naming and classifying newly discovered fossils, or revising previous classification schemes, and doing biostratigraphy to justify our existence, without any coherent, widely accepted species concept. Those (few) paleontologists looking for evolutionary patterns in the fossil record were concerned mostly with finding crucial "stepping stone" taxa (intermediate forms) and establishing the origins and relationships among higher taxonomic groups (Rudwick, 1985). In practice, fossil species were delimited based on skeletal features; they were grouped into genera and families based on shared anatomical characters, not dramatically different from the practice of 18th- and early 19th-century naturalists. Simpson (1943, 1944, 1951, 1953, 1961) was the first modern paleontologist to take up the "species problem" in a significant and influential way by emphasizing the time dimension in speciation theory and species ontology. And we have to wait until the introduction of punctuated equilibria (Eldredge and Gould, 1972; Gould and Eldredge, 1977) before interpretations of patterns in the fossil record begin to significantly bolster the view of species as stable, independent lineages (consisting of populations and comprising clades).

Dimensions of the "Species Problem"

When most evolutionary theorists, ecologists, and taxonomists discuss the "species problem," they are usually talking about the many competing *operational* concepts employed to delimit species. Although over two dozen concepts have been proposed, only a few are actually seriously discussed, but all of them can be reduced to three general categories (see below). There is in any case more to the problem than approaches to delimitation. The "species problem" also involves whether or not species are real things in nature, the issue of whether delimited taxa actually correspond to real evolutionary entities (a problem that gets mixed up in some instances with methods of delimitation), and the idea favored by some theorists that a single *universal* species concept is possible that would include all kinds of organisms—prokaryotes and eukaryotes, sexual and asexual, uni- and biparentals, fossil and Recent. And it should be remembered that the "species problem" impacts alpha taxonomy, the practical work of discovering and describing new species, leading on to the basic operation of making species inventories for different applications.

Adventures in alpha taxonomy

It is conceivable that one could discover and name new species, and use collections and monographic treatments of groups of organisms to identify specimens, without any strong theoretical commitments (indeed, generations of biostratigraphers and naturalists did essentially this). One could proceed simply by placing organisms in species based on the most obvious or well-preserved phenotypic traits without any conscientious adherence to a species concept—operational or ultimate—and without accepting any particular version of speciation theory. Many workers, at least since Darwin's time, have accepted that some form of evolution produced the patterns of taxonomic relatedness/distinctiveness, that the patterns consist of species, but that there is no pressing need for grand generalizations about the nature of species or about evolutionary theory. But even nuts-and-bolts taxonomists and taxonomy consumers who think they are theory-free really are not (recall "The Cloven Hoofprint of Theory" in Eldredge and Gould, 1972, 84–86). They have either developed personal procedures of discovery/identification based on experience and ingrained methodology, or were indoctrinated in such a method during their professional training (embodying a concept, stated or not); or they use literature produced by others who actually did have theoretical orientations and species concepts in mind.

Neo- and paleobiologists using (as opposed to producing) published classification systems make contact with the "species problem" indirectly and the results are obvious. If the author of a monograph on Cretaceous bivalves adheres to something close to the Dobzhansky-Mayr biologic species concept, species delimitation is likely to be relatively generalized, with similar forms or morphs regarded as examples of intraspecific variation, even if they come from different localities or stratigraphic levels. If the author wanted to coin a limited number of names, so the classification would not overwhelm the stratigraphers trying to use the fossils to estimate relative ages of formations and working out inter-regional correlations, something similar might result. If the author tried to apply some version of phylogenetic systematics, it is possible that every variant would have gotten its own species name, especially if specimens came from different locations or stratigraphic levels. If the literature is from the late 1800s–early 1900s, a typologic concept may have been followed, which could produce a similar result. And if neontological monographic work involved an effort to disclose cryptic species within a clade suspected of harboring them, a lot of specimens that look pretty much the same might be regarded as separate species—indicating that the author was hedging her bets for a good (environmental, ecologic) reason. The lumper-splitter tension is at play here, resulting from different motivations, and the people trying to identify specimens become the downstream recipients of someone else's convictions about how to discover species and possibly speciation theory. At the least, taxonomists discovering and naming new species should clearly state the operational (or ultimate) concept they favor. On the consumer side, paleoecologists, biostratigraphers, and others using published classifications and collections should be aware of the kind of concept that guided the taxonomists. But which operational approach works best? Do species taxa delimited using any of the methods now in use correspond to evolutionary entities?

Operationalism, parochialism, and a few decades of feuding

Since the 1970s, most of the noise generated about the "species problem" has resulted from clashes between different groups of neobiologists favoring different operational methods of species delimitation. An even narrower internecine conflict has erupted within the overall debate concerning different versions of the phylogenetic concept—a feud within a feud (see Hull, 1988; Wheeler and Meier, 2000). At least 26 ostensibly different species concepts have been proposed or applied in this interval (Mayden, 1997; Harrison, 1998; Brooks and McLennan, 1999; Hey, 2001; Wu, 2001; Coyne and Orr, 2004; Wilkins, 2009a; Hausdorf, 2011), although as previously noted most of them can be reduced to three general categories based on major criteria (Miller, 2001; Wilkins, 2009b). It is important to emphasize here that the brief review that follows is about operational approaches—methods of delimitation within clades sometimes containing many superficially similar species.

Category 1: Concepts based on reproductive unity, cohesion, recognition, or intergroup sterility. These concepts emphasize reproductive isolation or maintenance of reproductive networks. The most familiar concept in this category is the *biologic species concept*, having an old pedigree but formalized as an indispensable element of evolutionary theory in the Modern Synthesis (Mayr, 1942; Eldredge, 1985). In this view, species are groups of populations that consist of individual organisms capable of interbreeding to produce fertile offspring. Prezygotic or postzygotic "isolating mechanisms" prevent or limit successful out-crossing with other species. From the perspective of transspecific evolution, the establishment of reproductive isolation is the key step in the origin of a new species. From the point of view of delimitation, individual organisms that can mate to produce fertile offspring belong to the same species; organisms that fail this test (or lack the requisite isolating mechanisms) belong to separate species (see one of the most recent formulations in Mayr, 2000). Templeton's (1989, 25) *cohesion concept* is similar, emphasizing that species consist of the most inclusive group of individuals "having the potential for genetic and/or demographic exchangeability." The *recognition concept* of Paterson (1993) belongs in the same general category, as the defining characteristic of a fully formed species is a common system of fertilization, which includes mate recognition (anatomic compatibility, chemical signals, "hardwired" behavior). This is the modernized version of the biologic concept (Eldredge, 1995). All of these concepts stress some form of reproductive unity or exclusion as the major criterion; at least some phenotypic properties—those associated with isolation and recognition—have to be related to reproductive unity, cohesion, or recognition, but other adaptive morphologic features may be of secondary importance in delimitation.

Category 2: Concepts based on shared or exclusive properties of identified lineages. These are the approaches collected under the general category of the *phylogenetic species concept*. One version requires "dissolution of the stem species in a speciation event" (the *Hennigian concept*; Meier and Willman, 2000, 31), another specifies that organisms are "grouped into species because of evidence of monophyly" (*monophyletic concept*; Mishler and Theriot, 2000, 46–47), and a basic version simply requires that species taxa should be the "smallest aggregation of (sexual) populations or (asexual) lineages diagnosable by a unique combination of character states" (*diagnosable concept*; Wheeler and Platnick, 2000, 58). Beyond these requirements, all phylogenetic concepts emphasize the detection of ancestral characteristics (plesiomorphies) as opposed to new phenotypic properties (apomorphies) acquired during the process of divergence.

In most versions, species (as targets of delimitation methods) represent minimally separate lineages having initiations (speciation events), separate histories, and eventual terminations (usually lineage extinction). Some versions require or anticipate reproductive isolation, some do not; and only a few subtle features are required to discover separate lineages (Cracraft, 1989). The idea is that species can be delimited by (1) identifying unique phenotypic properties (diagnosable characters) and (2) determining the order of appearance of these properties during the development of a clade (represented by the familiar nodes and branches of cladograms) (for a lucid summary, see Futuyma, 1998). The primary goal is to construct hypothetical trees consisting of related organisms, the smallest branches of which are probable species taxa positioned with respect to each other based on phenotypic features, denoting ancestor-descendant relationships or degree of relatedness.

Category 3: Concepts based on the presence or absence of various properties (characters) of specimens. This is a large collection of operational concepts that group organisms into species if those organisms exhibit or contain the same features. These approaches may either be guided by a particular speciation theory, or could be free of any reliance on theory; and practitioners may view the clusters of organisms so delineated as actual genealogic entities, as approximations to evolutionary products, or simply as things that fit the criteria of a classification scheme having unknown or uncertain evolutionary significance (the tension between methodologic precision and taxonomic accuracy is lurking here, as it is in the methods of phylogenetic systematics; see Wiley and Mayden, 2000). This is the general category that includes the *morphologic concept* (based on the presence and form of key anatomic properties), the closely related *phenetic concept* ("variation in a set of characters is less within a group than between groups" is indicative of a separate species; Mayden, 1997, p. 404), and various forms of the *genetic concept* (delimitation based on properties of the genomes of organisms that can be detected and compared using the methods of molecular genetics). The recent applications of sequencing techniques and especially the *genotypic cluster concept* (Mallet, 1995) fall under this category. Many neobiologists involved in microevolutionary studies, taxonomic revisions and biodiversity assessments of living organisms employ some version of this approach coupled with morphometric analysis, and if feasible observations of ecologic and behavioral properties (for an interesting example, see Ishida et al., 2011 and references therein)—sources of information that may not be readily available to paleobiologists. Genetic analysis without connection to phylogenetic methods is conceptually rather close to a traditional phenetic approach.

Other kinds of operational species concepts. The foregoing is not an exhaustive survey. In the case of asexual organisms, some form of an *agamospecies concept* (acknowledging a special status for species of bacteria and protists) could be applied. If adaptive phenotypic features are regarded as key characters in delimitation, the *ecological species concept* might be favored. As mentioned above, paleontologists have divided what they regarded as anagenetic lineages into biostratigraphically useful segments referred to generally as *successional species* based on differences in skeletal features in specimens collected at different stratigraphic levels. For a more complete review of operational concepts, evaluations, and the important literature, Mayden's 1997 essay remains the best introduction.

The Potential Role of Paleobiology in the Resolution of the "Species Problem"

For at least a century, paleontologists have rehashed the "species problem" from their point of view, but I do not think we have had much of an impact in terms of potential resolution (Allmon, chapter 3, this volume)—until fairly recently. In terms of the practical problems associated with delimitation of species taxa, neobiologists have provided us with multiple approaches and concepts. And it turns out that the most effective approaches have involved combinations of these methods. In my view, however, the flow of ideas about the *reality of species* is starting to run in

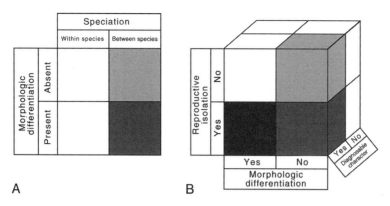

Figure 2.4 Variation among the products of speciation.
(A) Vrba's (1980, fig. 5) view, accounting for reproductively isolated + differentiated products and cryptic species; (B) a more comprehensive conception that includes minimally diagnosable taxa (having few or subtle traits that differ from parent/sister species, with or without reproductive isolation, yet representing a separate lineage—a separate evolutionary entity) (Miller, 2001, fig. 2).

the opposite direction, as a result of high-resolution studies of spatiotemporal distribution of fossils, application of phylogenetic methods to fossil data sets, and new macroevolutionary theories that place species-lineages at center stage. It is also becoming clear that the outcome of species-level evolution is varied (fig. 2.4), and that speciation can occur in different ways (although there may be dominant modes for different kinds of organisms or certain environmental situations), involving different geographic contexts and mechanisms (Otte and Endler, 1989) (some important modes are illustrated in fig. 2.5).

Speciation, species-lineages, and the proof of time

Species formation in all kinds of organisms involves (1) appearance and spread of new phenotypic traits (ranging from modest modifications [e. g., in clams, slight modification in shell ornamentation] to dramatic reorganization of phenotypes [e. g., involving a complex of anatomic, physiologic, and behavioral traits, as when developmental restructuring produces a progenetic paedomorph); (2) fragmentation of cohesion systems/reproductive communities/habitat clusters; and (3) especially the *formation of new, independent lineages*. Fully formed species-lineages are often stable over thousands to millions of generations; they are simultaneously parts

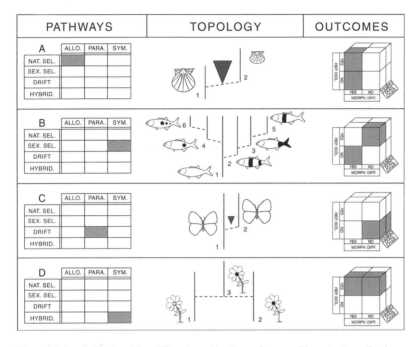

Figure 2.5 Speciation involving different combinations of geographic context, mechanisms, and outcomes.

Although four speciation *pathways* resulting in several different *outcomes* are depicted, all result in *separate lineages* having a beginning (speciation), a unique (phylogenetic, ecologic) history, and eventual termination (extinction) (see fig. 2.6). Out of the many possibilities, some examples are: (A) geographic separation + natural selection, resulting in either morphologically differentiated + reproductively isolated species or geographic variants; (B) differentiation in sympatry, possibly driven by subtle phenotypic differences under sexual selection, resulting in a flock of slightly different species; (C) intermittent geographic separation + drift, producing lineages that may or may not feature diagnosable characters but may interbreed when reunited; (D) hybridization, producing a morphologically differentiated + reproductively isolated species having traits derived from the two parent species. Miller, 2006, fig. 2.

and wholes, consisting of local population systems (demes) and comprising clades (groups of species-lineages sharing a common ancestor) (fig. 2.6). And in most cases, the interval of time required for species to completely part from an ancestor and achieve (phylogenetic, ecologic) autonomy is orders of magnitude shorter than the overall lifespan of the species-lineage (Gould, 2002). As things stand today in speciation theory, these are the only really safe generalizations about the speciation process that would apply to nearly all kinds of life forms.

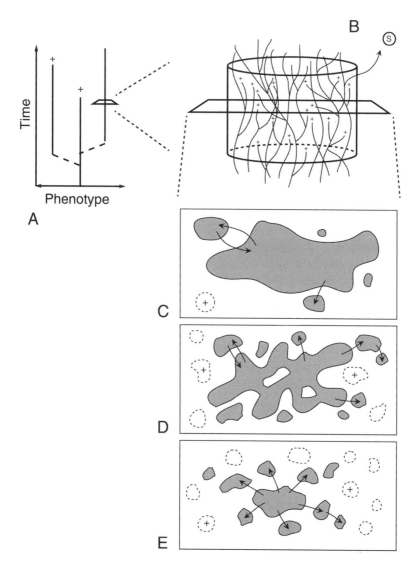

Figure 2.6 Species consist of population systems extended in time, and comprise clades of related species-lineages.

(A) Clade consisting of three species; (B) internal structure of a species-lineage made of demes that branch, fuse, go extinct, or diverge to form one or more possible new species; different possible metapopulation structures at one particular instance in the history of an established species include (C) broad geographic deployment with few barriers and outliers, (D) metapopulation consisting of a large "plasmodial" core with outliers that require re-plenishment/reestablishment after local extinction, and (E) structure consisting of a small core (or system of cores) that acts as a source for outliers that function as sinks requiring subsidies to avoid local extinction or that undergo repeated recolonization. Model E is the classic source-sink structure considered to be characteristic of many continental metapopula-tions (Pulliam, 1988); large marine metapopulations may be more like models C or D (see the discussions for varied taxa in Kritzer and Sale, 2006).

Harrison (1998) has discussed the steps in the speciation process, including appearance of novel characters, separation of populations, disruption in gene flow, and attainment of "exclusivity" (distinctive, stable properties). It is important to realize that these steps can come in different combinations or sequences in different instances of speciation. Not only can the products of the speciation process be strikingly different (for metazoans: fully reproductively isolated + anatomically differentiated, reproductively isolated but minimally differentiated [the usual meaning of cryptic or sibling species], and "minimally diagnosable taxa" [truly separate lineages, but neither reproductively isolated nor strongly differentiated from ancestors/ closely related species]), but the geographic contexts and mechanisms (selection, drift) may occur in different combinations (figs. 2.4, 2.5). I think this complex picture of species formation is one of the main sources of the conflict over ultimate nature of species. But to set the stage for speciation, involving any realistic pathway or order of events, some form of (habitat, geographic, reproductive) separation of population systems is required.

As I wrote in 2006 (565), "The general product of any speciation pathway, no matter which combination of geographic context and mechanism is involved, is the species-lineage, a historical entity having a beginning (speciation), a unique history and internal structure, and an eventual termination (extinction)." *Viewed in this way, species have a special status in evolutionary theory.* This is the level of biological organization and change where innovations are saved and injected into the history of life (Futuyma, 1987), where tokogenetic dynamics leave off for the most part, and where phylogenetic processes (definitive splitting) take over—the great "ratchet click" of species-level evolution (Gould, 2001).

Perspectives on speciation and the structure of species-lineages in neo- and paleobiology are understandably different owing to differences in spatiotemporal resolution. Operational concepts focusing on reproductive unity and phylogenetic patterns appear to resolve *stages* in the speciation process (Harrison, 1998); the fossil record mostly reflects the condensed history of species of mainly skeleton-bearing organisms—the *end-products* of speciation processes. In many cases in the modern biota, it is difficult to determine conclusively whether separated demes or demic complexes have become full-fledged species or are still varieties, subspecies, incipient species, etc. If neobiologists could fast-forward the history of an incipient species, the actual/eventual status could be confirmed. Although the same kind of problems (intergrading morphologies, slight variations in allochronic/allopatric specimens, possible cryptics) is encountered in the interpretation of fossil taxa, *paleobiologists have the advantage, in a sense, of seeing the finished products of speciation.* (Think about the main street of a small town, lined with bars and restaurants, all very busy on a Saturday night. You are strolling past a pub when suddenly a brawl breaks out on the sidewalk caused by a dozen or so patrons trying to beat each other to a pulp. When did the fight begin: at the uttering of the first insult, when the first beer glass was launched, when the first blows were struck, or when the police arrived? Every fistfight consists of a unique series of events and situated in the middle of a melee you might be able to identify the crucial developmental stages of any particular fight, but not the outcome. Now view the same brawl captured on video from across the street, but sped up. All of the events embedded within the fight are collapsed into a relatively short interval of entangled bodies, between much longer bracketing intervals of people moving in and out of the pub. The former is rather like the perspective of neobiologists; the latter is more like what paleobiologists detect in fossil data sets [e.g., fig. 2.7].)

Species-lineages as fundamental units of modern macroevolutionary theory

The laws of motion and thermodynamics, based on empirical experience and expressed mathematically, have broad application in astronomy, chemistry, geology, engineering, and biology (Silver, 1998). When generalizations about what nature contains and how it functions work this well and in varied, distantly related (or derivative) scientific endeavors, we accept the reliability (or provisional reality as *core concepts*) of the generalizations. The apparent dominance of allopatry (really any form of habitat, reproductive, or geographic separation) and natural or sexual selection as the main components of the speciation process (in terms of mode, or relative frequency) may deserve similar status in evolutionary biology (Lieberman et al. 2007). And I would argue that a lineage concept (Simpson, 1951, 1961; de Queiroz, 1998, 1999, 2007; Miller, 2001, 2006; Allmon, chapter 3, this volume) might qualify for the same status in macroevolution and macroecology. Punctuated speciation and lineage stasis, species selection, turnover pulses and persistence of regional ecosystems, "sloshing bucket" dynamics, and macroevolutionary consonance (Eldredge and Gould, 1972; Gould and Eldredge, 1977; Stanley, 1979; Eldredge, 1985, 1989, 2003; Vrba, 1980, 1985a, 1993, 2004; Lieberman et al., 1995, 2007; Benton and Pearson, 2001; Miller, 2002, 2003, 2004, 2005; Eldredge et al. 2005; Jablonski, 2008)

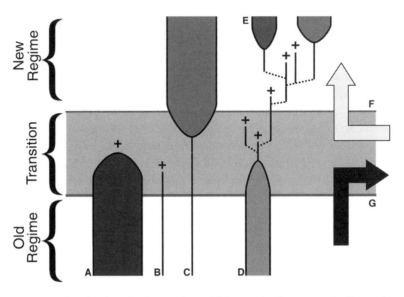

Figure 2.7 The role of species lineages in establishment, development, and collapse of regional ecologic systems: related to turnover pulses and coordinated stasis.
Based on Miller, 2006, fig. 1. (A) Previous dominant species driven to extinction by climate change or geologic processes affecting a large region; (B) rare species having the same fate; (C) rare species in a previous ecologic regime promoted to dominance in the succeeding system; (D) species in the old ecologic regime giving rise to numerous descendants (some becoming important players in the new system, some not attaining ecologic prominence [E]) in the new regime; (F and G) species invading and abandoning the region (see the recent discussion of connections between development of large scale ecologic systems and macroevolutionary patterns in Lieberman et al., 2007).

are plausible interpretations of recurrent large-scale patterns in the history of life *only* if species are viewed as historical entities or individuals (having births, life histories, and eventual deaths) uniquely positioned in the genealogic hierarchy (consisting of demic systems and making up clades of related lineages). In theories that attempt to connect recurrent macroevolutionary patterns to development of large ecologic systems, the lineage concept is particularly important, both in terms of description and interpretation (fig. 2.7).

Philosophers of science and language will see a lot to pick at here. For example, Stamos (2003) thinks that punctuated equilibria (which is a cornerstone of all of the theoretical movements mentioned above) is not a scientific theory but a metaphysical proposition, because of the impossibility of proving that the earliest and last members of a lineage belong to the same reproductive community. And citing patterns of punctuated speciation and subsequent phenotypic stasis as both a reason to accept the lineage concept (empirical experience) and a way to build macroevolutionary theory ("theoretical ammunition") may be getting too close to circularity. Punctuated equilibria is firmly established on empirical grounds (summarized in Gould, 2002; Eldredge et al., 2005, 2015) and functions logically as a working part or precondition of other recent macroevolutionary and macroecologic theories (e.g., Eldredge, 2003; Miller, 2004; Vrba, 2005; Lieberman and Vrba, 2005; Lieberman et al., 2007; Jablonski, 2008). If we do not trust theory, we ought to remember the experience of the earliest biostratigraphers, who two centuries ago discovered that fossils could be used to estimate relative ages of strata and to correlate formations from one region to another. The fossils they used had to be stratigraphically persistent and distinctive enough to do these things. High-resolution zonation and correlation are possible in the locally incomplete, mostly telescoped stratigraphic record only if the fossils being used represent stable, independent lineages with at least a few unique skeletal features. And this returns our attention to G. G. Simpson.

All the Way Back to Simpson

Eldredge (1993, 3) began an essay on the "species problem" with this not-so-upbeat comment: "Each generation, evolutionary biologists seem doomed to grapple with the species question. . . . The task is often seen as Sisyphean, with no final consensus on what really constitutes a 'species' ever emerging." I think the first part of this is actually a good thing—that we should continue to confront the problem and keep trying to achieve resolution. The last part of the quote might be too gloomy, because I think we are beginning to see a realistic picture of the true nature of species as evolutionary entities. In fact, many popular operational concepts seem to be beating separate paths to the same ultimate concept: some version of the lineage concept. This is not (yet) a consensus position, however, because some neo- and paleobiologists either don't believe in ultimate concepts or will not admit that taxa correspond to evolutionary units; because in the view of some geneticists the Darwinian view of continuous, gradual transitions precludes the notion of anything like species-as-individuals; and because some alpha taxonomists and applied biologists are not concerned with theory at all.

Yet recent developments in paleobiology are supporting the lineage concept of species in more than one way. One of the dominant modes of species-level evolution involves some form of separation or fracturing of a predecessor or ancestor species, producing buds or fragments that often become phenotypically stable entities featuring little in the way of directional change until terminated by extinction (allopatric speciation as a component of punctuated equilibria). And punctuated speciation and lineage stasis serve as theoretical scaffolding for the interpretation of a variety of related, recurrent macroevolutionary patterns. Things begin to look "real" when they keep recurring in nature and function well as working parts of theories (Blackburn, 2005). From my point of view, accepting that some form of lineage concept is the most accurate picture of species of all kinds of organisms does not require a leap of faith, just an acknowledgment of the significance of patterns preserved in the fossil record. Thus, I think Simpson (1961, 153) had it right all along: "An evolutionary species is a lineage (an ancestral-descendant sequence of populations) evolving separately from others with its own unitary evolutionary role and tendencies."

Acknowledgments

My sincere thanks go to Warren Allmon and Margaret Yacobucci for their invitation to contribute this chapter, and to Allmon and three anonymous reviewers for many thoughtful, useful suggestions. I remain responsible, however, for any missteps, misinterpretations, or overgeneralizations. Humboldt State University supported travel and preparation of the figures, produced with great skill by Andrew Schmidt. This essay is dedicated to my teacher, colleague, and friend, Ken McKinney, who started me thinking about species.

References

Allmon, W. A. 2013. Species, speciation, and paleontology up to the Modern Synthesis: persistent themes and unanswered questions. *Palaeontology* 56:1199–1223.

Bachman, K. 1998. Species as units of diversity: an outdated concept. *Theory in Biosciences* 117:213–230.

Benton, M. J., and P. N. Pearson. 2001. Speciation in the fossil record. *Trends in Ecology and Evolution* 16:405–411.

Blackburn, S. 2005. *Truth: A Guide*. New York: Oxford Univ. Press.

Brocchi, G. 1814. *Conchiologia Fossile Subapennina con Osservazioni geologiche sugli Apennini e sul suolo adiacente.* Milan: Stamperia Reale.

Brooks, D. R., and D. A. McLennan. 1999. Species: turning a conundrum into a research program. *Journal of Nematology* 31:117–133.

Claridge, M. F., H. A. Dawah, and M. R. Wilson. 1997. *Species: The Units of Biodiversity.* London: Chapman and Hall.

Coyne, J. A., and H. A. Orr. 2004. *Speciation.* Sunderland, MA: Sinauer Associates.

Cracraft, J. 1989. Speciation and its ontology: the empirical consequences of alternative species concepts for understanding patterns and processes of differentiation. In *Speciation and Its Consequences*, ed. D. Otte and J. A. Endler, 28–59. Sunderland, MA: Sinauer Associates.

Darwin, C. 1859. *On the Origin of Species by Means of Natural Selection, or the Preservation of Favoured Races in the Struggle for Life.* London: John Murray.

de Queiroz, K. 1998. The general lineage concept of species, species criteria, and the process of speciation: a conceptual unification and terminological recommendations. In *Endless Forms: Species and Speciation*, ed. D. J. Howard and S. H. Berlocher, 57–75. New York: Oxford Univ. Press.

de Queiroz, K. 1999. The general lineage concept of species and the defining properties of the species category. In *Species: New Interdisciplinary Essays*, ed. R. A. Wilson, 49–89. Cambridge: MIT Press.

de Queiroz, K. 2005. Ernst Mayr and the modern concept of species. *Proceedings of the National Academy of Sciences* 102:6600–6607.

de Queiroz, K. 2007. Species concepts and species delimitation. *Systematic Biology* 56:879–886.

Dobzhansky, T. 1935. A critique of the species concept in biology. *Philosophy of Science* 2:344–355.

Dobzhansky, T. 1937. *Genetics and the Origin of Species.* New York: Columbia Univ. Press.

Dominici, S., and N. Eldredge. 2010. Brocchi, Darwin, and transmutation: phylogenetics and paleontology at the dawn of evolutionary biology. *Evolution: Education and Outreach* 3:576–584.

Eldredge, N. 1985. *Unfinished Synthesis: Biological Hierarchies and Modern Evolutionary Thought.* New York: Oxford Univ. Press.

Eldredge, N. 1989. *Macroevolutionary Dynamics: Species, Niches, and Adaptive Peaks.* New York: McGraw-Hill.

Eldredge, N. 1993. What, if anything, is a species? In *Species, Species Concepts, and Primate Evolution*, ed. W. H. Kimbel and L. B. Martin, 3–20. New York: Plenum.

Eldredge, N. 1995. Species, selection, and Paterson's concept of the specific-mate recognition system. In *Speciation and the Recognition Concept: Theory and Application*, ed. D. M. Lambert and H. G. Spencer, 464–477. Baltimore: Johns Hopkins Univ. Press.

Eldredge, N. 2003. The sloshing bucket: how the physical realm controls evolution. In *Evolutionary Dynamics: Exploring the Interplay of Selection, Accident, Neutrality, and Function*, ed. J. P. Crutchfield and P. Schuster, 3–32. Oxford: Oxford Univ. Press.

Eldredge, N. 2009a. Experimenting with transmutation: Darwin, the *Beagle*, and evolution. *Evolution: Education and Outreach* 2:35–54.

Eldredge, N. 2009b. A question of individuality: Charles Darwin, George Gaylord Simpson, and transitional fossils. *Evolution: Education and Outreach* 2:150–155.

Eldredge, N. 2015. *Eternal Ephemera: Adaptation and the Origin of Species from the Nineteenth Century through Punctuated Equilibria and Beyond*. New York: Columbia Univ. Press.

Eldredge, N., and S. J. Gould. 1972. Punctuated equilibria: an alternative to phyletic gradualism. In *Models in Paleobiology*, ed. T. J. M. Schopf, 82–115. San Francisco: Freeman, Cooper and Co.

Eldredge, N., J. N. Thompson, P. M. Brakefield, S. Gavrilets, D. Jablonski, J. B. C. Jackson, R. E. Lenski, B. S. Lieberman, M. A. McPeek, and W. Miller III. 2005. The dynamics of evolutionary stasis. *Paleobiology* 31:133–145.

Erwin, D. H., and R. L. Anstey. 1995. *New Approaches to Speciation in the Fossil Record*. New York: Columbia Univ. Press.

Fisher, R. A. 1930. *The Genetical Theory of Natural Selection*. Oxford: Clarendon Press.

Fox, R. C. 1986. Species in paleontology. *Geoscience Canada* 13:73–84.

Futuyma, D. J. 1987. On the role of species in anagenesis. *American Naturalist* 130:465–473.

Futuyma, D. J. 1998. *Evolutionary Biology* (3rd ed.). Sunderland, MA: Sinauer Associates.

Gardner, J. 1944. Mollusca from the Miocene and lower Pliocene of Virginia and North Carolina. Part 1. Pelecypoda. *U.S. Geological Survey Professional Paper* 199-A: 19–178 + 23 plates.

Gardner, J. 1948. Mollusca from the Miocene and lower Pliocene of Virginia and North Carolina. Part 2. Scaphopoda and Gastropoda. *U.S. Geological Survey Professional Paper* 199-B: 179–310 + 38 plates.

Gould, S. J. 1977. Eternal metaphors of palaeontology. In *Patterns of Evolution, as Illustrated by the Fossil Record*, ed. A. Hallam, 1–26. Amsterdam: Elsevier.

Gould, S. J. 2001. The interrelationship of speciation and punctuated equilibrium. In *Evolutionary Patterns: Growth, Form, and Tempo in the Fossil Record*, ed. J. B. C. Jackson, S. Lidgard, and F. K. McKinney, 196–217. Chicago: Univ. of Chicago Press.

Gould, S. J. 2002. *The Structure of Evolutionary Theory*. Cambridge: Belknap Press.

Gould, S. J., and N. Eldredge. 1977. Punctuated equilibria: the tempo and mode of evolution reconsidered. *Paleobiology* 3:115–151.

Harrison, R. G. 1998. Linking evolutionary pattern and process: the relevance of species concepts for the study of speciation. In *Endless Forms: Species and Speciation*, ed. D. J. Howard and S. H. Berlocher, 19–31. New York: Oxford Univ. Press.

Hausdorf, B. 2011. Progress toward a general species concept. *Evolution* 65: 923–931.

Hey, J. 2001. The mind of the species problem. *Trends in Ecology and Evolution* 16:326–329.

Hey, J. 2006. On the failure of modern species concepts. *Trends in Ecology and Evolution* 21:447–450.

Hull, D. L. 1988. *Science as a Process: An Evolutionary Account of the Social and Conceptual Development of Science*. Chicago: Univ. of Chicago Press.

Hull, D. L. 1997. The ideal species concept—and why we can't get it. In *Species: The Units of Biodiversity*, ed. M. F. Claridge, H. A. Dawah, and M. R. Wilson, 357–380. London: Chapman and Hall.

Huxley, J. 1943. *Evolution: The Modern Synthesis*. New York: Harper and Brothers.

Imbrie, J. 1957. The species problem with fossil animals. In *The Species Problem*, ed. E. Mayr, 125–153. Washington: American Association for the Advancement of Science Publication 50.

Ivany, L. C., and K. M. Schopf (eds.). 1996. New perspectives on faunal stability in the fossil record. *Palaeogeography, Palaeoclimatology, Palaeoecology* 127:1–59.

Ishida, Y., Y. Demeke, P. J. van Coeverden de Groot, N. J. Georgiadis, K. E. A. Leggett, V. E. Fox, and A. L. Roca. 2011. Distinguishing forest and savanna African elephants using short nuclear DNA sequences. *Journal of Heredity* 102:610–616.

Jablonski, D. 2008. Species selection: theory and data. *Annual Review of Ecology, Evolution, and Systematics* 39:501–524.

King, M. 1993. *Species Evolution: The Role of Chromosome Change*. Cambridge: Cambridge Univ. Press.

Kritzer, J. P., and P. F. Sale (eds.). 2006. *Marine Metapopulations*. Amsterdam: Elsevier Academic Press.

Lambert, D. M., and H. G. Spencer (eds.). 1995. *Speciation and the Recognition Concept: Theory and Application*. Baltimore: Johns Hopkins Univ. Press.

Lieberman, B. S. and E. S. Vrba. 2005. Stephen Jay Gould on species selection: 30 years of insight. *Paleobiology* 31:113–121.

Lieberman, B. S., C. E. Brett, and N. Eldredge. 1995. A study of stasis and change in two species lineages from the Middle Devonian of New York State. *Paleobiology* 21:15–27.

Lieberman, B. S., W. Miller III, and N. Eldredge. 2007. Paleontological patterns, macroecological dynamics, and the evolutionary process. *Evolutionary Biology* 34:28–48.

Lyell, C. 1832. *Principles of Geology, Being an Attempt to Explain the Former*

Changes of the Earth's Surface, by Reference to Causes Now in Operation (vol. 2). London: John Murray.

Mallet, J. 1995. A species definition for the Modern Synthesis. *Trends in Ecology and Evolution* 10:294–299.

Mallet, J. 2001. The speciation revolution. *Journal of Evolutionary Biology* 14: 887–888.

Mayden, R. L. 1997. A hierarchy of species concepts: the denouement in the saga of the species problem. In *Species: The Units of Biodiversity*, ed. M. F. Claridge, H. A. Dawah, and M. R. Wilson, 381–424. London: Chapman and Hall.

Mayr, E. 1940. Speciation phenomena in birds. *American Naturalist* 74:249–278.

Mayr, E. 1942. *Systematics and the Origin of Species*. New York: Columbia Univ. Press.

Mayr, E. 1982. *The Growth of Biological Thought: Diversity, Evolution, and Inheritance*. Cambridge: Belknap Press.

Mayr, E. 1993. What was the evolutionary synthesis? *Trends in Ecology and Evolution* 8:31–34.

Mayr, E. 2000. The biological species concept. In *Species Concepts and Phylogenetic Theory: A Debate*, ed. Q. D. Wheeler and R. Meier, 17–29. New York: Columbia Univ. Press.

Meier, R., and R. Willman. 2000. The Hennigian species concept. In *Species Concepts and Phylogenetic Theory: A Debate*, ed. Q. D. Wheeler and R. Meier, 30–43. New York: Columbia Univ. Press.

Miller, W., III. 2001. The structure of species, outcomes of speciation, and the 'species problem': ideas for paleobiology. *Palaeogeography, Palaeoclimatology, Palaeoecology* 176:1–10.

Miller, W., III. 2002. Regional ecosystems and the origin of species. *Neues Jahrbuch für Geologie und Paläontologie, Abhandlungen* 225:137–156.

Miller, W., III. 2003. A place for phyletic evolution within the theory of punctuated equilibria: Eldredge pathways. *Neues Jahrbuch für Geologie und Paläontologie, Monatshefte* 2003:463–476.

Miller, W., III. 2004. Assembly of large ecologic systems: macroevolutionary connections. *Neues Jahrbuch für Geologie und Paläontologie, Monatshefte* 2004: 629–640.

Miller, W., III. 2005. The paleobiology of rarity: some new ideas. *Neues Jahrbuch für Geologie und Paläontologie, Monatshefte* 2005:683–693.

Miller, W., III. 2006. What every paleontologist should know about species: new concepts and questions. *Neues Jahrbuch für Geologie und Paläontologie, Monatshefte* 2006:557–576.

Mishler, B. D. and E. C. Theriot. 2000. The phylogenetic species concept (*sensu* Mishler and Theriot): monophyly, apomorphy, and phylogenetic species concepts. In *Species Concepts and Phylogenetic Theory: A Debate*, ed. Q. D. Wheeler and R. Meier, 44–54. New York: Columbia Univ. Press.

Otte, D., and J. A. Endler. 1989. *Speciation and Its Consequences*. Sunderland, MA: Sinauer.

Paterson, H. E. H. 1993. *Evolution and the Recognition Concept of Species: Collected Writings*. Baltimore: Johns Hopkins Univ. Press.

Pulliam, H. R. 1988. Sources, sinks, and population regulation. *American Naturalist* 132:652–661.

Rudwick, M. J. S. 1985. *The Meaning of Fossils: Episodes in the History of Palaeontology*. Chicago: Univ. of Chicago Press.

Sepkoski, D. 2012. *Rereading the Fossil Record: The Growth of Paleobiology as an Evolutionary Discipline*. Chicago: Univ. of Chicago Press.

Silver, B. L. 1998. *The Ascent of Science*. New York: Oxford Univ. Press.

Simpson, G. G. 1943. Criteria for genera, species, and subspecies in zoology and paleozoology. *Annals of the New York Academy of Sciences* 44:145–178.

Simpson, G. G. 1944. *Tempo and Mode in Evolution*. New York: Columbia Univ. Press.

Simpson, G. G. 1951. The species concept. *Evolution* 5:285–298.

Simpson, G. G. 1953. *The Major Features of Evolution*. New York: Columbia Univ. Press.

Simpson, G. G. 1961. *Principles of Animal Taxonomy*. New York: Columbia Univ. Press.

Stamos, D. N. 1996. Was Darwin really a species nominalist? *Journal of the History of Biology* 29:127–144.

Stamos, D. N. 2003. *The Species Problem: Biological Species, Ontology, and the Metaphysics of Biology*. Lanham, MD: Lexington Books.

Stanley, S. M. 1979. *Macroevolution: Pattern and Process*. San Francisco: W. H. Freeman.

Sylvester-Bradley, P. C. 1951. The subspecies in palaeontology. *Geological Magazine* 88:88–102.

Sylvester-Bradley, P. C., ed. 1956. *The Species Concept in Palaeontology*. London: Systematics Association Publication 2.

Templeton, A. R. 1989. The meaning of species and speciation: a genetic perspective. In *Speciation and its Consequences*, ed. D. Otte and J. A. Endler, 3–27. Sunderland, MA: Sinauer Associates.

Trueman, A. E. 1924. The species concept in palaeontology. *Geological Magazine* 61:355–360.

Vrba, E. S. 1980. Evolution, species, and fossils: how does life evolve? *South African Journal of Science* 76:61–84.

Vrba, E. S. 1985a. Evolution and the environment: alternative causes of the temporal distribution of evolutionary events. *South African Journal of Science* 81:229–236.

Vrba, E. S. 1985b. Introductory comments on species and speciation. In *Species and Speciation*, ed. E. S. Vrba, ix–xviii. Pretoria: Transvaal Museum Monograph 4.

Vrba, E. S. 1993. Turnover-pulses, the Red Queen, and related topics. *American Journal of Science* 293-A:418–452.

Vrba, E. S. 2004. Ecology, development, and evolution: perspectives from the fossil record. In *Environment, Development, and Evolution: Toward a Synthesis*, ed. B. K. Hall, R. D. Pearson, and G. B. Müller, 85–105. Cambridge: MIT Press.

Vrba, E. S. 2005. Mass turnover and heterochrony events in response to physical change. *Paleobiology* 31:157–174.

Vrba, E. S., and N. Eldredge. 2005. *Macroevolution: Diversity, Disparity, Contingency.* Supplement to *Paleobiology* 31.

Wheeler, Q. D., and R. Meier. 2000, eds. *Species Concepts and Phylogenetic Theory: A Debate.* New York: Columbia Univ. Press.

Wheeler, Q. D., and N. I. Platnick. 2000. The phylogenetic species concept (*sensu* Wheeler and Platnick). In *Species Concepts and Phylogenetic Theory: A Debate*, ed. Q. D. Wheeler and R. Meier, 55–69. New York: Columbia Univ. Press.

Wiley, E. O., and R. L. Mayden. 2000. The evolutionary species concept. In *Species Concepts and Phylogenetic Theory: A Debate*, ed. Q. D. Wheeler and R. Meier, 70–89. New York: Columbia Univ. Press.

Wilkins, J. S. 2009a. *Defining Species: A Sourcebook from Antiquity to Today.* New York: Peter Lang.

Wilkins, J. S. 2009b. *Species: A History of the Idea.* Berkeley: Univ. of California Press.

Wu, C. 2001. The genic view of the process of speciation. *Journal of Evolutionary Biology* 14:851–865.

CHAPTER THREE

Studying Species in the Fossil Record: A Review and Recommendations for a More Unified Approach

Warren D. Allmon

There is no doubt that considerable differences in outlook exist between palaeontologists and neontologists, due to a mutual failure to understand each other's concepts of species.
—Thomas, 1956: 17

Species recognition is at the core of the paleontological enterprise and is an essential component in building an accurate understanding of evolution.—White, 2003: 1997

Introduction

The theoretical literature on species and speciation has frequently been viewed as chaotic, onerous, and unenlightening (e.g., Hull 1997; Morrison 2011; Harrison 2012), but it remains unclear the degree to which this apparent disorder indicates nature's complexity or our ignorance, or both. For example, "The argument on the appropriate way(s) to define a 'species' is still unsettled" (Gavrilets 2003, 2197); "There is no consensus on what exactly a species is" (Barton 2001, 325); and "No binding rules exist for the classification of individuals into species, nor is it clear whether species exist at all" (Kunz 2012, 1). The study of speciation has been called the "continuing search for the unknown and unknowable" (Paterson 1981, 113). Speciation, says Hunt, is "elusive—too slow to observe in the present day, too rapid to capture in the geological record" (2010, S69). "The problem of speciation," wrote Provine, "exemplifies . . . a sequence in which the relative assurance of earlier explanations unravels in the light of greater knowledge, leaving confusion and disunity" (1991, 201). Arguments about species and speciation are often heated and philosophical, and the response of many evolutionary biologists "has been to ignore (if not disparage) these discussions" (Harrison 1998, 19). For many biologists, this "lack of interest reflects discomfort with a debate that appears to have no end and seems to have become increasingly muddled" (Howard 1998, 439). Such controversy "seems to suggest that there is no general agreement about what species are, and if this is the case, then the possibility of understanding how species come into being also seems unlikely" (de Queiroz 1998, 57).

The situation in paleontology has frequently seemed, if anything, even murkier (Allmon and Yacobucci, introduction, this volume). In their survey of 20 years of paleontological studies of speciation, for example, Erwin and Anstey (1995, 12) suggested that the diversity of neontological species discussions "leads to the seemingly inescapable conclusion that there is neither a unified definition of species nor a unified explanation of the mechanisms of speciation; nor does either seem likely in the near future." This lack of consensus—combined with the inherent incompleteness of the fossil record—seems to have discouraged most modern systematic paleontologists from paying much more than perfunctory attention to the details of what species are and how they should be recognized.

Despite this seemingly widespread ennui, however, there has recently in fact emerged something of a working consensus on a number of important aspects of our understanding of what species are and how they arise, both in living forms and in the fossil record (e.g., de Queiroz 1998, 2007; Miller 2001, 2006, chapter 2 of this volume; Harrison 2014). This often unrecognized common ground surely does not mean that there remains no unexplained complexity in nature around these issues. It does, however, appear to provide an opportunity for at least some standardization of methods for recognizing and discussing species, especially as they are perceived in fossils.

This chapter and the one following (Allmon and Sampson, chapter 4, this volume) attempt to lay out what appear to be the commonalities in our understanding of the nature of species and speciation, and to put these into an analytical framework that can be applied across the fossil record (at least of animals) in order to develop generalizations about the causes of speciation for evolutionary paleobiology. I have attempted a thorough survey of the literature, and included a large number of citations to it, because discussion of these topics is scattered widely and has not been elsewhere comprehensively reviewed. My treatment is both an exercise in methodological simplification and an endorsement of the idea that at least *some* substantive aspects of nature may actually be simpler than they might at first appear. As I try to make clear, I am under no illusions that all of the questions around species and speciation can be addressed, much less solved, by this discussion. My goal is, rather, to advocate for a more unified approach that can generate new data and conclusions useful for addressing some of the "big questions" that we want answered about the history and diversity of life. In both paleontology and neontology, there are significant signs that the biological phenomena of species and speciation are becoming more rather than less clear. In some important respects, however, evolutionary paleobiology cannot expect to be taken seriously by neontological evolutionary biology—at least on discussions of species—unless it pays explicit attention to these signs.

More particularly, this chapter has two main objectives. First, I argue for adoption by paleontologists of a single, "unified" species concept, of the sort that is becoming widely accepted among neontologists. I limit my discussion to biparental animals, and there is an emphasis on marine invertebrates (although I have tried to adequately represent vertebrates as well). I recommend that paleontologists formally embrace the general lineage concept (GLC) of species as the best description of both what species are and how they can be understood in the fossil record.

Second, I provide an overview of the current operational status of species in studies of fossil animals. Despite the widespread perception that recognition and delimitation of species in fossils are arbitrary, futile, and biologically meaningless activities, there is in fact something of a working consensus on how to carry out these basic and essential tasks. Yet this consensus is, at best, irregularly applied in the technical literature. I review this consensus and recommend that it be more formally recognized and adopted.

The species problem

Since the term was coined by paleontologist E. R. Trueman in 1924, the "species problem" has had various neontological characterizations (Mayr 1957; O'Hara 1993; Ruse 1995; Brooks and McLennan 1999; Isaac and Purvis 2004; de Queiroz 2005a; Richards 2010), and the "species problem" in paleontology has seemed even more complex (Allmon 2013). As mentioned above, discussion of such issues is seen by many evolutionary

TABLE 3.1 **Decomposition of the challenge of studying species in the fossil record into three component questions. Based on Allmon (2013).**

I. Species nature problem. What are species?
A. Are species real?
B. Are they one thing or more than one thing?
II. Species recognition problem. How can and should species be recognized and delimited?
A. How are living species delimited?
B. To what degree can "species" as recognized in living organisms be recognized in the fossil record?
III. Fossil species study problem. What can we learn about species (and speciation) from fossils?
A. To what degree can fossil species be studied as modern species are?
B. What can we learn about species (and speciation) from fossils that we would not otherwise know?

---

biologists and paleontologists as interminable and unproductive. Yet the frequently noted magnitude and apparent chaos of the literature on species concepts and definitions (e.g., Hull 1997; Mayden 1997, 1999; Harrison 1998, 2010, 2012; Brooks and McLennan 1999; Coyne and Orr 2004; Reydon 2004; Mallet 2007; Hausdorf 2011; Kunz 2012) easily obscures the two fundamental biological questions that lie at its core: What are species? And how can/should we recognize them? Adding fossils adds more questions to this twofold inquiry.

Seen in this way, the species problem in evolutionary paleontology actually consists of at least three separate but closely related issues (Allmon 2013): I. What are species in living organisms? (the "species nature problem"); II. How are living species recognized or delimited, and to what degree can "species" as recognized among living organisms be recognized in the fossil record? (the "species recognition problem"); and III. To the degree that species can be so recognized, to what degree can fossil species be studied as modern species are? That is, what can we learn about the origin and evolution of species from fossils that we could not otherwise learn? (the "fossil species study problem") (see table 3.1). This classification forms the framework for this and the following chapter.

The first question, what species are, is among the oldest questions in biology (Mayr 1957, 1982; Stamos 2003; Wilkins 2009; Allmon 2013). This question has at least two major components (table 3.1). These are, as Mayr (1957, 6; 1982, 285) suggested: (IA) Are species real things in nature, or only constructs of human thinking/perception? This is "the question of whether biological nature really is discontinuous. Do species exist as discrete, objective entities?" (Coyne and Orr 2004, 9). (IB) If species exist, what if any features do they all have in common, and are these commonalities sufficient to refer to all of them by a single term, concept, or definition?

Question I—the reality and nature of species—has been repeatedly confused with question II—species delimitation, and (as pointed out by many others, e.g., Eldredge and Cracraft, 1980, 94; Adams 2001; Ghiselin 2002; Coyne and Orr 2004; de Queiroz 1998, 2005a, 2007; Hey 2006; Wiens 2007; Kunz 2012) this conflation has produced much of the persistent and frustrating muddlement in the literature. The criteria by which living species are delimited or recognized (question II) constitutes its own valid and extensive topic (e.g., Wiens and Servedio 2000; Sites and Marshall 2003, 2004; Marshall et al. 2006; de Queiroz 2007; Knowles and Carstens 2007; Wiens 2007; Hausdorf and Hennig 2010; Ross et al. 2010; Lim et al. 2011; Sauer and Hausdorf 2012), certain aspects of which are paleontologically important. They are not, however, germane to the issue of what species *are*—what has come to be called the "ontology" of species (question I; e.g., Eldredge 1985a,c; Mayr 1987; Stamos 2003).

The Species Nature Problem: What Are Species in Living Animals?

Although they describe the literature on species concepts as "vast and stupefying," Coyne and Orr (2004, 6) admit that "it seems wise to decide what species are before considering how they arise" and that "deriving a species concept is important because it frames one's entire research program on the origin of species" (2004, 10). As Rosenberg (1985, 193) puts it, when we talk about species "we have got to know what we are talking about." Similarly, discussions of evolution must, writes Kimbel (1991, 368) "be preceded, both logically and operationally, by delineation of species, the irreducible units of evolutionary analysis."

Do species exist? Or are they merely human mental constructs? Some authors continue to maintain the latter view (e.g., Bachmann 1998; Hendry et al. 2000; Mallet 2005, 2007; Hart 2010; Mishler 1999, 2010). The majority of biologists, however, appear to regard the living world (at least the biparental animal part of it) as divided into naturally occurring discontinuous clusters of phenotypes and genotypes that are independent of human perception and maintain their separateness. These clusters have conventionally been called "species" (e.g., Mayr 1963; Wiley 1978, 1981, 2002; Ghiselin 1997; Adams 2001; Stamos 2003, 17; Coyne and Orr 2004, 10ff.;

de Queiroz 2007; LaPorte 2007; Wiley and Lieberman 2011; Kunz 2012; Marie Curie Speciation Network 2012, 27; Harrison 2014). For example:

> The living world is quite obviously not a continuum. Instead, it is somehow "packaged" into discrete groups: groups that we clearly have to define before we can begin to consider their histories and interactions. This is not simply an artifact of human perception. It is self-evident fact, and if we cannot agree on this, we have no basis whatever for any rational discussion of the hierarchically organized living world around us. (Tattersall and Mowbray 2005, 371, 372)

> Species . . . are self-delimiting and exist independent of our observations. They are not merely groups of organisms lumped or split to suit our fancy . . . Species are spatiotemporally limited and internally cohesive. They maintain their independence from other entities over time and space. . . . A species has a unique origin and a unique historical fate. The same species does not arise multiple times, nor does it share its fate with any other entities . . ." (Adams 2001, 156)

If species really exist, what are they, and are they the same for all taxa (table 3.1)? Consensus appears to be less robust on these questions. Controversy persists about which among the bewildering array of formally proposed "species concepts" (at least 26 by some counts: Mayden 1997; Wilkins 2006) is preferable and under what circumstances. It is also not clear whether "species" in all groups (again, mainly animals) are the same, or whether a single species concept or definition can be applied across higher taxa. For example, is a "species" of bivalve the "same" as a "species" of mammal? Are all bivalve or mammal species "the same"? That is, can they be delimited in the same way, and/or are they equivalent in a functional or ontological way?

One persistently influential school of thought, sometimes known as "species pluralism," argues that the only single generalization possible about species is that there isn't one (e.g., Mishler and Donoghue 1982; Ereshefsky 1992, 1998; Dupré 1993, 1999; Rosenberg 1994; Stanford 1995; Sluys and Hazevoet 1999; Hendry et al. 2000; Brigandt 2003; Mallet 2007; Gourbière and Mallet 2009; Mishler 2010). As Coyne and Orr (2004, 26) put it, "Evolutionists now appreciate that no single species concept can encompass sexual taxa, asexual taxa, and taxa having mixed modes of reproduction." They quote Kitcher (1984, 309), who writes that "there is no unique relation which is privileged in that the species taxa it generates will answer the needs of all biologists and will be applicable to all groups of organisms."

The alternative view—which has been called the "monist" position (for this terminology, see Hull 1997; Dupré 1999; Ereshefsky 2010)—maintains that, despite clear differences among taxa, "species" have more in common than not; that is, there can and should be a single species concept—a "unified" (de Queiroz 2005c), "ultimate" (Miller 2001, 2006, chapter 2 of this volume), "unit" (Wheeler 2007), or "universal" (Pauers 2010) concept. Just such a concept has, in fact, emerged over the past 15 years or so, in the idea that "all modern species definitions describe variants of a single general concept of species"—the general lineage concept (GLC) (de Queiroz 1998, 57, 60; 1999, 2005a,b,c, 2007).

The general lineage concept of species

In proposing the GLC, de Queiroz (1998) argued that "despite the diversity of alternative species definitions, there really is only one general species concept in modern systematic and evolutionary biology—species are segments of population level evolutionary lineages" (1998, 63). Most, if not all, of the multitude of species-related phenomena emphasized by various other species concepts are, de Queiroz suggested, related not to what species are, but to the criteria by which we recognize them at various stages in the process of speciation (see also Harrison 2014; Allmon and Sampson, chapter 4, this volume). The GLC, according to de Queiroz, "is not a new species concept but simply the clear separation of the theoretical concept of species (as separately evolving metapopulation lineages) from operational criteria (lines of evidence) that are used for its empirical application" (2007, 883).

The GLC is in many ways similar to the "evolutionary species," first articulated by Simpson (1951), and later elaborated by him and others (Simpson 1961; Wiley 1978, 1981; Eldredge 1985c, 1989; Frost and Kluge 1994; Brooks and McLennan 1999; Wiley and Mayden 2000a,b,c; Wiens 2004a,b; de Queiroz 2007, 883; see also Naomi 2010). Simpson's original definition—"a phyletic lineage (ancestral-descendent sequence of interbreeding populations) evolving independently of others, with its own separate and unitary evolutionary role and tendencies" (1951, 289–290)—emphasized the status of species as "evolutionary units" which have "been evolving separately, or which will do so, or, as a rule, both." Genetic difference was both a result and evidence of this status, and reproductive isolation, in Simpson's view, "makes the prediction of separate evolutionary roles certain." Wiley modified Simpson's language by "not implying that species must change (evolve)", thereby accommodating species in stasis (Wiley also christened it the "evolutionary species concept," or ESC):

> A species is a single lineage of ancestral descendant populations of organisms which maintains its identity from other such lineages and which has its own evolutionary tendencies and historical fate. (Wiley 1978, 18)

The GLC is a generalization and "reformulation of the evolutionary species concept"; both are "lineage-based concepts" (Hausdorf 2011, 924) and "recognize historically distinct evolutionary lineages that are likely to remain distinct" (Rissler and Apodaca 2007, 925). Wiley and Lieberman (2011, 34) argue that the GLC is simply a new name for the ESC.

In addition to the ESC, the GLC includes important elements of other species concepts (see also Miller, chapter 2, this volume), such as the phylogenetic species concept (PSC; Eldredge and Cracraft 1980; Cracraft 1983, 1987, 1989), which defines a species as "an irreducible (basal) cluster of organisms, diagnosably distinct from other such clusters, and within which there is a parental pattern of ancestry and descent" (Cracraft 1989, 34– 35). In common with the PSC, the GLC "acknowledges the biological discreteness of species and partitions natural variation in a way that equates taxonomic species with evolutionary (phylogenetic) units." It recognizes species as "spatiotemporally discrete entities that constitute particular segments of the 'genealogical nexus' (Ghiselin 1974; Hull 1976)" and "delineates taxa in the fossil record that are perceived to be units of evolution. It thus creates equivalence between the taxonomic species and the evolutionary species, and establishes a framework of consensus on phylogenetic pattern" (Kimbel 1991, 361, 366–368).

As noted by Bruner (2004, 98), "the ESC and the PSC are . . . particular formulations of a more general individualistic thesis, for which species are approached as 'individuals', with a personal ontogenetic pathway which involves birth, growth, ageing, death and, eventually, genealogy." Bruner acknowledges that this idea may be impractical for recognizing species in paleontology, but he says "it represents one of the best conceptual and romantic [sic] approaches to the matter, telling exactly what a species is or, better, the way we should have to look at it." In seeing species as "lineage segments" (de Queiroz 1998), the GLC is also consistent with the internodal species concept (Kornet 1993; Baum 1998; Polly 1997).

Despite skepticism from some neontologists (e.g., Mallet 2005, 2007; Hausdorf 2011) and paleontologists (e.g., Smith 1994; Tattersall and Mowbray 2005), the idea that something like the GLC/ESC comes close to a universal species concept has become increasingly accepted (e.g., Adams 2001; Hey 2001, 2006; Miller 2001, 2006, chapter 2 of this volume; Wiens 2004a,b, 2007; Pauers 2010; Reeves and Richards 2011; Harrison 2014). As Wiens (2007, 875) puts it, "There has been real progress made in thinking about species concepts, which now makes some general agreement seem possible." There seems to be general consensus, he continues, "that species are lineages. . . ." Adams (2001, 156) phrases the same idea more colorfully (expanding on the description of Baum 1998, 644), calling species "space-time worms." Speciation, argues Wiens, is "the origin of new lineages . . . specifically, the largest lineages that are connected by gene flow" (2004a, 915). Species are thus "independent lineages, having unique internal characteristics and histories," and formed via a "transition from processes producing patterns of tokogenesis [see below] among populations to those producing phylogenesis (definitive division) within clades" (Miller 2006, 557–558).

The GLC is based on the notion that species "usually consist of multiple populations geographically and temporally separated from one another; that these subdivisions bud off from other populations, divide, fuse or go extinct; and therefore the structure of species is like the bundle of fibers making up a hemp rope" (Miller 2001, 2). This image—of species as a "braided stream," in which "large channels represent lineages; the smaller channels represent gene flow between them" (Holliday 2003, 655)—has frequently been depicted with a particular style of diagram (fig. 3.1) that suggests both the internal complexity and coherence of species.

Within such a plexus, a number of processes may keep the populations from diverging from one another. The most frequently cited is gene flow, or *tokogenesis*, a term coined by Hennig (1966, 20) and applied subsequently by, e.g., Eldredge (1993); Holliday (2003, 653); Anstey and Pachut (2004); Bruner (2004); Fitzhugh (2005, 2009); Schwartz (2009); Wiley and Lieberman (2011, 29–34); and Mayden (2013). At least for sexual organisms, tokogenesis means that "organism lineages continually anastomose as a result of sexual reproduction to create a higher level lineage whose component organism lineages are unified by that very process" (de Queiroz 1998, 60). Speciation in this view is a disruption of tokogenetic relationships. This conception has much in common with other ideas that species are held together by various factors, and speciation is the result of the breakdown of these factors (e.g., the cohesion species concept of Templeton 1989; and the specific-mate recognition system, or SMRS, of Paterson 1982, 1985).

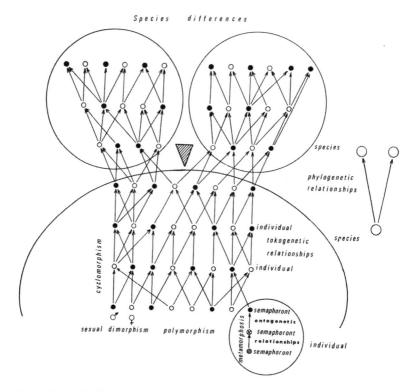

Figure 3.1 "Braided" or "plexus" diagrams showing relationships of tokogenesis (interbreeding) within species, and their breakdown during speciation.
A. From Hennig (1966). B. From Baum and Shaw (1995). C. From Adams (2001). D. Modified from Miller (2001, 4). i. A species "lineage consisting of a bundle of demes that oscillate through the phenotypic/habitat space of an established species but does not fill up the space." ii, iv. Species lineages "in which the demes explore most of the phenotypic/habitat possibilities of a species or define those possibilities through time." iii. A "single, well-mixed population filling essentially the entire phenotypic/habitat space of an established species." "Small crosses are demic or habitat cluster extinctions. Speciation associated with pronounced phenotypic differentiation . . . is indicated with sm; speciation without this kind of differentiation is indicated with s." E. From Fitzhugh (2005). F. From Baum (2009).

Species might also be held together by factors other than, or in addition to, gene flow, such as stabilizing selection or selective extinction of divergent populations (Ehrlich and Raven 1969; Wiley 1981; Beurton 1995; Wood and Mayden 2002; Eldredge et al. 2005; Barker 2007). For example, genetic exchange can still occur between more or less independent lineages without the remerging of these lineages (Nosil 2008; Feder et al. 2012). The unity of a species might also be a consequence of lack of survival of any potentially separate branches (M. Bell, personal communication,

January 2014). A species lineage might persist in one environment, occasionally giving rise to divergent and rapidly evolving isolates in adjacent different environments, but these isolates do not persist for long, largely because the habitat on which they depend is ephemeral (see also discussion of "isolate persistence" in Allmon and Sampson, chapter 4, this volume). Williams (1992) called this a "phylogenetic raceme" (see also Bell and Foster 1994; Perdices et al. 2000). In such cases, something very like

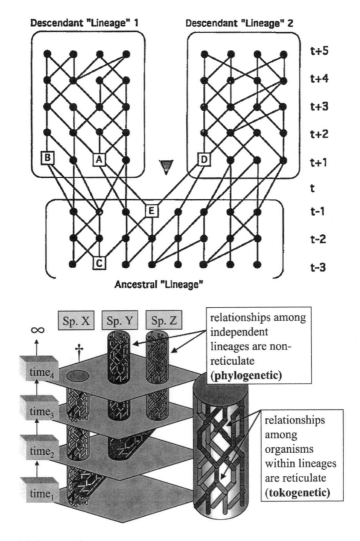

Figure 3.1 (*continued*)

the GLC could still apply, but the mechanisms for it would go beyond gene flow.

In this conception, the GLC is potentially "universal" in at least four respects: (1) it purports to apply to *all* species (or at least all sexually reproducing ones) (de Queiroz 2007); (2) it focuses on *what* species are (cf., Eldredge 1993); (3) it offers one (perhaps not the only; see below) explanation for *why* species are what they are; and (4) it is applicable to fossils, as discussed below. According to the GLC, whatever else they are

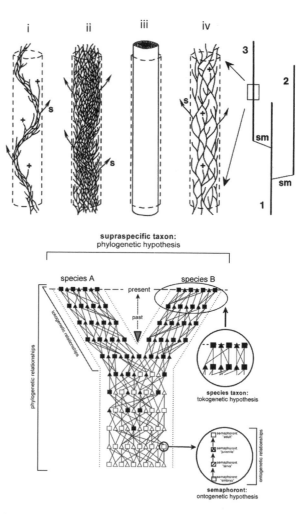

Figure 3.1 (*continued*)

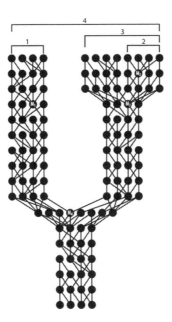

Figure 3.1 (*continued*)

and however they arise, species are separate branches of the tree of life characterized by a range of more or less discrete phenotypes. If they did not exist as such, life would be less of a diversity of recognizable entities and more a variably lumpy smear of variation. Species are the means by which life not only separates itself into kinds, but also how life maintains the extraordinary range of phenotypic diversity by which it survives and reproduces (Futuyma 1987, 1989). The GLC is in the monist tradition, but it does not require that all species or instances of speciation are the same. It says that all species are exemplars of the same basic phenomenon—separate evolutionary lineages—but these lineages can vary enormously in a variety of ways. This variety, however, "attaches to mechanisms, pathways and products, not to the topology of independent lineages and to their position in the genealogic hierarchy" (Miller 2006, 566).

The Fossil Species Recognition Problem

"Species" have been recognized and named based on fossils since before the time of Linnaeus, but views have always been diverse about what such designations really mean. In practice, paleontologists have taken two different approaches to this question (Cain 1954, 107ff.; Sylvester-Bradley 1956b; Allmon 2013). The first—driven mainly by the close connection between invertebrate paleontology and biostratigraphy—has been to regard "species" of fossils as purely utilitarian tools for geological correlation: a fossil species is any morphotype circumscribed so as to make it more recognizable and therefore biostratigraphy easier. Arkell and Moy-Thomas (1940, 395), for example, stated that a species is no more than "a practical and convenient unit by which fossils are distinguished." Such definitions imply a major role for subjective personal opinions, and some paleontologists (mostly in an older tradition) made this subjectivity an explicit part of their notion of species. Gilmore (1940, 469) wrote that a species is "a group of individuals which in the sum total of their attributes resemble each other to a degree usually accepted as specific, the exact degree being ultimately determined by the more or less arbitrary judgment of taxonomists." Davies (1961, 263) similarly said that

> a species is a collection of individuals so nearly alike that they may conveniently be denoted by the same name. This definition leaves the decision to the judgment of every palaeontologist, as to whether there is sufficient likeness, and judgments will inevitably differ: as such differences constantly occur, this definition recognizes the facts of the case.

In their influential textbook, Raup and Stanley (1978) recalled this tradition by paraphrasing Regan (1926):

> The fact that species discrimination depends largely on the experience of the person making the discrimination has led to an informal definition of the species that is invoked with surprising frequency: "A species is a species if a competent specialist says it is." (Raup and Stanley 1978, 108; see Trewavas 1973 and Allmon 2013 for further discussion of Regan's original text)

The prioritization of stratigraphic utility also led in many, if not most, cases to application of narrow species definitions and thus highly "split" taxonomies (e.g., Shaw and Lespérance 1994; Hughes and Labandeira 1995; Nardin et al. 2005).

With the coming of the Modern Synthesis in the 1940s, many (although certainly not all) paleontologists began to adjust their species concepts and definitions to reflect the contemporary neontological focus on populational thinking (e.g., Newell 1947; see Sepkoski 2012, chapter 1 of this volume).

In another influential textbook, for example, Moore et al. (1953) defined species as "assemblages of individuals having identity or near identity of form and anatomical features, except for sex differences, and measurable distinctness from other assemblages" (1953, 9), and reprinted Newell's 1947 figure showing a species as a variable population. Much paleontological attention then turned to "the species problem" in fossils (sensu Trueman 1924)—how to divide up a continuously evolving lineage into separate nameable pieces—referred to variously as paleospecies, successional species, or chronospecies[1] (e.g., Sylvester-Bradley 1956a; Imbrie 1957; Sepkoski, chapter 1, this volume). For examples in invertebrates, see Stenzel (1949) and Rodda and Fisher (1964); for vertebrates see Gingerich (1976, 1985); Rose and Bown (1986); and Jordana and Köhler (2011). The answer to this conundrum was (and remains) that such a decision is completely arbitrary (e.g., Cain 1954; Thomas 1956; Simpson 1961, 165; Mayr 1963, 24; Mayr and Ashlock 1991, 106).

In practice, however, it was widely suggested that this anagenetic continuity was seldom a serious problem because of the incompleteness of the stratigraphic record; indeed, evolutionists frequently expressed what seemed like grudging gratitude for gaps in the record, because they served as convenient boundaries for naming segments of continuous lineages (e.g., Mayr 1942, 153–154; Simpson 1951, 291; Cain 1954, 107; Clark 1956; George 1956, 133; Newell 1956; see Eldredge and Cracraft 1980, 116). This ironic, even perverse, celebration of the incompleteness of the fossil record at least partly justified its characterization of Eldredge and Gould (1972, 92) as "a theoretical debate unsurpassed in the annals of paleontology for its ponderous emptiness. . . ." This critique, of course, was also based on the contention that anagenesis is rare to nonexistent in the fossil record (Eldredge 1993; Gould 2002, 775–776, 784). When anagenesis is detected, however, in either biostratigraphic or paleobiologic studies, such discussions cannot be avoided (e.g., Newell 1947; Maglio 1971; Sykes and Callomon 1979; Hallam 1982; Bell et al. 1985; Rose and Bown 1986; Allmon 1990; Geary 1995; Levinton 2001, 314–316; Dzik 2005; Waller 2011, 16). The anagenesis vs. cladogenesis dichotomy leads to the problem of the relative frequency of different tempos and modes of evolution in the fossil record (sensu Gould

1. The term "chronospecies" was apparently first used by George (1956, 129). For more recent usage, see, e.g., Stanley (1979), Wheeler and Meier (2000), Dzik (2005), and White (2013). The term "successional species" was apparently first used by Imbrie (1957), and then also by, among others, Simpson (1961, 167) and Rose and Bown (1986).

1991), which, despite decades of debate and study, remains poorly under-stood (Barnosky 1987; Erwin and Anstey 1995; Jackson and Cheetham 1999; Benton and Pearson 2001; Geary 2009; Hunt 2008, 2010, 2013).

Morphospecies and reproductive isolation

All species recognized solely from fossil material ("paleospecies")[2] are "morphospecies," defined by Cain (1954, 121) as "species which have been established solely on morphological evidence" (see also George 1956; Cronquist 1978). (This is effectively equivalent to the "phenetic species concept" of Sokal and Crovello [1970]; see Smith 1994, 10ff.) This defini-tion was intended to distinguish such forms from species defined by re-production via the biological species concept; such species are sometimes termed "biospecies" (e.g., Cain 1954, 121; Gould 2002, 785) or "good spe-cies," meaning species which are both reproductively isolated and differ in morphology (e.g., Cain 1954, 49, 73, 110; Coyne and Orr 2004, 60, 75; Mal-let 2005, 109; Descimon and Mallet 2009; see further discussion below). Reproductive isolation is presumed to be in principle demonstrable among biospecies, with or without morphological differences (fig. 3.2).

Although the prevailing custom in modern paleobiology is to assume (frequently without explicitly acknowledging it) that fossil morphospe-cies are similar enough to living biospecies for most evolutionary purposes (Allmon and Yacobucci, introduction, this volume), a number of authors have over the years explicitly declared that it is in principle misleading to try to equate species based on fossils with those based on living organisms. For example, although he is a staunch advocate of species-level paleonto-logical studies, Stanley (1979, 8) writes that "there is no question that fos-sil entities recognized as species are not strictly comparable to species in the living world." Other paleontologists have suggested that at least some, and perhaps many, fossil morphospecies are more morphologically compa-rable to modern species groups or genera than to individual living species

2. The term "paleospecies" was apparently first used by Wilmott (1949) to refer to seg-ments of continuously evolving lineages, a usage followed by Cain (1954, 107) and Simpson (1961, 155). This is essentially the same meaning as the term "chronospecies," coined by George (1956) (see note 1). See Bonde (1981) for additional synonyms. More recently, "pa-leospecies" has been used to refer to any species based solely on fossils (e.g., Neige et al. 1997; Gould 2002, 784ff.; Barnosky et al. 2005; Marco 2007).

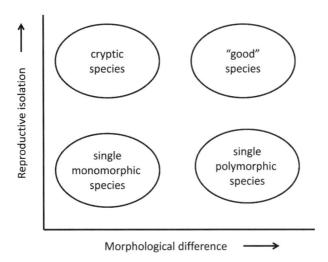

Figure 3.2 The relationship between morphological differentiation and reproductive isolation in sister species or populations.
Modified from Vrba (1980, 68). On "good species" see Shaw (1996), Mallet (1996), and Descimon and Mallet (2009).

(McCune 1987, 3; Chaline 1990, 7; see also Fox 1986, 81; Levinton 2001; Forey et al. 2004; Hills et al. 2012).

In recognition of these potential differences between Recent and fossil species, some authors have suggested various ways to formally distinguish species based on fossils from those based on living material. It has, for example, been suggested that fossil organisms be classified separately from living taxa (Crowson 1970), or that taxa based on fossils be explicitly labeled (Collinson 1986, cited in Forey et al. 2004, 650). It is widespread practice among vertebrate systematists to mark extinct taxa with a dagger (†) (e.g., Patterson and Rosen 1977; Lauder and Liem 1983; Grande and Bemis 1998; Bemis, chapter 11, this volume).

Other designations for fossil species include "stem species" (an extinct taxon that is more closely related to its crown group than to any other group, but may lack some of the defining apomorphies of the crown group), which is increasingly widespread (e.g., Ax 1985; Donoghue 2005; Wiley and Lieberman 2011, 83, 242). The term "metaspecies" was coined by Donoghue (1985) to refer to a diagnosable assemblage of organisms that lacks a unique apomorphy; for example, an ancestral or stem species of a daughter species that retains its defining apomorphy (see Anstey and Pachut 2004, 652–653; Pachut and Anstey 2009 for its paleontological application). Donoghue (1985) suggested that metaspecies be distinguished in print with an asterisk (*).

Morphospecies, however, are by no means restricted to fossils. Many modern species in many higher taxa (e.g., mollusks, crustaceans, insects, fishes, snakes, frogs) have been, and continue to be, described solely on the basis of dead, preserved material (e.g., Claridge et al. 1997; Clarkson 1998, 8; Benton and Pearson 2001; Forey et al. 2004, 643; Wiley and Lieberman 2011, 55–56), although this practice is slowly changing as new genetic techniques become more widely available. Indeed, as paleontologists frequently point out (e.g., Newell 1956, 70–71; Allmon 1990, 8; Kimbel 1991, 361–362; Gould 2002, 785; Forey et al. 2004, 643; Saupe and Selden 2011, 184–185), most living species in most major groups of invertebrates with fossilizable hard parts are described on the basis of skeletal morphology, not reproduction, soft anatomy, or genetics. (The situation appears to be different for vertebrates, in which nonskeletal characters are also commonly used for species description and discrimination.) Indeed, very few rigorous studies of reproductive isolation in extant species have even been done, so while one may *infer* that an extant species is a "good" biospecies, rarely is this actually tested (Benton and Pearson 2001; Sangster 2014).

Whether or not nonskeletal characters are used in defining or describing modern species, skeletal characters may at least sometimes be highly reliable in identifying them. In modern echinoderms, for example, "studies of recent species confirm the overall usefulness of the skeleton for differentiating species as well as higher taxa" (Guensburg 1984, 19–20). The same applies to teeth in distinguishing species of primates (Gingerich 1974a) and rodents (Escudé et al. 2008; Vianey-Liaud et al. 2011) (but this may not always apply to the teeth of other mammals: e.g., Roth 1992; Gould 2001; Dayan et al. 2002, 523; Tattersall 2007, 140).

Many, perhaps most, paleospecies are therefore no more poorly characterized than many, perhaps most, modern species. (Ultimately, of course, this is not a very useful argument for their validity: it is not a good defense of paleontology to point out "that ordinary practice with fossils follows the worst habits (majoritarian though they may be) of neontological taxonomy" [Gould 2002, 785].) What are the implications of this state of affairs?

What is known about the process of speciation based on living species clearly indicates that there is no necessary connection between morphological difference and reproductive isolation (Vrba 1980; Eldredge 1989, 1993; Coyne and Orr 2004). In some taxa, there may be a priori reasons for expecting potentially fossilizable morphology to accurately reflect reproductive isolation, particularly in groups in which vision and/or sexual selection play major roles. Perhaps the most well-known example is horns in African bovid mammals, as explicated by Vrba (1980). Primates may be another instance (Albrecht and Miller 1993). Lack of correlation between reproductive isolation or genetic difference and morphological difference can take two forms (fig. 3.2): populations may be reproductively isolated but morphologically similar (sibling or cryptic species), or morphologically differentiated but able to reproduce (intraspecific polymorphism). Each of these can pose major challenges for paleontological recognition of species.

Mayr (1942, 151) coined the term "sibling species" for reproductively isolated populations that were morphologically distinguished by no or very few morphological differences (see also Mayr 1963, 31ff.). The term "cryptic species" has become a more accepted alternate because it does not imply that the species in question are each other's close relatives (Henry 1985; Knowlton 1986). When they mention cryptic species at all, paleontologists usually do so only to acknowledge that they can say little about them (e.g., Vrba 1980, 68; Chaline 1990, 7; Gould 2002, 785ff.; Hunt 2013, 720; Prothero 2013, 51). In living animals, cryptic species have been reported from almost all higher taxa (e.g., Mayr 1963, 37ff.), and the rate of discovery has accelerated dramatically with the increasing use of genetic sequencing (Knowlton 1986, 1993, 2000; Knowlton et al. 1992; Knowlton and Weigt 1997; Sáez and Lozano 2005; Stuart et al. 2006; Beheregaray and Caccone 2007; Bickford et al. 2007; Pfenninger and Schwenk 2007; Castelin et al. 2010; Perez-Ponce de Leon and Nadler 2010; Lohman et al. 2010; Allmon and Smith 2011; Budd et al. 2012; Tepper et al. 2012; Brandão 2013; Demos et al. 2014; Highton 2014). Given these results, two major questions of paleontological importance remain: what proportion of all speciation events produce cryptic species? and does this proportion vary in any regular way among higher taxa?

Some authors have argued that cryptic speciation is, in principle, highly likely, perhaps even more likely than speciation that produces "good species" (i.e., those that are both reproductively isolated and differ in morphology), and that this is a fundamental obstacle to studying species and speciation in fossils. Schopf (1982), for example, suggested that cryptic species might comprise 10–20 % of all species. Levinton (1988, 182) went even further, positing that "the vast majority of speciation events probably beget no significant [morphological] change" (see also Levinton and Simon 1980; Hoffman and Reif 1990; Levinton 2001, 312ff.). Neither of these

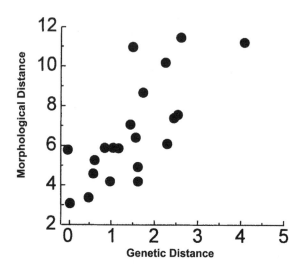

Figure 3.3 Comparison of genetic and morphological distances between populations of the Recent bryozoan *Stylopoma*. Modified from Jackson and Cheetham (1994).

suggestions, however, seems to have been based on abundant empirical evidence.

The traditional paleontological defense of the morphological species concept is that, when careful studies of genetic versus morphological differences between living species have been done, the results suggest that morphospecies *are* frequently correlated closely to biospecies (for the general argument, see, e.g., Eldredge 1989, 108; Gould 2002, 785; for specific examples, see, e.g., Michaux 1987, 1989a,b; Dorit 1990; Jackson and Cheetham 1990, 1994; Knowlton et al. 1992; Budd et al. 1994; Chiba 2007; Pilbrow 2010; Hills et al. 2012; see also Coyne and Orr 2004, 45–46) (fig. 3.3).

The apparently high frequency of cryptic species nevertheless leaves unclear what either the frequency or relative proportion of cryptic species among different higher taxa actually is. There are at least three potential situations, each with a different implication for the application of morphospecies in paleontology:

(1) The proportion of cryptic species may be approximately equal

across higher taxa (e.g., Pfenninger and Schwenk 2007). If this is true, then, at least for macroevolutionary comparisons among taxa, cryptic species could be essentially ignored (although we still would not know what the actual absolute frequency is in a particular taxon).

(2) The proportion of cryptic species may differ—in some regular and knowable pattern—among higher taxa (e.g., Mayr 1963, 37; Bickford et al. 2007; Trontelj and Fišer 2009; Poulin 2010). If this is true, then it might in principle be possible to assess the relative frequency in different taxa, and then apply the results to their respective fossil representatives using uniformitarian reasoning.

(3) The proportion of cryptic species might vary among taxa or over space or time in such a way that it is difficult or impossible to measure and/or predict.

Data currently available are not sufficient to determine which of these situations applies. It is clear (from the literature cited above and more), however, that cryptic species are numerous and occur widely among many living higher taxa of paleontological significance. This implies that estimates of species numbers and speciation rates based on fossils (e.g., Sepkoski 1998; Lieberman 2001a,b) are *minimum* estimates. As Tattersall and Mowbray (2005, 377–378) put it:

> Where aggregations of comparable fossil specimens can be consistently distinguished from one another on the basis of morphology (that is to say, if consistent morphs are recognizable), we may have reasonable confidence that at least two species are present. The error, if any, will be on the conservative side: fewer species will tend to be recognized than are actually represented . . . while such systematic error will inevitably produce an oversimplified picture, that picture will not be actually distorted in its essentials . . . the bias toward simplicity involved in equating readily recognizable morphs with species is an acceptable one, and preferable to all other possible alternatives when dealing with the inevitably limited data of the fossil record.

It is possible that more careful assessment of morphological differences among fossils may reveal examples of very subtle patterns of morphological variation, yielding paleontological recognition of "pseudo-cryptic" species (sensu Knowlton 1993), at least in certain taxa. Several studies have reported such patterns (e.g., Gili and Martinell 2000; Herbert and Portell 2004; Adrain and Westrop 2005, 2006; Westrop and Adrain 2007; Budd 2010; see also Shaw and Lespérance 1994, 820).

Intraspecific variation or polymorphism can also be a problem for species recognition, in a direction opposite from that of cryptic species. The polytypic species concept—which allowed for sometimes wide phenotypic variability within reproductively connected groups—was a major component of the population-based approach to recognizing species in the Modern Synthesis (Mayr 1942, 1963). A number of instances of species-level morphological differences within genetically undifferentiated groups have been documented in diverse taxa (e.g., the famous case of the Cuatro Cienegas cichlid fishes [Sage and Selander 1975; Kornfield et al. 1982]; see also Wilke and Falniowski [2001] for an example from gastropods). As with cryptic species, paleontologists are clearly at a disadvantage here. Difficulties with recognizing intraspecific polymorphism in fossils mean that paleontological species counts may be overestimates, while difficulties with recognizing cryptic species in fossils mean that counts may be underestimates. It is possible that these two factors will cancel each other out, but there does not appear to be any rigorous quantitative way to test this suggestion.

In sum, the relationship between morphological and reproductive difference—at least in animals—is a two-dimensional spectrum, with four recognizable end-members (fig. 3.2). Most paleontological researchers assume, conclude, or hope that most species in their group of interest fall into the category of "good species" (upper right oval in fig. 3.2). It is clear, however, that this is not always true in living taxa. Both cryptic species (upper left oval in fig. 3.2) and polymorphic species (lower right oval) are present in many, perhaps most, higher taxa, but we do not know their relative frequency, as a proportion of all species or among taxa.

Does this mean that paleontological analysis of species is impossible or completely incommensurable with neontological analysis? No. But it does mean that for those taxa containing cryptic species (which may be most or all of them), the relationship of morphological and reproductive difference with the phenomenon of "separate and unitary evolutionary role and tendencies" described by the ESC/GLC is somewhat different from that shown in fig 3.2. As shown in fig. 3.4, a morphospecies that includes multiple cryptic species contains its own cladogenetic structure (and therefore violates one of the main criteria of the ESC: Wiley [1978]; Anstey and Pachut [2004, 652]). Yet such a morphospecies still maintains its own "separate and unitary evolutionary role and tendency," and so at least with respect to morphology is consistent with an ESC/GLC conception of species.

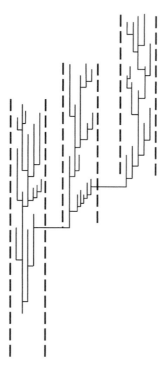

Figure 3.4 Diagram of a clade of morphospecies (each outlined with vertical dashed lines), each of which includes numerous cryptic species (vertical solid lines).

Species in paleontological practice

Most contemporary paleontological papers that discuss, describe, and analyze "species" of fossil organisms do not explicitly address how exactly the authors recognize or delimit their species. For example, examination of a sample of early twenty-first-century literature shows that of 109 papers describing new species, only 8 included explicit discussion of what the authors thought a "species" is, other than a list of distinguishing characteristics or autapomorphies.[3] Some modern paleontological authors do,

3. I examined all papers published in *Palaeontology* in 2008, *Journal of Paleontology* in 2012 and 2013, and *Journal of Vertebrate Paleontology* in 2012 and 2013. Relevant papers included only those naming new species of animals, excluding new species described in new genera. None of the papers describing new species of vertebrates included explicit discussion of species concepts; only three included explicit comparison of modern ranges of variation (Brochu and Storrs 2012; Lambert and De Muizon 2013; Bates et al. 2014).

TABLE 3.2 **Examples of explicit species concepts from recent systematic paleontological papers.**

Higher taxon	Operational species definition	Reference
Insects	The hypothesis that a new specimen belongs to a known species must be rejected for the new specimen to be assigned to a new species. . . . Rejection requires a sample demonstrating a bi- or multi-modal distribution of some trait, or data on closely related species demonstrating that the observed differences exceed the range of known intra-specific variability.	Gu et al. (2011:311)
Trilobites	A group of specimens that are more closely related mor-phologically to each other than to any other specimens with consistent and persistent diagnostic traits distinguishable by "ordinary means" . . . delimit the boundaries of extinct species.	Crônier et al. (2011:152)
Trilobites	The species concept employed in this study is "the smallest aggregation of populations . . . diagnosable by a unique combination of character states in comparable individuals" (Nixon and Wheeler, 1990:218). Such phylogenetic species cannot be further subdivided even if they show considerable continuous variation within the lineages (Wiley, 1978).	Hughes (1994:6)
Trilobites	We take species to be consistently distinguishable (usually morphologically) biologic entities. Paleontology should try to identify such entities, even though much of the stratigraphic, geographic, and biologic record is absent.	Shaw and Lespérance (1994:813)
Trilobites	We define a species as the smallest aggregation of comparable individuals diagnosable by a unique combination of characters.	Hopkins and Webster (2009: 524; see also Webster, 2009, 2011)
Bivalves	The specific-level taxonomic approach used here is essen-tially a quasi-population method in which co-varying individuals from the same bedding plane are considered to belong to a single population. In order to recognize and define species, an emphasis has been placed on the use of discrete character traits that are unique and, in species for which there are sufficient numbers of individuals, to establish quantitatively morphologic variation within the fossil sample.	McRoberts (2011: 622)
Osteichthyes	Taking into account that it is an operational necessity to use a morphological species concept . . . and recognizing that certain characters will always be lacking, every attempt has been made to interpret variation in these fossils as one would if they were living fishes . . . I view these species descriptions as hypotheses of biological species as inferred from the available morphology.	McCune (1987: 3)

however, lay out the conception of species they use. A selection is pre-sented in table 3.2.

Lack of explicit species concepts is not unique to paleontology. Most neontological species have long been, and continue to be, described without reference to their authors' operational species concepts (Sangster 2014).

Most neobotanical monographs, for example, are neither explicit nor consistent about their species concepts (Luckow 1985; McDade 1995). Neither neontological nor paleontological journals typically require authors to do otherwise. From 1981 to 2007, however, the journals *Bulletins of American Paleontology* and *Palaeontographica Americana*, published by the Paleontological Research Institution, which include mostly longer systematic papers (and which I edited from 1992 to 2007), required or strongly encouraged its authors to include in their manuscripts a section called "Introduction to Systematic Paleontology," which was to include discussion of "philosophical considerations." Here, authors were asked to answer a number of questions, including: "What is a fossil species of the group you are describing? How does this differ from an equivalent-level taxonomic entity of a related living organism, whatever that may be? What is your personal taxonomic philosophy?" (*Instructions to Authors*, PRI Publications, 1999). A sampling of the responses of authors to this request is presented in table 3.3.

Although the statements of "species concepts" given in tables 3.2 and 3.3 vary (e.g., some make more or less explicit reference to trying to be equivalent to biospecies), most are actually representative of a rough working consensus about the proper approach to fossil species that has existed in paleontology for many decades (e.g., Simpson 1951, 291; Raup and Stanley 1978, 102; Tasch 1980, 7–8; Nadachowski 1993; Miller 2001, 2006; Foote and Miller 2006, 79–80; McGowran 2008, 372ff; Benton and Harper 2009, 121).

This consensus approach consists of at least two basic components: (1) examination of the closest living relative(s) of the fossil taxon of interest to determine typical ranges of variation within versus among species; and (2) application of this information via uniformitarian reasoning to the fossils by testing for morphological clusters of fossil specimens separated by gaps.

In order to establish a valid species it should be necessary to show characters in the available fossil material which purport to be of the same magnitude as those which separate related living species. (Simons and Pilbeam 1965; quoted in Tattersall 1986, 165)

The usual "rule of thumb" is to include within a species a range of morphological variation comparable to that in the same features in related living species, and to distinguish between species by gaps similar in dimension to those separating their living relatives; if none are known, comparison can be made to gaps between species in living analogues. (Fox 1986, 79)

TABLE 3.3 **Explicit species concepts of authors publishing in** *Bulletins of American Paleontology*, **1981–2007. (See also Jung 1986, 9.)**

Taxon	Age	Species concept	Reference
brachiopods	Permian	Species names are simply handles for convenient discussion of time segments of evolving lineages.	Hoover (1981,38)
brachiopods	Devonian	The species names [used here] are merely conveniences for discussing clusters of morphologic variability within what are, for the most part, well-established generic concepts.	Cooper and Dutro (1982, 30)
corals (Rugosa)	Ordovician	The species recognized herein are considered valid because they are separated from all others by morphologic "gaps" . . . Assignment of species to these genera serves to indicate particular gross morphologic features, although it is uncertain if they were determined by environment, genotype, or both.	Elias (1982, 52)
crinoids	Ordovician	Modern relatives aid in development of criteria for recognition of fossil species. The species concept in paleontology is therefore somewhat arbitrary but nevertheless does offer a reasonable approximation of biological species boundaries.	Guensburg (1984, 19)
bivalves	Neogene	It is obvious that the determination of the validity of a species on shell morphology alone is impossible . . . All we can do is try to compare the morphology of spine patterns, numbers and relative sizes of ribs, and overall shell shape. Thus, the determination of which name is correct for any particular fossil is tentative at best.	Vokes and Vokes (1992, 7)
gastropods	Paleogene	Fossil species are groups of morphologically distinct populations within which variation is of the magnitude displayed by closely related, or presumably analogous, living species and their local populations, and between which the differences are of the kind and degree expected to result from reproductive isolation of populations in such related or analogous species.	Allmon (1996, 13)
corals (Rugosa)	Devonian	Species . . . are treated as comprising morphologically cohesive intergrading populations of fossil corals that were restricted in their time span and morphological limits by an unknown but variable amount . . . students of the Rugosa should expect variation within populations of fossil coral species approximating that seen within modern and fossil species of the Scleractinia.	Sorauf (1998, 24)
crinoids	Silurian	We utilize a morphospecies approach with explicit recognition that biological species are impossible to define with fossils. Also, many morphospecies recognized by paleontologists, on the basis of specimens distributed through both time and space, may well represent groups of closely related species.	Eckert and Brett (2001,19)

The usual inferential procedure is to employ some quantitative measure of morphological variation . . . to test the null hypothesis that a fossil sample under consideration is analogous to a single, extant biological species. If the fossil species is shown to be no more variable than single, extant species, the null hypothesis has successfully resisted refutation, whereas if it is revealed to be significantly more variable than extant analogs, the null hypothesis has been refuted and the inference is drawn that the sample is composed of different biological species. (Kimbel 1991, 362)

Based on this logic, I previously proposed (Allmon 1996, 13) a definition for a species when working with fossils (modified from Beerbower 1968, 80–81; Waller 1969, 8; Gould 1969, 459ff.; Raup and Stanley 1978, 108ff.): *fossil species are groups of morphologically distinct populations within which variation is of the magnitude displayed by closely related, or presumably analogous, living species and their local populations, and between which the differences are of the kind and degree expected to result from reproductive isolation of populations in such related or analogous species.*
As mentioned above, the great majority of current systematic paleontological papers do not make clear, or even mention, what they think a species is or how it is to be recognized in the fossil record. Albrecht and Miller (1993, 155), however, urged their paleoanthropological colleagues to "strengthen their research designs" with respect to fossil species via more explicit attention to selecting both the fossil sample to be studied, and the appropriate samples of modern relatives to serve as comparative analogs. They wrote that

comparative studies of fossils should differ little from experimental science in their ideal research design; both require careful controls on the different sources of variation involved in the problem . . . this may involve more time, more travel, more museums, more measuring, and more attention to locality data. The result, however, will be an appropriate foundation for rigorous analysis that will be based on larger, more numerous samples that more accurately reflect the nature of morphometric variation within and among species of both living and fossil primates. (Albrecht and Miller 1993, 156)

Following these recommendations of Albrecht and Miller, each of the steps involved in applying this consensus approach to fossil species is discussed further below.

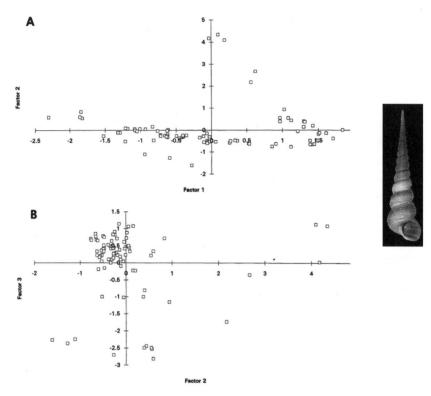

Figure 3.5 Comparison of Recent and fossil gastropods for purposes of species discrimination. A, B. Results of factor analysis of specimens of the Recent gastropod species *Turritella terebra* (inset). C, D. Results of factor analysis of specimens of the fossil turritellid gastropod *Kapalmerella mortoni sensu lato* (inset). A and C show first- and second-factor axes, and B and D show results for second- and third-factor axes. Xs represent specimens from Virginia (*K. mortoni sensu stricto*) (inset). Open squares represent specimens from Alabama, usually referred to a separate species, *K. postmortoni*. The analysis concluded that these should be recognized as two geographic subspecies. Modified from Allmon (1996).

STEP I. COMPARISON WITH MODERN RELATIVES OR ANALOGS. Comparison of fossils with living organisms is at the center of all paleontology. Accordingly, "there is a general consensus," write Plavcan and Cope (2001, 204), "that species in the fossil record should be comparable in some way to living species, at least as far as morphological variability is concerned." As Foote and Miller note, "the approaches of biologists and paleontologists are often rather similar: One typically starts by determining whether the phenotypic difference between two populations is large relative to the variation within the population" (2006, 78). Examples of such an explicit uniformitarian approach include Waller (1969), Chaline (1990, 7),

Jackson and Cheetham (1990), Nehm and Geary (1994), Allmon (1996), Knowlton and Budd (2001), and Anstey and Pachut (2004) (fig. 3.5).

It is not always clear, however, exactly *which* living species within a group should be compared with which fossil(s). This is a major issue in dealing with the relationship between reproductive isolation and morphological difference, and has been discussed with particular vigor by paleoanthropologists (e.g., Tattersall 1986, 1992, 2007; Turner 1986; Foley 1991, 2005; Kimbel 1991; Wood 1992, 1993; Albrecht and Miller 1993; Kelley 1993; Plavcan 1993; Rose and Bown 1993; Szalay 1993; Howell 1999; Plavcan and Cope 2001; Holliday 2003; Hunt 2003; White 2003; Bruner 2004; Tattersall and Mowbray 2005; Groves 2007; Rightmire 2008; Wood and Lonergan 2008; Quintyn 2009; Lordkipanidze et al. 2013). "What is the most appropriate choice of [Recent] reference sample to compare to the fossil sample?" ask Plavcan and Cope; "Unfortunately, there are no absolute answers . . ." (2001, 205). As Tattersall notes, "Taxonomic decisions are critically affected by the

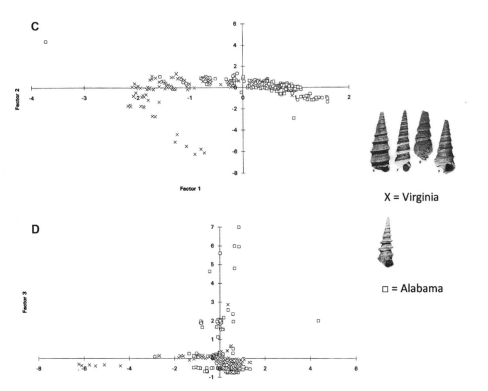

Figure 3.5 (*continued*)

choice of which living species—or pair of related species—is to be taken as arbiter of the variability permissible in a fossil assemblage recognized as a species" (1986, 166). Comparisons of fossil and living forms, writes Kelley (1993, 451),

> are intended to facilitate the taxonomic process by providing some guiding operational limits for allowable amounts of variation . . . within species. However [such studies assume] . . . a methodology that seemingly precludes a demonstration of greater variation . . . in the past, whether or not such are considered possible or even likely. . . . Such approaches, while unquestionably bringing more rigour to the analysis of variation, also preclude certain kinds of evolutionary novelty.

Kelley therefore reasonably argues that rather than slavishly applying the present to the past, alternative taxonomic hypotheses should be "discussed in a comparative fashion," with the strengths and weaknesses of alternatives used to rank them according to explicit criteria:

> There will not necessarily be agreement about the ranking, but at least the nature of the disagreements will be explicit. These criteria, and the premises and analyses upon which the taxonomies are based, will always be based one way or the other on the present, but hopefully on a present that is recognized as being a single evolutionary instant that does not encompass all known evolutionary phenomena. (Kelley 1993, 451)

The "closest living relative" may be obvious enough if the fossil and living forms are morphologically identical (or almost so) and therefore judged to be conspecifics (e.g., Michaux 1995). Even here, however, the possibility of invisible anagenetic change cannot be excluded. If the fossil and living form are judged to be different species (which is usually the case), then conclusions about the fossil taxon will have to be drawn from a separate taxon, with congenerics assumed to be more similar than confamilials in different genera, and so on up the Linnaean hierarchy. At any of these taxonomic levels, the assumption of similarity between fossil and living forms may be in error (i.e., taxonomic uniformitarianism may not be valid; e.g., Allmon 2007). A phylogenetic analysis might identify the closest living relative in a cladistic sense, but such an analysis may not exist.

In discussing fossil hominids, for example, Tattersall (1986, 170) first used all mammals and then all primates as a basis for comparison (see also Kimbel 1991, 362):

In any group other than Hominidae the presence of several clearly recognizable morphs in the record of the middle to upper Pleistocene would suggest (indeed, demonstrate) the involvement of several species. Any mammalian paleontologist seeing morphological differences on the order of those separating modern humans from their precursors, and the latter from each other, would have no difficulty in recognizing a number of separate species . . . [while some fossils show similarities with others] they also show differences of a magnitude that in any other primate family would be accepted without demur as demarcating separate species.

For extinct higher taxa (e.g., ammonoids, trilobites, graptolites, archaeocyathids), it seems widely accepted that the comparison becomes even more tenuous, although this is not frequently explicitly stated (e.g., Néraudeau 2011). In trilobites, for example, some authors have adopted species concepts that make little or no reference to any living arthropods (see, e.g., table 3.2). Similarly, in conodonts, Girard and Renaud argue for a species definition "that takes into account the variation that might have been encompassed into the biological species." They admit, however, that "whatever the definition of species, one should admit that any attempt to define a conodont species is doomed to remain hypothetical for such ancient fossils devoid of any unambiguous modern relatives" (Girard and Renaud 2011, 113).

STEP 2A. RECOGNITION OF OTUS.   All systematics begins with the attempt to sort individual organisms into groups, based on some kind of characteristics. For fossils, this means recognizing groups of individuals within which there is less variation than among them. These groups have been referred to by many terms; I will here call them "operational taxonomic units" (OTUs). Although this term has a long and varied history (e.g., Sneath and Sokal 1973), I use it here in a theoretically agnostic sense to refer to groups of organisms that, based on whatever character is used, are more similar within than among. Other similar uses of "OTU" for fossils include Michaux (1989a,b) and Parham et al. (2012) (but see Riedel 1978 for a narrower usage). In paleontology, such groups have previously been called "phena" (Sylvester-Bradley 1958; Mayr 1969, 36; Hoffman and Reif 1990), "paleo-demes" (Howell 1999), and "paleophena" (Dzik 1990).

Recognition of such OTUs may be accomplished (at least initially) by eye, or may require morphometric and statistical analyses of varied sophistication (e.g., McCune 1987, 19; Hageman 1991; Plavcan and Cope

2001; Foote and Miller 2006, 75ff.; Hunt 2013). Parsimony analysis of dinosaur specimens (rather than taxa) has been used to cluster them into possible species (e.g. Yates 2003; Upchurch et al. 2004).

STEP 2B. SEARCHING FOR MORPHOLOGICAL GAPS. In practical terms, the correlation between morphology and reproductive isolation assumed in the application of morphospecies occurs via a criterion of "morphological gaps" (fig. 3.6; see additional examples in Foote and Miller 2006), which again is not that different than the usual approach to modern species (see, e.g., Zapata and Jiménez 2012). If a phylogeny is available, "gap size" is equivalent to the number of autapomorphies shown by a putative species compared to its putative sister species. While autapomorphies may be objectively defined, the number that justifies species recognition is arbitrary and raises the perennial topic of "splitters" versus "lumpers" in taxonomy (Teichert 1949, 29; Simpson 1961, 135; Mayr 1969, 238ff.). This is very much an active concern in modern paleontological systematics (e.g., Alroy 2002; Smith 2007; Benton 2010; Allmon 2011). As noted by Carpenter (2010), for example, a number of dinosaur species have been defined based on single autapomorphies. Shaw and Lespérance (1994, 808) argue that "informed 'lumping' will usually tell more about the history of life than enthusiastic, typological splitting" (see also Hendricks 2009). On the other hand, Tattersall (1986, 168) states that "where distinct morphs can readily be identified it would seem most productive to assume that they represent species unless there is compelling evidence to believe otherwise. To brush morphological diversity under the rug of an all-encompassing species is simply to blind oneself to the complex realities of phylogeny."

Whatever the size of the gap, it is preferable that it be quantified rather than just eyeballed or estimated qualitatively. Methods range from simple bivariate plots and relatively straightforward multivariate approaches (e.g., Allmon 1990, 1996; Adrain and Westrop 2006; Grey et al. 2008; Tapanila and Pruitt 2013), to more complex and elaborate multivariate techniques (e.g., Van Bocxlaer and Schultheiß 2010; Zapata and Jiménez 2012; Hunt 2013).

STEP 3. GEOGRAPHY. Paleontologists are not limited to morphological information about fossils; they may also have data about geography, sometimes equivalent in resolution to that used by neontologists (e.g., Newell 1947, 1956; Imbrie 1957; Boucot 1981, 27). This will vary widely depending on circumstances; geographic ranges of fossil taxa are, of course, incomplete, but this is not unique to fossils. As Albrecht and Miller (1993, 127) put it, incomplete information will always

> make it difficult to interpret the meaning of geographic differences. Do distinct morphs from two fossil sites represent valid species that were sympatric at some geographically intermediate, [but] nonfossiliferous locality, or do the morphs represent demes of a single species that has undiscovered, geographically intermediate populations forming a continuum of morphological variation?

Geographic information has nevertheless long played a role in both species recognition and phylogenetic reconstruction in paleontology. For example, if attention is focused only on a single sedimentary basin, and there is no evidence of biogeographic dispersal into or out of it, then it is common practice to assign ancestral or stem status to the earliest member of a clade (e.g., Geary 1990a). Even when the geographic setting is broader than a single basin, if some assumptions are made about dispersal, geography has been used as a taxonomic character to identify likely ancestor or stem taxa (e.g., Allmon 1996, 16–19). Geographic distribution may thus be useful for judging whether a morphologically distinct form in a fossil assemblage is more likely to be a separate species. Subspecies as neontologically defined are geographically separate, and so a geographically separate fossil variant might be judged to be a distinct subspecies, whereas a co-occurring form would not (e.g., Newell 1947; McKerrow 1952; Allmon 1990).

STEP 4. ECOPHENOTYPY.   Evolution involves only variation that is inherited. Nonheritable phenotypic variation that is due to variation in the environment (ecophenotypy or phenotypic plasticity) can be difficult to distinguish from heritable or genetically determined variation, even in living taxa. This has long been acknowledged to be a challenge for studies of fossils (e.g., Boucot 1981, 27; 1982; Reif 1985, 279ff.; Foote and Miller 2006, 62; Prothero 2013, 38), and has been addressed in in a number of studies (e.g., Johnson 1981; Seilacher et al. 1985; Scrutton 1996; Samadi et al. 2000; Stewart 2002; Webber and Hunda 2007; Grey et al. 2008; Freiheit and Geary 2009; Schneider et al. 2010; Dynowski and Nebelsick 2011; Hageman et al. 2011; Maillet et al. 2013; Hageman and Todd 2014). As noted by Reif (1985, 279), however, "To prove ecophenotypy beyond doubt it is necessary to breed the animals with known genetic constitution under different conditions," and so there is no general paleontological solution to the problem.

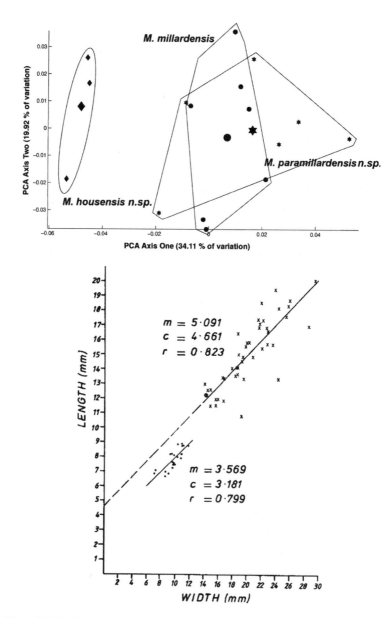

Figure 3.6 Morphological discrimination.
A. Morphological discrimination of species of the Ordovician trilobite genus *Millardicurus* from Utah. From Adrain and Westrop (2006). B. Morphological discrimination of two species of Jurassic brachiopods of the genus *Homoeorhynchia* from Yorkshire, England. From Ager (1983). C., D. (next page). Morphological discrimination of species of the bivalve genus *Buchia* from the Upper Jurassic–Lower Cretaceous of British Columbia. From Grey et al. (2008).

Two approaches to ecophenotypy have been discussed by paleontologists. The first is to examine the pattern of morphology and paleoenvironment; if similar morphologies occur in different lithologies, a genetic basis for the traits is supported (e.g., Grey et al. 2008). Yet the converse (different morphology in different environments) does not convincingly demonstrate that the variation is heritable, but only that there was local adaptation to environment. Other authors have emphasized uniformitarian comparison with living relatives or analogues (e.g., Samadi et al. 2000; Maillet et al. 2013; Hageman and Todd 2014). For example, since modern oysters are well-known to show ecophenotypy (Carriker 1996), fossil oysters might be expected to do the same (e.g., Seilacher et al. 1985; Haglund 1998).

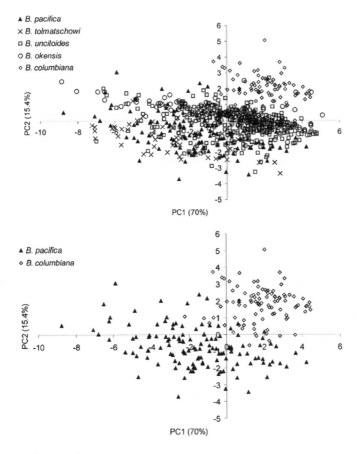

Figure 3.6 (*continued*)

Applying the GLC to Fossils

Even though we might like to understand species and speciation in all taxa represented in the fossil record, it is clear that not all fossils are suitable for recognizing species or studying speciation (i.e., it will not be possible to learn what their patterns/processes of speciation were). Hallam (1982, 355) provided a good summary of the requirements for successful study of species and speciation in the fossil record:

(1) "The fossils should be abundant and easy to collect and should have a high preservation potential so that they will give a fair reflection of the original communities."

(2) "There should be good biostratigraphic control, allowing correlation over large areas and a minimizing of the potential complication of hiatuses in the stratal sequence."

(3) "There should be good geographic control so that, in alliance with biostratigraphy, one can determine the role and extent of species migrations."

To this might be added (M. Bell, pers. comm., January 2014):

(4) close phylogenetic relationship and phenetic similarity to extant analogues, and

(5) fine microstratigraphic resolution.

Assuming that these conditions are present, the task of applying the GLC, as described above, to a set of fossils involves two major conceptual steps (e.g., Kimbel 1991, 365): perceiving the separate "identity" or "unitary evolutionary role" (i.e., distinguishing species), and the "assumption of lineage, and hence, reproductive continuity through time" (recognizing and tracing the lineage containing the species through time and space). As noted by Roopnarine (2001), the second step in many respects is equivalent to the approach to phylogenetic analysis known as "stratophenetics" (Gingerich 1974b, 1979) in its assumption that stratigraphic and geographic proximity can be taken as evidence (when combined with morphology) of genetic continuity. There is, of course, "no 'silver bullet' that will infallibly tell you that you are dealing with an historically individuated entity. And this is why we are obliged to look at the preponderance of the evidence" (Tattersall 2007, 140). Kimbel similarly admits:

> Imprecise knowledge of past variation, the vagaries of the fossil record, and, not least, descent with modification itself will surely render it impossible to attribute unequivocally each and every specimen to a species. (1991, 368)

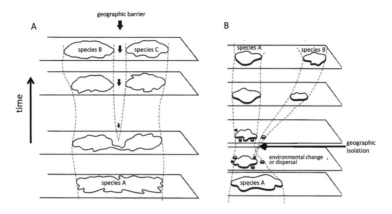

Figure 3.7 Allopatric speciation through geological time.
These figures show by implication a species concept equivalent to the ESC/GLC discussed here. In each case, ancestor-descendant continuity is inferred via continuity through time and space. A. Division of the parental species ("species A") by vicariance (aka "dumbbell allopatry"), resulting in divergence into two daughter species ("species B" and "species C"). B. Formation of small peripheral daughter populations of the parental species ("species A") by vicariance or dispersal, resulting in origin of one daughter species ("species B") with persistence of the parent. Modified from Benton and Harper (2009).

As described above, inclusion of the increasingly reasonable assumption that cryptic species are common in most if not all higher taxa requires some modification of the application of the GLC to the paleontological record (fig. 3.4). In this context, textbook diagrams of "speciation" should perhaps be viewed in a subtly different light (fig. 3.7): they are not necessarily representations of species as potentially recognizable to the neontologist, but— using the GLC conception of species—they can still be viewed in much the same way evolutionarily (as monophyletic lineages of cryptic species).

Conclusions

Although, in principle, many species based on fossils may be similar to those based on modern material, in practice many fossil species will unavoidably differ in important ways from modern species. Given the difficulties with recognizing intraspecific variation and ecophenotypy in fossils, and especially the increasing recognition of cryptic species in many higher taxa of living animals, the assumption that fossil morphospecies are "usually" effectively the same as reproductively isolated biospecies (e.g., Gould

2002, 785) is, at best, wishful thinking. It is surely valid in many cases, but it is just as surely invalid in many others.

Yet this uncertainty does not make paleontologically recognized species evolutionarily meaningless or completely noncomparable to modern species. (It has the same implications for all those Recent morphospecies as it does for paleontology: e.g., Knowlton 1993; Sáez and Lozano 2005; Bickford et al. 2007). The growing number of and controversy about species concepts has distracted both neontologists and paleontologists from the conclusion articulated by G. G. Simpson half a century ago: species are separately evolving lineages (Simpson 1951, 1961; de Queiroz 1998; Miller 2001, 2006, chapter 2 in this volume; Harrison 2014). This means that when either neontologists or paleontologists talk about species, they *are* in some important respects talking about the same thing.

Morphospecies, whether fossil or living, are more or less spatiotemporally discrete packages of morphology that may or may not consist of single genetically and evolutionarily separate units. Paleontologists can frequently detect morphospecies and their origination. Operationally, however, paleontologists in general do a poor job of being clear on what we think the "species" we are describing really are (or were). We can and should do much better. We can clarify what we think our species are and/or are not, which will surely improve the potential for "species-level data" from fossils to contribute to improved understanding of evolution.

Some modest recommendations:

(1) Paleozoologists should explicitly adopt something like the ESC/ GLC for all species-level work, and journal editors should hold them to it in their submitted papers. At the very least, authors should explain why they think their species do *not* fit the ESC/GLC concept.

(2) All descriptions of fossil species should make explicit reference to the standard by which morphospecies are being recognized, whether it is relatively close modern relatives or more distant modern analogs. This practice should apply to both quantitative and qualitative species discriminations.

(3) The nature of fossil species—both what they do and do not share with modern species—should be made clearer to both producers and consumers of paleontological species descriptions. Toward this end, conventions for distinguishing modern and fossil species—in print and verbally— should be standardized. The current mix of symbols and terms for "fossil," "extinct," or "stem" species should be reduced to one system, and paleontologically defined species should be referred to by a uniform term. For the sake of putting forth a proposal for discussion, I suggest that species

described solely on the basis of fossil material be called "paleospecies." This will at least allow readers to recognize immediately that species so marked are based exclusively on hard part morphology, and to distinguish them from species that are based on both modern and fossil data, or solely on modern material.

Acknowledgments

I am indebted to many colleagues for thought-provoking discussion of these ideas over many years. Most recently, I have benefitted from conversations with Willy Bemis, Dana Geary, Gene Hunt, Amy McCune, and Dolph Schluter. Thanks also to Brendan Anderson, Michael Foote, Dana Friend, Dana Geary, Jon Hendricks, William Miller, Chuck Mitchell, Scott Sampson, and especially Mike Bell and Peg Yacobucci, for comments on previous drafts, and to Andrew Matthiessen for assistance with figures and references. Preparation of this paper was supported in part by NSF grants EAR-0719642 and EAR-1053517.

References

Adams, B. J. 2001. "The species delimitation uncertainty principle." *Journal of Nematology*. 33(4): 153–160.

Adams, C. E., and F. A. Huntingford. 2004. "Incipient speciation driven by phenotypic plasticity? Evidence from sympatric populations of Arctic char." *Biological Journal of the Linnean Society*. 81: 611–618.

Adrain, J. M., and S. R. Westrop. 2005. "Late Cambrian ptychaspidid trilobites from western Utah: Implications for trilobite systematics and biostratigraphy." *Geological Magazine*. 142(4): 377–398.

Adrain, J. M., and S. R. Westrop. 2006. "New earliest Ordovician trilobite genus *Millardicurus*: The oldest known hystricurid." *Journal of Paleontology*. 80(4): 650–671.

Ager, D. V. 1983. "Allopatric speciation—an example from the Mesozoic Brachiopoda." *Palaeontology*. 26(3): 555–565.

Albrecht, G. H., and J. M. A. Miller. 1993. "Geographic variation in primates: A review with implications for interpreting fossils." In *Species, species concepts, and primate evolution*, W. H. Kimbel and L. B. Martin, eds., 123–161. New York: Plenum Press.

Allmon, W. D. 1990. "A review of the *Bullia* group (Neogastropoda: Nassariidae), with comments on its evolution, biogeography and phylogeny." *Bulletins of American Paleontology* 99(335): 1–179.

Allmon, W. D. 1996. "Evolution and systematics of Cenozoic American Turritellidae (Gastropoda). I. Paleocene and Eocene species related to *"Turritella mortoni* Conrad" and *"Turritella humerosa* Conrad" from the U.S. Gulf and Atlantic Coastal Plains." *Palaeontographica Americana.* 59: 1–134.

Allmon, W. D., 2007. "Cretaceous marine nutrients, greenhouse carbonates, and the abundance of turritelline gastropods." *Journal of Geology.* 115(5): 509–524.

Allmon, W. D. 2011. Review of "Molluscan paleontology of the Chesapeake Miocene" by Edward J. Petuch and Mardie Drolshagen." *Palaios* (doi: 10.2110/palo. 2011.BR64).

Allmon, W. D. 2013. "Species, speciation, and paleontology up to the Modern Synthesis: Persistent themes and unanswered questions." *Palaeontology.* 56(4): 1199–1223.

Allmon, W. D., and U. E. Smith. 2011. "What, if anything, can we learn from the fossil record about speciation in marine gastropods? Biological and geological considerations." *American Malacological Bulletin.* 29(1): 247–276.

Alroy, J. 2002. "How many named species are valid?" *Proceedings of the National Academy of Sciences of the USA.* 99(6): 3706–3711.

Anstey, R. L., and J. F. Pachut. 2004. "Cladistic and phenetic recognition of species in the Ordovician bryozoan genus *Peronopora.*" *Journal of Paleontology.* 78(4): 651–674.

Arkell, W. J., and J. A. Moy-Thomas. 1940. "Palaeontology and the taxonomic problem." In *The new systematics,* J. Huxley, ed., 395–410. Oxford, UK: Clarendon Press.

Ax, P. 1985. "Stem species and the stem lineage concept." *Cladistics.* 1(3): 279–287.

Bachmann, K. 1998. "Species as units of diversity: an outdated concept." *Theory in Biosciences.* 117(3): 213–230.

Barker, M. J. 2007. "The empirical inadequacy of species cohesion by gene flow." *Philosophy of Science.* 74: 654–665.

Barnosky, A. D. 1987. "Punctuated equilibrium and phyletic gradualism: Some facts from the Quaternary mammalian record." *Current Mammalogy.* 11: 109–147.

Barnosky, A., M. A. Carrasco, and E. B. Davis. 2005. "The impact of the species-area relationship on estimates of paleodiversity." *PLoS Biology.* 3(8): e266.

Barton, N. H. 2001. "Speciation." *Trends in Ecology and Evolution.* 16(7): 325.

Bates, H., K. J. Travouillon, B. Cooke, R. M. D. Beck, S. J. Hand, and M. Archer. 2014. "Three new Miocene species of musky rat-kangaroos (Hypsiprymnodontidae, Macropodoidea): Description, phylogenetics, and paleoecology." *Journal of Vertebrate Paleontology.* 34(2): 383–396.

Baum, D. A. 1998. "Individuality and the existence of species through time." *Systematic Biology.* 47(4): 641–653.

Baum, D. A. 2009. "Species as ranked taxa." *Systematic Biology.* 58(1): 74–86.

Baum, D. A., and K. L. Shaw. 1995. "Genealogical perspectives on the species problem." *Monographs in Systematic Botany from the Missouri Botanical Garden*. 53: 289–303.

Beerbower, J. 1968. *Search for the past: An introduction to paleontology*. 2nd ed. Englewood Cliffs, NJ: Prentice- Hall.

Beheregaray, L. B., and A. Caccone. 2007. "Cryptic biodiversity in a changing world." *Journal of Biology* 6:9 (doi:10.1186/jbiol60).

Bell, M. A., and S. A. Foster. 1994. "Introduction to the evolutionary biology of the threespine stickleback." In *The evolutionary biology of the threespine stickleback*, M. A. Bell and S. A. Foster, eds., 1–27. Oxford: Oxford University Press.

Bell, M. A., J. V. Baumgartner, and E. C. Olson. 1985. "Patterns of temporal change in single morphological characters of a Miocene stickleback fish." *Paleobiology*. 11: 258–271.

Benton, M. J. 2010. "Naming dinosaur species: The performance of prolific authors." *Journal of Vertebrate Paleontology*. 30(5): 1478–1485.

Benton, M. J., and D. A. T. Harper. 2009. *Introduction to paleobiology and the fossil record*. Chichester, UK: Wiley-Blackwell.

Benton, M. J., and P. N. Pearson. 2001. "Speciation in the fossil record." *Trends in Ecology and Evolution*. 16(7): 405–411.

Beurton, P. 1995. "How is a species kept together?" *Biology and Philosophy*. 10: 181–196.

Bickford, D., D. J. Lohman, N. S. Sohdi, P. K. L. Ng, R. Meier, K. Winker, K. K. Ingram, and I. Das. 2007. "Cryptic species as a window on diversity and conservation." *Trends in Ecology and Evolution*. 22: 148–155.

Bonde, N. 1981. "Problems of species concepts in paleontology." International symposium on concept and method in paleontology, J. Martinell, ed., 19–34. Barcelona: Universitat de Barcelona.

Boucot, A. J. 1981. *Principles of benthic marine paleoecology*. New York: Academic Press.

Boucot, A. J. 1982. "Ecophenotypic or genotypic?" *Nature*. 296: 609–610.

Brandão, S. N. 2013. "Challenging cosmopolitanism in the deep sea: The case of 'Cythere acanthoderma Brady, 1880' (Crustacea, Ostracoda)." *Revue de Micropaleontogie*. 56: 2–19.

Brigandt, I. 2003. "Species pluralism does not imply species eliminativism." *Philosophy of Science*. 70(5): 1305–1316.

Brochu, C. A., and G. W. Storrs. 2012. "A giant crocodile from the Plio-Pleistocene of Kenya, the phylogenetic relationships of Neogene African crocodylines, and the antiquity of *Crocodylus* in Africa." *Journal of Vertebrate Paleontology*. 32(3): 587–602.

Brooks, D. R., and D. A. McLennan. 1999. "Species: Turning a conundrum into a research program." *Journal of Nematology*. 31(2): 117–133.

Bruner, E. 2004. "Evolution, actuality, and species concept: A need for a palaeontological tool." *Human Evolution*. 19(2): 93–112.

Budd, A. F. 2010. "Tracing the long-term evolution of a species complex: Examples from the Montastraea 'annularis' complex." *Palaeoworld*. 19: 348–356.

Budd, A. F., F. L. D. Nunes, E. Weil, and J. M. Pandolfi. 2012. "Polymorphism in a common Atlantic reef coral (*Montastrea cavernosa*) and its long-term evolutionary implications." *Evolutionary Ecology*. 26: 265–290.

Budd, A. F., K. G. Johnson, and D. C. Potts. 1994. "Recognizing morphospecies in colonial corals. I. Landmark-based methods." *Paleobiology*. 20: 484–505.

Cain, A. J. 1954. *Animal species and their evolution*. London: Hutchinson.

Carpenter, K. 2010. "Species concept in North American stegosaurs." *Swiss Journal of Geosciences*. 103(2): 155–162.

Carpenter, K., and P. J. Currie. 1990. "Introduction: On systematic and morphological variation." In *Dinosaur systematics: Approaches and perspectives*, K. Carpenter and P. J. Currie, eds., 1–8. Cambridge: Cambridge University Press.

Carriker, M. R. 1996. "The shell and ligament." In *The eastern oyster Crassostrea virginica*, V. S. Kennedy, R. I. E. Newell, and A. F. Eble, eds., 75–168. College Park: Maryland Sea Grant College.

Castelin, M., J. Lambourdiere, M.-C. Boisselier, P. Lozouet, A. Couloux, C. Cruaud, and S. Samadi. 2010. "Hidden diversity and endemism on seamounts: Focus on poorly dispersive neogastropods." *Biological Journal of the Linnean Society*. 100: 420–438.

Chaline, J. 1990. *Paleontology of vertebrates*. Berlin: Springer-Verlag.

Chiba, S. 2007. "Taxonomic revision of the fossil land snail species of the genus *Mandarina* in the Ogasawara Islands." *Paleontological Research*. 11(4): 317–329.

Claridge, M. F., H. A. Dawah, and M. R. Wilson, eds. 1997. *Species: The units of biodiversity*. London: Chapman and Hall.

Clark, R. B. 1956. "Species and systematics." *Systematic Zoology*. 5:1–10.

Clarkson, E. N. K. 1998. Invertebrate palaeontology and evolution. 4th ed. Oxford, UK: Blackwell Science.

Collinson, M. E. 1986. "The use of generic names for fossil plants." In *Systematic and taxonomic approaches in palaeobotany*, R. A. Spicer and B. A. Thomas, eds., 91–104. Oxford: Clarendon.

Cooper, G. A., and J. T. Dutro, Jr., 1982. "Devonian brachiopods of New Mexico." *Bulletins of American Paleontology*. 82–83: 1–215.

Coyne, J. A., and H. A. Orr. 2004. *Speciation*. Sunderland, MA: Sinauer Associates.

Cracraft, J. 1983. "Species concepts and speciation analysis." *Current Ornithology*. 1: 159–187.

Cracraft, J. 1987. "Species concepts and the ontology of evolution." *Biology and Philosophy*. 2: 329–346.

Crônier, C., A. Bignon, and A. François. 2011. "Morphological and ontogenetic criteria for defining a trilobite species: The example of Siluro-Devonian Phacopidae." *Comptes Rendus Palevol.* 10(2): 143–153.

Cronquist A. 1978. *The evolution and classification of flowering plants.* 2nd ed. New York: New York Botanical Garden.

Crowson, R. A. 1970, *Classification in biology.* New York: Atherton Press.

Davies, A. M. 1961. *An introduction to palaeontology.* 3rd ed. Revised and partly rewritten by J. Stubblefield. London: Thomas Morby & Co.

Dayan, T., D. Wool, and D. Simberloff. 2002. "Variation and covariation of skulls and teeth: Modern carnivores and the interpretation of fossil mammals." *Paleobiology.* 28(4): 508–526.

de Queiroz, K. 1998. "The general lineage concept of species, species criteria, and the process of speciation: A conceptual unification and terminological recommendations." In *Endless forms: Species and speciation,* D. J. Howard and S. H. Berlocher, eds., 57–75. Oxford: Oxford University Press.

de Queiroz, K. 1999. "The general lineage concept of species and the defining properties of the species category." In *Species: New interdisciplinary essays,* R. Wilson, ed., 49–89. Cambridge, MA: MIT Press.

de Queiroz, K. 2005a. "Different species problems and their resolution." *BioEssays.* 27: 1263–1269.

de Queiroz, K. 2005b. "Ernst Mayr and the modern concept of species." *Proceedings of the National Academy of Sciences.* 102 (supplement 1): 6600–6607.

de Queiroz, K. 2005c. "A unified concept of species and its consequences for the future of taxonomy." *Proceedings of the California Academy of Sciences.* Ser. 4, 56 (supplement 1), 18: 196–215.

de Queiroz, K. 2007. "Species concepts and species delimitation." *Systematic Biology.* 56(6): 879–886.

Demos, T. C., J. C. K. Peterhans, B. Agwanda, and M. J. Hickerson. 2014. "Uncovering cryptic diversity and refugial persistence among small mammal lineages across the Eastern Afromontane biodiversity hotspot." *Molecular Phylogenetics and Evolution.* 71: 41–54.

Descimon, H., and J. Mallet. 2009. "Bad species." In *Ecology of butterflies in Europe,* J. Settele, M. Konvicka, T. Shreeve, and H. Van Dyck, eds., 219–249. Cambridge, UK: Cambridge University Press.

Donoghue, M. J. 1985. "A critique of the biological species concept and recommendations for a phylogenetic alternative." *Bryologist.* 88: 171–181.

Donoghue, P. C. J. 2005. "Saving the stem group—a contradiction in terms?" *Paleobiology.* 31(4):553–558.

Dorit, R. L. 1990. "The correlates of high diversity in Lake Victoria haplochromine cichlids: A neontological perspective." In *Causes of evolution: A paleontological perspective,* R. M. Ross and W. D. Allmon, eds., 322–353. Chicago: University of Chicago Press.

Dupré, J. 1993. *The disorder of things: Metaphysical foundations of the disunity of science*. Cambridge, MA: Harvard University Press.

Dupré, J. 1999. "On the impossibility of a monistic account of species." In *Species*, R. A. Wilson, ed., 3–22. Cambridge, MA: MIT Press.

Dynowski, J. F., and J. H. Nebelsick. 2011. "Ecophenotypic variations of *Encrinus liliiformis* (Echinodermata: Crinoidea) from the middle Triassic Muschelkalk of Southwest Germany." *Swiss Journal of Palaeontology*. 130(1): 53–67.

Dzik, J. 1990. The concept of chronospecies in ammonites. In *Atti del secondo convegno internazionale Fossili Evoluzione Ambiente*, Pergola, 25–30 October 1987, G. Pallini, F. Cecca, S. Cresta, and M. Santantonio, eds., 273–289.

Dzik, J. 2005. "The chronophyletic approach: Stratophenetics facing an incomplete fossil record." *Special Papers in Palaeontology*. 73: 159–183

Eckert, J., and C. E. Brett. 2001. "Early Silurian (Llandovery) crinoids from the Lower Clinton Group, western New York State." *Bulletins of American Paleontology*. 360:1–88.

Ehrlich, P. R., and P. H. Raven. 1969. "Differentiation of populations." *Science*. 165: 1228–1232.

Eldredge, N. 1985a. *Unfinished synthesis: Biological hierarchies and modern evolutionary thought*. Oxford, UK: Oxford University Press.

Eldredge, N. 1985b. *Time frames: The rethinking of Darwinian evolution and the theory of punctuated equilibria*. New York: Simon & Schuster.

Eldredge, N. 1985c. "The ontology of species." In *Species and speciation*, E. S. Vrba, ed., 17–20. Pretoria: Transvaal Museum Monograph no. 4.

Eldredge, N. 1989. *Macroevolutionary dynamics*. New York: McGraw-Hill.

Eldredge, N. 1993. "What, if anything, is a species?" In *Species, species concepts, and primate evolution*, W. H. Kimbel and L. B. Martin, eds., 3–20. New York: Plenum Press.

Eldredge, N., and J. Cracraft. 1980. *Phylogenetic patterns and the evolutionary process*. New York: Columbia University Press.

Eldredge, N., and S. J. Gould. 1972. "Punctuated equilibria: An alternative to phyletic gradualism." In *Models in paleobiology*, T. J. M. Schopf, ed., 82–115. San Francisco: Freeman, Cooper & Co.

Eldredge, N., J. N. Thompson, P. M. Brakefield, S. Gavrilets, D. Jablonski, J. B. C. Jackson, R. E. Lenski, B. S. Lieberman, M. A. McPeek, and W. Miller III. 2005. "The dynamics of evolutionary stasis." *Paleobiology*. 31(sp5): 133–145.

Elias, R. J. 1982. "Latest Ordovician solitary rugose corals of eastern North America." *Bulletins of American Paleontology*. 314: 1–116.

Ereshefsky, M. 1992. "Eliminative pluralism." *Philosophy of Science*. 59: 671–690.

Ereshefsky, M. 1998. "Species pluralism and anti-realism." *Philosophy of Science*. 65: 103–120.

Ereshefsky, M. 2010. "Darwin's solution to the species problem." *Synthese*. 175: 405–425.

Erwin, D. H., and R. L. Anstey. 1995. "Speciation in the fossil record." In *New approaches to speciation in the fossil record*, D. Erwin and R. Anstey, eds., 11–38. New York: Columbia University Press.

Escudé, E., S. Montuire, E. Desclaux, J.-P. Quéré, E. Renvoisé, and M. Jeannet. 2008. "Reappraisal of 'chronospecies' and the use of *Arvicola* (Rodentia, Mammalia) for biochronology." *Journal of Archaeological Science*. 35: 1867–1879.

Feder, J. L., S. P. Egan, and P. Nosil. 2012. "The genomics of speciation-with-geneflow." *Trends in Genetics*. 28(7): 342–350.

Fitzhugh, K. 2005. "The inferential basis of species hypotheses: The solution to defining the term 'species.'" *Marine Ecology*. 26: 155–165.

Fitzhugh, K. 2009. "Species as explanatory hypotheses: Refinements and implications." *Acta Biotheoretica*. 57: 201–248.

Foley, R. 1991. "How many hominid species should there be?" *Journal of Human Evolution*. 20: 413–427.

Foley, R. 2005. "Species diversity in human evolution: Challenges and opportunities." *Transactions of the Royal Society of South Africa*. 60(2): 67–72.

Foote, M., and A. I. Miller. 2006. *Principles of paleontology*. 3rd ed. New York: W. H. Freeman and Company.

Forey, P. L., R. A. Fortey, P. Kenrick, and A. B. Smith. 2004. "Taxonomy and fossils: A critical appraisal." *Philosophical Transactions of the Royal Society of London I*. 359: 639–653.

Fox, R. C. 1986. "Species in paleontology." *Geoscience Canada*. 13(2): 73–84.

Freiheit, J. R., and D. H. Geary. 2009. "Neogene paleontology of the northern Dominican Republic. 23. Strombid gastropods (genera *Strombus* and *Lobatus*; Mollusca: Gastropoda: Strombidae) of the Cibao Valley." *Bulletins of American Paleontology*. 376: 1–54.

Frost, D. R., and A. G. Kluge. 1994. "A consideration of epistemology in systematic biology, with special reference to species." *Cladistics*. 10: 259–294.

Futuyma, D. J. 1987. "On the role of species in anagenesis." *American Naturalist*. 130: 465–473.

Futuyma, D. J. 1989. "Speciational trends and the role of species in macroevolution." *American Naturalist*. 134(2): 318–321.

Gavrilets, S. 2003. "Models of speciation: What have we learned in 40 years?" *Evolution*. 57(10): 2197–2213.

Geary, D. H. 1990a. "Patterns of evolutionary tempo and mode in the radiation of *Melanopsis* (Gastropoda; Melanopsidae)." *Paleobiology*. 16: 492–511.

Geary, D. H. 1990b. "Exploring the roles of intrinsic and extrinsic factors in the evolutionary radiation of *Melanopsis*." In *Causes of evolution*, R. M. Ross and W. D. Allmon, eds., 305–321. Chicago: University of Chicago Press.

Geary, D. H. 1995. "The importance of gradual change in species-level transitions." In *New approaches to speciation in the fossil record*, D. H. Erwin and R. L. Anstey, eds., 67–86. New York: Columbia University Press.

Geary, D. H. 2009. "The legacy of punctuated equilibrium." In *Stephen Jay Gould: Reflections on his view of life*, W. D. Allmon, P. H. Kelley, and R. M. Ross, eds., 127–145. New York: Oxford University Press.

George, T. N. 1956. "Biospecies, chronospecies, and morphospecies." In *The species concept in palaeontology*, P. C. Sylvester-Bradley, ed., 123–137. London: Systematics Association.

Ghiselin, M. T. 1974. "A radical solution to the species problem." *Systematic Zoology*. 23: 536–544.

Ghiselin, M. T. 1997. *Metaphysics and the origin of species*. Albany: SUNY Press.

Ghiselin, M. T. 2002. "Species concepts: The basis for controversy and reconciliation." *Fish and Fisheries*. 3: 151–160.

Gili, C., and J. Martinell. 2000. "Phylogeny, speciation, and species turnover: The case of the Mediterranean gastropods of genus *Cyclope* Risso, 1826." *Lethaia*. 33(3): 236–250.

Gilmore, J. S. L. 1940. "Taxonomy and philosophy." In *The new systematics*, J. S. Huxley, ed., 461–474. Oxford: Oxford University Press.

Gingerich, P. D. 1974a. "Stratigraphic record of Early Eocene *Hyopsodus* and the geometry of mammalian phylogeny." *Nature*. 248: 107–109.

Gingerich, P. D. 1974b. "Size variability of the teeth in living mammals and the diagnosis of closely related sympatric fossil species." *Journal of Paleontology*. 48(5): 895–903.

Gingerich, P. D. 1976. "Paleontology and phylogeny: Patterns of evolution at the species level in early Tertiary mammals." *American Journal of Science*. 276: 1–28.

Gingerich, P. D. 1979. "Stratophenetic approach to phylogeny reconstruction in vertebrate paleontology." In *Phylogenetic analysis and paleontology*, J. Cracraft and N. Eldredge, eds., 41–78. New York: Columbia University Press.

Gingerich, P. D. 1985. "Species in the fossil record: Concepts, trends, and transitions." *Paleobiology*. 11(1): 27–41.

Girard, C., and S. Renaud. 2011. "The species concept in a long-extinct fossil group, the conodonts." *Comptes Rendus Palevol*. 10: 107–115.

Gould, G. C. 2001. "The phylogenetic resolving power of discrete dental morphology among extant hedgehogs and the implications for their fossil record." *American Museum Novitates*. 3340: 1–52.

Gould, S. J. 1969. "An evolutionary microcosm: Pleistocene and Recent history of the land snail *P.* (*Poecilozonites*) in Bermuda." *Bulletin of the Museum of Comparative Zoology*. 138: 407–532.

Gould, S. J. 1991. "Exaptation: A crucial tool for an evolutionary psychology." *Journal of Social Issues*. 47(3): 43–65.

Gould, S. J. 2002. *The structure of evolutionary theory*. Cambridge, MA: Harvard University Press.

Gourbière, S., and J. Mallet. 2009. "Are species real? The shape of the species boundary with exponential failure, reinforcement, and the 'missing snowball'." *Evolution*. 64(1): 1–24.

Grande, L., and W. E. Bemis. 1998. "A comprehensive phylogenetic study of amiid fishes (Amiidae) based on comparative skeletal anatomy: An empirical search for interconnected patterns of natural history." *Society of Verterbrate Paleontology Memoir.* 4:1–690.

Grey, M., J. W. Haggart, and P. L. Smith. 2008. "Species discrimination and evolutionary mode of *Buchia* (Bivalvia: Buchiidae) from Upper Jurassic-Lower Cretaceous strata of Grassy Island, British Columbia, Canada." *Palaeontology.* 51(3): 583–595.

Groves, C. 2007. "Species concepts and speciation: Facts and fantasies." In *Handbook of paleoanthropology.* Vol. 3. *Phylogeny of hominids,* W. Henke and I. Tattersall, eds., 1861–1879. Berlin: Springer-Verlag.

Gu, J.-J., O. Béthoux, and D. Ren. 2011. *Longzhua loculata* n. gen. n. sp., one of the most completely documented Pennsylvanian Archaeorthoptera (Insecta; Ningxia, China)." *Journal of Paleontology.* 85(2): 303–314.

Guensburg, T. E. 1984. "Echinodermata of the Middle Ordovician Lebanon Limestone." *Bulletins of American Paleontology.* 86: 1–100.

Hageman, S. J. 1991. "Approaches to systematic and evolutionary studies of perplexing groups: An example using fenestrate Bryozoa." *Journal of Paleontology.* 65(4): 630–647.

Hageman, S. J., and C. D. Todd. 2014. "Hierarchical (mm- to km-scale) environmental variation affecting skeletal phenotype of a marine invertebrate (*Electra pilosa,* Bryozoa): Implications for fossil species concepts." *Palaeogeography, Palaeoclimatology, Palaeoecology.* 396: 213–226.

Hageman, S. J., P. N. Wyse Jackson, A. R. Abernethy, and M. Steinthorsdottir. 2011. "Calendar scale, environmental variation preserved in the skeletal phenotype of a fossil bryozoan (*Rhombopora blakei* n.sp.), from the Mississippian of Ireland." *Journal of Paleontology.* 85(5): 853–870.

Haglund, W. M. 1998. "Ecophenotypes of the late Cretaceous oyster *Crassostrea subtrigonalis* (Evans and Shumard, 1857), Central Alberta, Canada." In *Bivalves: An eon of evolution.* P. Johnston, ed., 295–304. Calgary: University of Calgary Press.

Hallam, A. 1982. "Patterns of speciation in Jurassic *Gryphaea*." *Paleobiology.* 8(4): 354–366.

Hallam, A. 1988. "The contribution of palaeontology to systematics and evolution." In *Prospects in systematics,* D. L. Hawksworth, ed., 128–147. Oxford: Systematics Association and Clarendon Press.

Harrison, R. G. 1998. "Linking evolutionary pattern and process: The relevance of species concepts for the study of speciation." In *Endless forms: Species and speciation,* D. J. Howard and S. H. Berlocher, eds., 19–31. New York: Oxford University Press.

Harrison, R. G. 2010. "Understanding the origin of species: Where have we been? Where are we going?." In *Evolution since Darwin: The first 150 years,* M. A. Bell, D. J. Futuyma, W. F. Eanes, and J. S. Levinton, eds., 319–346. Sunderland, MA: Sinauer Associates.

Harrison, R. G. 2012. "The language of speciation." *Evolution*. 66 (12): 3643–3657.

Harrison, R. G., 2014, "Species and speciation." In *The Princeton guide to evolution*, J. B. Losos, ed., 489–495. Princeton, NJ: Princeton University Press.

Hart, M. W. 2010. "The species concept as an emergent property of population biology." *Evolution*. 65(3): 613–616.

Hausdorf, B. 2011. "Progress toward a general species concept." *Evolution*. 65 (4): 923–931.

Hausdorf, B., and C. Hennig. 2010. "Species delimitation using dominant and codominant multilocus markers." *Systematic Biology*. 59(5): 491–503.

Hendricks, J. R. 2009. "The genus *Conus* (Mollusca: Neogastropoda) in the Plio-Pleistocene of the southeastern United States." *Bulletins of American Paleontology*. 375: 1–178.

Hendry, A. P., S. M. Vamosi, S. J. Latham, J. C. Heilbuth, and T. Day. 2000. "Questioning species realities." *Conservation Genetics*. 1: 67–76.

Hennig, W. 1966. *Phylogenetic systematics*. Urbana: University of Illinois Press.

Henry, C. S. 1985. "Sibling species, call differences, and speciation in green lacewings (Neuroptera: Chrysopidae: *Chrysoperla*)." *Evolution*. 39(5): 965–984.

Herbert, G. S., and R. W. Portell. 2004. "First paleontological record of larval brooding in the calyptraeid gastropod genus Crepidula Lamarck, 1799." *Journal of Paleontology*. 78: 424–429.

Hey, J. 2001. *Genes, categories, and species: The evolutionary and cognitive cause of the species problem*. New York: Oxford University Press.

Hey, J. 2006. "Recent advances in assessing gene flow between diverging populations and species." *Current Opinion in Genetics and Development*. 16(6): 592–596.

Highton, R. 2014. "Detecting cryptic species in phylogeographic studies: Speciation in the California slender salamander, *Batrachoseps attenuatus*." *Molecular Phylogenetics and Evolution*. 71: 127–141.

Hills, S. F. K., J. S. Crampton, S. A. Trewick, and M. Morgan-Richards. 2012. "DNA and morphology unite two species and 10 million year old fossils." *PloS One*. 7(12): e52083.

Hoffman, A., and W.-E. Reif. 1990. "On the study of evolution in species-level lineages in the fossil record: Controlled methodological sloppiness." *Paläontologische Zeitschrift*. 64(1/2): 5–14.

Holliday, T. W. 2003. "Species concepts, reticulation, and human evolution." *Current Anthropology*. 44(5): 653–673.

Hoover, P. 1981. "Paleontology, taphonomy, and paleoecology of Palmarito Formation (Permian of Venezuela)." *Bulletins of American Paleontology*. 80: 1–313.

Hopkins, M. P., and M. Webster. 2009. "Ontogeny and geographic variation of a new species of the corynexochine trilobite *Zacanthopsis* (Dyeran, Cambrian)." *Journal of Paleontology*. 83(4): 524–547.

Howard, D. J. 1998. "Unanswered questions and future directions in the study of speciation." In *Endless Forms: Species and Speciation*, D. J. Howard and S. H. Berlocher, eds., 439–448. New York: Oxford University Press.

Howell, F. C. 1999. "Paleo-demes, species clades, and extinctions in the Pleistocene hominin record." *Journal of Anthropological Research*. 55(2): 191–243.

Hughes, N. C. 1994. "Ontogeny, intraspecific variation, and systematics of the Late Cambrian trilobite *Dikelocephalus*." *Smithsonian Contributions to Paleobiology*. 79:1–89.

Hughes, N., and C. Labandeira. 1995. "The stability of species in taxonomy." *Paleobiology*. 21(4): 401–403.

Hull, D. L. 1976. "Are species really individuals?" *Systematic Zoology*. 25:174–191.

Hull, D. L. 1997. "The ideal species concept—and why we can't get it." In *Species: The units of biodiversity*, M. F. Claridge, H. A. Dawah, and M. R. Wilson, eds., 357–380. London: Chapman & Hall.

Hunt, G. 2008. "Gradual or pulsed evolution: When should punctuational explanations be preferred?" *Paleobiology*. 34: 360–377.

Hunt, G. 2010. "Evolution in fossil lineages: Paleontology and the origin of species." *American Naturalist*. 176: S61–S76.

Hunt, G. 2012. "Measuring rates of phenotypic evolution and the inseparability of tempo and mode." *Paleobiology*. 38(3): 351–373.

Hunt, G. 2013. "Testing the link between phenotypic evolution and speciation: An integrated palaeontological and phylogenetic analysis." *Methods in Ecology and Evolution*. 4: 714–723.

Hunt, G., and D. L. Rabosky. 2014. "Phenotypic evolution in fossil species: Pattern and process." *Annual Review of Earth and Planetary Sciences*. 42:421–441.

Hunt, K. D. 2003. "The single species hypothesis: Truly dead and pushing up bushes, or still twitching and ripe for resuscitation?" *Human Biology*. 75: 485–502.

Imbrie, J. 1957. "The species problem with fossil animals." In *The species problem*, E. Mayr, ed., 125–153. AAAS Publ. no. 50.

Isaac, N. J. B., and A. Purvis. 2004. "The 'species problem' and testing macroevolutionary hypotheses." *Diversity and Distributions*. 10: 275–281.

Jackson, J. B. C., and A. H. Cheetham. 1990. "Evolutionary significance of morphospecies: A test with cheilostome Bryozoa." *Science*. 248: 579–583.

Jackson, J. B. C., and A. H. Cheetham. 1994. "Phylogeny reconstruction and the tempo of speciation in cheilostome Bryozoa." *Paleobiology*. 20: 407–423.

Jackson, J. B. C., and A. H. Cheetham. 1999. "Tempo and mode of speciation in the sea." *Trends in Ecology and Evolution*. 14(2): 72–77.

Johnson, A. L. A. 1981. "Detection of ecophenotypic variation in fossils and its application to a Jurassic scallop." *Lethaia*. 14: 277–285.

Johnson, J. G. 1975. "Allopatric speciation in fossil brachiopods." *Journal of Paleontology*. 49(4): 646–661.

Jordana, X., and M. Köhler. 2011. "Enamel microstructure in the fossil bovid

Myotragus balearicus (Majorca, Spain): Implications for life-history evolution of dwarf mammals in insular ecosystems." *Palaeogeography, Palaeoclimatology, Palaeoecology*. 300(1): 59–66.

Jung, P. 1986. "Neogene paleontology in the northern Dominican Republic. 2. The genus *Strombina* (Gastropoda: Columbellidae)." *Bulletins of American Paleontology*. 90(324): 1–42.

Kelley, J. 1993. "Taxonomic implications of sexual dimorphism in *Lufengpithecus*." In *Species, species concepts, and primate evolution*, W. H. Kimbel and L. Martin, eds., 429–458. New York: Plenum Press.

Kimbel, W. H. 1991. "Species, species concepts, and hominid evolution." *Journal of Human Evolution*. 20: 355–371.

Kitcher, P. 1984. "Species." *Philosophy of Science*. 51: 308–333.

Knowles, L. L., and B. C. Carstens. 2007. "Delimiting species without monophyletic gene trees." *Systematic Biology*. 56(6): 887–895.

Knowlton, N. 1986. "Cryptic and sibling species among the decapod Crustacea." *Journal of Crustacean Biology*. 6(3): 356–363.

Knowlton, N. 1993. "Sibling species in the sea." *Annual Review of Ecology and Systematics*. 24: 189–216.

Knowlton, N. 2000. "Molecular genetic analyses of species boundaries in the sea." *Hydrobiologia*. 420: 73–90.

Knowlton, N., and A. F. Budd. 2001. "Recognizing coral species present and past." In *Evolutionary patterns: Growth, form, and tempo in the fossil record*, J. B. C. Jackson, S. Lidgard, and F. K. McKinney, eds., 97–119. Chicago: University of Chicago Press.

Knowlton, N., and L. A. Weigt. 1997. "Species of marine invertebrates: A comparison of the biological and phylogenetic species concepts." *In Species: The units of biodiversity*, M. F. Claridge, H. A. Dawah, and M. R. Wilson, eds., 199–219. London: Chapman and Hall.

Knowlton, N., E. Weil, L. A. Weight, and H. M. Guzman. 1992. "Sibling species in *Montastrea annularis*, coral bleaching and the coral climate record." *Science*. 255: 330–333.

Kornfield, I. L., D. C. Smith, P. S. Gagnon, and J. N. Taylor. 1982. "The cichlid fish of Cuatro Cienegas, Mexico: Direct evidence of conspecificity among distinct trophic morphs." *Evolution*. 36: 658–664.

Kornet, D. J. 1993. "Permanent splits as speciation events: A formal reconstruction of the internodal species concept." *Journal of Theoretical Biology*. 164: 407–435.

Kristensen, N. P., M. J. Scoble, and O. Karsholt. 2007. "Lepidoptera phylogeny and systematics: The state of inventorying moth and butterfly diversity." *Zootaxa*. 1668: 699–747.

Kunz, W. 2012. *Do species exist? Principles of taxonomic classification*. Weinheim, Germany: Wiley-Blackwell/Wiley-VCH Verlag & Co.

Lambert, O., and C. De Muizon. 2013. "A new long-snouted species of the

Miocene pontoporiid dolphin *Brachydelphis* and a review of the Mio-Pliocene marine mammal levels in the Sacaco Basin, Peru." *Journal of Vertebrate Paleontology.* 33(3): 709–721.

Lauder, G. V., and K. F. Liem. 1983. "The evolution and interrelationships of the actinopterygian fishes." *Bulletin of the Museum of Comparative Zoology.* 150:95–197.

LaPorte, J. 2007. "In defense of species." *Studies in History and Philosophy of Biological and Biomedical Sciences.* 38: 255–269.

Levinton, J. S. 1988. *Genetics, paleontology, and macroevolution.* New York: Cambridge University Press.

Levinton, J. S. 2001. *Genetics, paleontology, and macroevolution.* 2nd ed. New York: Cambridge University Press.

Levinton, J. S., and C. M. Simon. 1980. "A critique of the punctuated equilibria model and implications for the detection of speciation in the fossil record." *Systematic Zoology.* 29(2): 130–142.

Lieberman, B. S. 2001a. "A test of whether rates of speciation were unusually high during the Cambrian radiation." *Proceedings of the Royal Society B.* 268: 1707–1714.

Lieberman, B. S. 2001b. "Analyzing speciation rates in macroevolutionary studies." In *Fossils, phylogeny, and form: An analytical approach*, J. M. Adrain, G. D. Edgecombe, and B. S. Lieberman, eds., 340–358. New York: Plenum Press.

Lim, G. S., M. Balke, and R. Meier. 2011. "Determining species boundaries in a world full of rarity: Singletons, species delimitation methods." *Systematic Biology.* 61(1): 165–169.

Lohman, D. J., K. K. Ingram, D. M. Prawiradilaga, K. Winker, F. H. Sheldon, R. G. Moyle, P. K. L. Ng, P. S. Ong, L. K. Wang, T. M. Braile. 2010. "Cryptic genetic diversity in "widespread' Southeast Asian bird species suggests that Philippine avian endemism is gravely underestimated." *Biological Conservation.* 143: 1885–1890.

Lordkipanidze, D., M. S. Ponce de León, A. Margvelashvili, Y. Rak, G. P. Rightmire, A. Vekua, C. P. E. Zollikofer. 2013. "A complete skull from Dmanisi, Georgia, and the evolutionary biology of *Homo*." *Science.* 342: 326–331.

Luckow, M. 1995. "Species concepts: Assumptions, methods, and applications." *Systematic Botany.* 20(4): 589–605.

Maglio, V. J. 1971. "The nomenclature of intermediate forms: An opinion." *Systematic Zoology.* 20: 370–373.

Maillet, S., B. Milhau, M. Vreulx, T. Danelian, C. Monnet, and J. P. Nicollin, 2013. "Ecophenotypic variation of the Devonian benthic ostracod species *Cavellina rhenana* Krömmelbein, 1954: A paleoenvironmental proxy for the Ardenne (France-Belgium) and Rheno-Hercynian realm." *Palaeogeography, Palaeoclimatology, Palaeoecology.* 392: 324–334.

Mallet, J. 2005. "Speciation in the 21st century. (Review of J. A. Coyne and H. A. Orr, 2004, Speciation)." *Heredity*. 95: 105–109.

Mallet, J. 2007. "Species, Concepts of." In *Encyclopedia of biodiversity*, vol. 5, S. Levin et al., eds., 427–440. Oxford: Elsevier.

Marco, A. S. 2007. "New occurrences of the extinct vulture *Gyps melitensis* (Falconiformes, Aves) and a reappraisal of the paleospecies." *Journal of Vertebrate Paleontology*. 27(4): 1057–1061.

Marie Curie Speciation Network. 2012. "What do we need to know about speciation?" *Trends in Ecology and Evolution*. 27(1): 27–39.

Marshall, J. C., E. Arevalo, E. Benavides, J. L. Sites, and J. W. Sites. 2006. "Delimiting species: Comparing methods for Mendelian characters using lizards of the *Sceloporus grammicus* (Squamata: Phrynosomatidae) complex." *Evolution*. 60: 1050–1065.

Mayden, R. L. 1997. "A hierarchy of species concepts: The denoument in the saga of the species problem." In *Species: The Units of Biodiversity*, M. F. Claridge, H. A. Dawah, and M. R. Wilson, eds., 381–423. London: Chapman & Hall.

Mayden, R. L. 1999. "Consilience and a hierarchy of species concepts: Advances toward closure of the species puzzle." *Journal of Nematology*. 31(2): 95–116.

Mayden, R. L. 2013. "Species, trees, characters, and concepts: Ongoing issues, diverse ideologies, and a time for reflection and change." In *The Species Problem—Ongoing Issues*, I. Y. Pavlinov, ed., 171–192. Rijeka, Croatia: InTech. http://www.intechopen.com/books/the-species-problem-ongoing-issues.

Mayr, E. 1942. *Systematics and the origin of species*. New York: Columbia University Press.

Mayr, E. 1957. "Species concepts and definitions." In *The species problem*, E. Mayr, ed., 1–22. American Association for the Advancement of Science, publication no. 50.

Mayr, E. 1963. *Animal species and evolution*. Cambridge, MA: Harvard University Press.

Mayr, E. 1969. *Principles of systematic zoology*. New York: McGraw-Hill.

Mayr, E. 1982. *The growth of biological thought*. Cambridge, MA: Harvard University Press.

Mayr, E. 1987. "The ontological status of species: Scientific progress and philosophical terminology." *Biology and Philosophy*. 2: 145–166.

Mayr, E., and P. D. Ashlock. 1991. *Principles of systematic zoology*. 2nd ed. New York: McGraw Hill.

McCune, A. R. 1987. "Toward the phylogeny of a fossil species flock: Semionotid fishes from a lake deposit in the Early Jurassic Towaco Formation, Newark Basin." *Peabody Museum of Natural History Bulletin* 43.

McCune, A. R., K. S. Thompson, and P. E. Olsen. 1984. "Semionotid fishes from the Mesozoic Great Lakes of North America." In *Evolution of fish species flocks*, A. A. Echelle and I. Kornfield, eds., 27–44. Orono: University of Maine Press.

McDade, L. A. 1995. "Species concepts and problems in practice: Insights from botanical monographs." *Systematic Botany*. 10(4): 606–622.

McGowran, B. 2008. *Biostratigraphy: Microfossils and geological time*. Cambridge: Cambridge University Press.

McKerrow, W. S. 1952. "Notes on the species and subspecies in palaeontology." *Geological Magazine*. 89: 148–151.

McRoberts, C. A. 2011. "Late Triassic Bivalvia (chiefly Halobiidae and Monotidae) from the Pardonet Formation, Williston Lake Area, Northeastern British Columbia, Canada." *Journal of Paleontology*. 85(4): 613–664.

Meier, R., and R. Willmann. 2000a. "The Hennigian species concept." In *Species concepts and phylogenetic theory: A debate*, Q. D. Wheeler and R. Meier, eds., 30–43. New York: Columbia University Press.

Michaux, B. 1987. "An analysis of allozymic characters of four species of New Zealand *Amalda* (Gastropoda: Olividae: Ancillinae)." *New Zealand Journal of Zoology*. 14: 359–366.

Michaux, B. 1989a. "Cladograms can reconstruct phylogenies: An example from the fossil record." *Alcheringa*. 13: 21–36.

Michaux, B. 1989b. "Morphological variation of species through time." *Biological Journal of the Linnean Society*. 38: 239–255.

Michaux, B. 1995. "Species concepts and the interpretation of fossil data." In *Speciation and the recognition concept*, D. M. Lambert and H. G. Spencer, eds., 45–56. Baltimore: Johns Hopkins University Press.

Miller, W., III. 2001. "The structure of species, outcomes of speciation, and the 'species problem': Ideas for paleobiology." *Palaeogeography, Palaeoclimatology, Palaeoecology*. 176: 1–10.

Miller, W., III. 2006. "What every paleontologist should know about species: New concepts and questions." *Neues Jahrbuch Geologie Paläontologie Mohatshefte*. 2006(9): 557–576.

Mishler, B. D. 1999. "Getting rid of species?" In *Species*, R. A. Wilson, ed., 307–315. Cambridge: MIT Press.

Mishler, B. D. 2010. "Species are not uniquely real biological entities." In *Contemporary debates in philosophy of biology*, F. J. Ayala and R. Arp, eds., 110–122. Chichester, UK: Wiley-Blackwell.

Mishler, B. D., and M. J. Donoghue. 1982. "Species concepts: A case for pluralism." *Systematic Zoology*. 31: 491–503.

Moore, R. C., C. G. Lalicker, and A. G. Fischer. 1953. *Invertebrate fossils*. New York: McGraw Hill.

Morrison, D. A. 2011. "Review of 'Species: a history of the idea' by J. S. Wilkins." *Systematic Biology*. 60: 239–241.

Nadachowski, A. 1993. "The species concept and Quaternary mammals." *Quaternary International*. 19: 9–11.

Naomi, S.-I. 2010. "On the integrated frameworks of species concepts: Mayden's hierarchy of species concepts and de Queiroz's unified concept of species." *Journal of Zoological Systematics and Evolutionary Research*. 49(3): 177–184.

Nardin, E., I. Rouget, and P. Neige. 2005. "Tendencies in paleontological practice when defining species, and consequences on biodiversity studies." *Geology*. 33(12): 969–972.

Nehm, R., and D. H. Geary. 1994. "A gradual morphologic transition during a rapid speciation event in marginellid gastropods (Neogene: Dominican Republic)." *Journal of Paleontology*. 68(4): 787–795.

Neige, P., D. Marchand, and B. Laurin. 1997. "Heterochronic differentiation of sexual dimorphs among Jurassic ammonite species." *Lethaia*. 30(2): 145–155.

Néraudeau, D. 2011. "The species concept in palaeontology: Ontogeny, variability, evolution." *Comptes Rendus Palevol*. 10: 71–75.

Newell, N. D. 1947. "Infraspecific categories in invertebrate paleontology." *Evolution*. 1: 163–171.

Newell, N. D. 1956. "Fossil populations." In *The species concept in palaeontology: A symposium*, P. C. Sylvester-Bradley, ed., 63–82. London: Systematics Association.

Nosil, P. 2008. "Speciation with gene flow could be common." *Molecular Ecology*. 17(9): 2103–2106.

O'Hara, R. J. 1993. "Systematic generalization, historical fate, and the species problem." *Systematic Biology*. 42(3): 231–246.

Pachut, J. F., and R. L. Anstey. 2009. "Inferring evolutionary modes in a fossil lineage (Bryozoa: *Peronopora*) from the Middle and Late Ordovician." *Paleobiology*. 35(2): 209–230.

Parham, J. F., P. C. J. Donoghue, C. J. Bell, T. D. Calway, J. J. Head, P. A. Holroyd, J. G. Inoue, R. B. Irmis, W. G. Joyce, D. T. Ksepka, J. S. L. Patané, N. D. Smith, J. E. Tarver, M. van Tuinen, Z. Yang, K. D. Angielczyk, J. M. Greenwood, C. A. Hipsley, L. Jacobs, P. J. Makovicky, J. Müller, K. T. Smith, J. M. Theodor, R. C. M. Warnock, and M. J. Benton. 2012. "Best practices for justifying fossil calibrations." *Systematic Biology*. 61(2): 346–359.

Paterson, H. E. H. 1981. "The continuing search for the unknown and unknowable: A critique of contemporary ideas on speciation." *South African Journal of Science*. 77: 113–119.

Paterson, H. E. H. 1982. "Perspective on speciation by reinforcement." *South African Journal of Science*. 78: 53–57.

Paterson, H. E. H. 1985. "The recognition concept of species." In *Species and speciation*, E. S. Vrba, ed., 21–30. Pretoria: Transvaal Museum.

Patterson, C., and D. E. Rosen. 1977. "Review of ichthyodectiform and other Mesozoic teleost fishes, and the theory and practice of classifying fossils." *Bulletin of the American Museum of Natural History*. 158(2): 81–172.

Pauers, M. J. 2010. "Species concepts, speciation, and taxonomic change in the Lake Malawi mbuna, with special reference to the genus *Labeotropheus* Ahl

1927 (Perciformes: Cichlidae)." *Reviews in Fish Biology and Fisheries.* 20: 187–202.

Paul, G. S. 2008. "A revised taxonomy of the iguanodont dinosaur genera and species." *Cretaceous Research.* 29: 192–216.

Pearson, P. N., and K. G. Harcourt-Brown. 2001. "Speciation and the fossil record." In *Encyclopedia of Life Sciences*, 1–5. London: Nature Publishing Group; West Sussex: John Wiley and Sons.

Perdices, A., I. Doadrio, I. M. Côté, A. Machordom, P. Economidis, and J. D. Reynolds. 2000. "Genetic divergence and origin of Mediterranean populations of the river blenny *Salaria fluviatilis* (Teleostei: Blenniidae)." *Copeia.* 2000(3): 723–731.

Perez-Ponce de Leon, G., and S. A. Nadler. 2010. "What we don't recognize can hurt us: A plea for awareness about cryptic species." *Journal of Parasitology.* 96: 453–464.

Pfennig, D. W., M. A. Wund, E. C. Snell-Rood, T. Cruickshank, C. D. Schlichting, and A. P. Moczek. 2010. "Phenotypic plasticity's impact on diversification and speciation." *Trends in Ecology and Evolution.* 25: 459–467.

Pfenninger, M., and K. Schwenk. 2007. "Cryptic animal species are homogeneously distributed among taxa and biogeographical regions." *BMC Evolutionary Biology.* 7: 121. doi:10.1186/1471-2148-7-121.

Pilbrow, V. 2010. "Dental and phylogeographic patterns of variation in gorillas." *Journal of Human Evolution.* 59(1): 16–34.

Plavcan, J. M. 1993. "Catarrhine dental variability and species recognition in the fossil record." In *Species, species concepts, and primate evolution*, W. H. Kimbel and L. Martin, eds., 239–263. New York: Plenum Press.

Plavcan, J. M., and D. A. Cope. 2001. "Metric variation and species recognition in the fossil record." *Evolutionary Anthropology.* 10: 204–222.

Polly, P. D. 1997. *Ancestry and species definition in paleontology: A stratocladistic analysis of Paleocene-Eocene Viverravidae (Mammalia, Carnivora) from Wyoming.* Ann Arbor: Museum of Paleontology, University of Michigan.

Poulin, R. 2010. "Uneven distribution of cryptic diversity among higher taxa of parasitic worms." *Biology Letters.* 7(2): 241–244.

Prothero, D. R. 2013. *Bringing fossils to life: An introduction to paleobiology.* 3rd ed. New York: Columbia University Press.

Provine, W. B. 1991. "Mechanisms of speciation: A review." In *Evolution of life. Fossils, molecules, and culture.* S. Osawa and T. Honjo, eds., 201–214. Toyko: Springer-Verlag.

Quintyn, C. 2009. "The naming of new species in hominin evolution: A radical proposal—A temporary cessation in assigning new names." *Homo: Journal of Comparative Human Biology.* 60(4): 307–341.

Raup, D. M., and S. M. Stanley. 1978. *Principles of paleontology.* 2nd ed. San Francisco: W. H. Freeman.

Reeves, P. A., and C. M. Richards. 2011. "Species delimitation under the general lineage concept: An empirical example using wild North American hops (Cannabaceae: *Humulus lupulus*)." *Systematic Biology.* 60(1): 45–59.

Regan, C. T. 1926. "Organic evolution." *Reports of the British Association for the Advancement of Science.* 1925: 75–86.

Reif, W.-E. 1985. "Endemic evolution of *Gyraulus kleini* in the Steinheim Basin (Planorbid snails, Miocene, southern Germany)." In *Sedimentary and evolutionary cycles*, U. Bayer and A. Seilacher, eds., 256–294. Berlin: Springer-Verlag.

Reydon, T. A. C. 2004. "Why does the species problem still persist?" *Bioessays.* 26(3): 300–305.

Richards, R. A. 2010. *The species problem: A philosophical analysis.* Cambridge, UK: Cambridge University Press.

Riedel, W. R. 1978. "Systems of morphologic descriptors in paleontology." *Journal of Paleontology.* 52(1): 1–7.

Rightmire, G. P. 2008. "*Homo* in the Middle Pleistocene: Hypodigms, variation, and species recognition." *Evolutionary Anthropology.* 17(1): 8–21.

Rissler, L. J., and J. J. Apodacta. 2007. "Adding more ecology into species delimitation: Ecological niche models and phylogeography help define cryptic species in the black salamander (*Aneides flavipunctatus*)." *Systematic Biology.* 56(6): 924–942.

Rodda, P. U., and W. L. Fisher. 1964. "Evolutionary features of *Athleta* (Eocene, Gastropoda) from the Gulf Coastal Plain." *Evolution.* 18(2): 235–244.

Roopnarine, P. D. 2001. "The description and classification of evolutionary mode: A computational approach." *Paleobiology.* 27(3): 446–465.

Rose, K. D., and T. M. Bown. 1986. "Gradual evolution and species discrimination in the fossil record." *Contributions to Geology, University of Wyoming*, Special Paper 3, pp. 119–130.

Rose, K. D., and T. M. Bown. 1993. "Species concepts and species recognition in fossil primates." In *Species, species concepts, and primate evolution*, W. H. Kimbel and L. Martin, eds., 299–330. New York: Plenum Press.

Rosenberg, A., 1985. *The structure of biological science.* Cambridge: Cambridge University Press.

Rosenberg, A. 1987. "Why does the nature of species matter? Comments on Ghiselin and Mayr." *Biology and philosophy.* 2(2): 192–197.

Rosenberg, A. 1994. *Instrumental biology or the disunity of science.* Chicago: University of Chicago Press.

Ross, K. G., D. Gotzek, M. S. Ascunce, and D. D. Shoemaker. 2010. "Species delimitation: A case study in a problematic ant taxon." *Systematic Biology.* 59(2): 162–184.

Roth, V. L. 1992. "Quantitative variation in elephant dentitions: Implications for the delimitation of fossil species." *Paleobiology.* 18: 184–202.

Ruse, M. 1995. "The species problem." In *Concepts, theories, and rationality in the biological sciences*, G. Wolters and J. Lennox eds., 171–193. Pittsburgh: University of Pittsburgh Press.

Sáez, A. G., and E. Lozano. 2005. "Body doubles." *Nature*. 433: 111.

Sage, R. D., and R. K. Selander. 1975. "Trophic radiation through polymorphism in cichlid fishes." *Proceedings of the National Academy of Sciences*. 72:4669–4673.

Samadi, S., P. David, and P. Jarne. 2000. "Variation of shell shape in the clonal snail *Melanoides tuberculata* and its consequences for the interpretation of fossil series." *Evolution*. 54: 492–502.

Sangster, G. 2014. "The application of species criteria in avian taxonomy and its implications for debate over species concepts." *Biological Reviews*. 89: 199–214.

Sauer, J., and B. Hausdorf. 2012. "A comparison of DNA-based methods for delimiting species in a Cretan land snail radiation reveals shortcomings of exclusively molecular taxonomy." *Cladistics*. 28(3): 300–316.

Saupe, E., and P. Selden. 2011. "The study of fossil spider species." *Comptes Rendus Palevol*. 10: 181–188.

Schneider, S., F. T. Fürsich, T. Schulz-Mirbach, and W. Werner. 2010. "Ecophenotypic plasticity versus evolutionary trends—morphological variability in Upper Jurassic bivalve shells from Portugal." *Acta Palaeontologica Polonica*. 55(4): 701–732.

Schopf, T. J. M. 1982. "A critical assessment of punctuated equilibrium. 1. Duration of taxa." *Evolution*. 36(6): 1144–1157.

Schwartz, J. H. 2009. "Reflections on systematics and phylogenetic reconstruction." *Acta Biotheoretica*. 57: 295–305.

Scrutton, C. T. 1996. "Ecophenotypic variation in the Early Silurian rugose coral *Palaeocyclus porpita*." *Proceedings of the Yorkshire Geological and Polytechnic Society*. 51(1): 1–8.

Seilacher, A., B. A. Matyja, and A. Wierzbowski. 1985. "Oyster beds: Morphologic response to changing substrate conditions." In *Sedimentary and evolutionary cycles*, U. Bayer and A. Seilacher, eds., 421–435. Berlin: Springer-Verlag.

Sepkoski, D. 2012. *Rereading the fossil record: The growth of paleobiology as an evolutionary discipline*. Chicago: University of Chicago Press.

Sepkoski, J. J., Jr. 1998. "Rates of speciation in the fossil record." *Philosophical Transactions of the Royal Society of London. Series B, Biological Sciences*. 353: 315–326.

Shaw, F. C., and P. J. Lespérance. 1994. "North American biogeography and taxonomy of *Cryptolithus* (Trilobita, Ordovician)." *Journal of Paleontology*. 68(4): 808–823.

Simons, E. L., and D. R. Pilbeam. 1965. "Preliminary revision of the Dryopithecinae Pongidae Anthropoidea." *Folia Primatologia*. 3: 81–152.

Simpson, G. G. 1951. "The species concept." *Evolution*. 5: 285–298.

Simpson, G. G. 1953. *Major features of evolution*. New York: Columbia University Press.

Simpson, G. G. 1961. *Principles of animal taxonomy.* New York: Columbia University Press.

Sites, J. W., and J. C. Marshall. 2003. "Delimiting species: A renaissance issue in systematic biology." *Trends in Ecology and Evolution.* 18: 462–470.

Sites, J. W., and J. C. Marshall. 2004. "Operational criteria for delimiting species." *Annual Review of Ecology and Systematics.* 35: 199–227.

Sluys, R., and C. J. Hazevoet. 1999. "Pluralism in species concepts: Dividing nature at its diverse joints." *Species Diversity.* 4: 242–256.

Smith, A. B. 1994. *Systematics and the fossil record: Documenting evolutionary patterns.* Oxford: Blackwell.

Smith, A. B. 2007. "Marine diversity through the Phanerozoic: Problems and prospects." *Journal of the Geological Society.* 164: 731–745.

Sneath, P. H. A., and R. R. Sokal. 1973. *Numerical taxonomy.* San Francisco: W. H. Freeman.

Sokal, R. R., and T. J. Crovello. 1970. "The biological species concept: A critical evaluation." *American Naturalist.* 104:127–153.

Sorauf, J. E. 1998. "Frasnian (Upper Devonian) rugose corals from the Lime Creek and Shell Rock Formations of Iowa." *Bulletins of American Paleontology.* 355: 1–159.

Stamos, D. N. 2003. *The species problem: Biological species, ontology, and the metaphysics of biology.* Lanham, MD: Lexington Books.

Stanford, P. K. 1995. "For pluralism and against realism about species." *Philosophy of Science.* 62:70–91.

Stanley, S. M. 1979. *Macroevolution: Pattern and process.* San Francisco: W. H. Freeman.

Stanley, S. M., and X. Yang. 1987. "Approximate evolutionary stasis for bivalve morphology over millions of years: A multivariate, multilineage study." *Paleobiology.* 13(2): 113–139.

Stearn, C. W., and R. L. Carroll. 1989. *Paleontology: The record of life.* New York: Wiley.

Stenzel, H. B. 1949. "Successional speciation in paleontology: The case of the oysters of the *sellaeformis* stock." *Evolution.* 3(1): 34–50.

Stewart, J. R. 2002. "The evidence for the timing of speciation of modern continental birds and the taxonomic ambiguity of the Quaternary fossil record." In *Proceedings of the 5th Symposium of the Society of Avian Paleontology and Evolution*, Beijing, 1–4 June 2000, Z. Zhou and F. Zhang, eds., 259–280.. Beijing: Science Press.

Stuart, B. L., R. F. Inger, and H. K. Voris. 2006. "High level of cryptic species diversity revealed by sympatric lineages of Southeast Asian forest frogs." *Biology Letters.* 2: 470–474.

Sykes, R. M., and J. H. Callomon. 1979. "The *Amoeboceras* zonation of the Boreal Upper Oxfordian." *Palaeontology.* 22(4): 839–903.

Sylvester-Bradley, P. C., ed. 1956a. *The species concept in paleontology.* Systematics Association Publication no. 2. London: Systematics Association.

Sylvester-Bradley, P. C. 1956b. "The new palaeontology." In *The species concept in palaeontology*, P. C. Sylvester-Bradley, ed., 1–8. London: Systematics Association.

Sylvester-Bradley, P. C. 1958. "The description of fossil populations." *Journal of Paleontology.* 32(1): 214–235.

Szalay, F. S. 1993. "Species concepts: The tested, the untestable, and the redundant." In *Species, species concepts, and primate evolution*, W. H. Kimbel and L. B. Martin, eds., 21–42. New York: Plenum Press.

Tapanila, L., and J. Pruitt. 2013. "Unraveling species concepts for the *Helicoprion* tooth whorl." *Journal of Paleontology*, 87(6): 965–983.

Tasch, P. 1980. *Paleobiology of the invertebrates: Data retrieval from the fossil record.* 2nd ed. New York: John Wiley & Sons.

Tattersall, I. 1986. "Species recognition in human paleontology." *Journal of Human Evolution.* 15: 165–175.

Tattersall, I. 1992. "Species concepts and species identification in human evolution." *Journal of Human Evolution.* 22: 341–349.

Tattersall, I. 2007. "Neanderthals, *Homo sapiens*, and the question of species in paleoanthropology." *Journal of Anthropological Sciences.* 85: 139–146.

Tattersall, I., and K. Mowbray. 2005. "Species and paleoanthropology." *Theory in Biosciences.* 123: 371–379.

Teichert, C. 1949. *Permian crinoid Calceolispongia.* New York: Geological Society of America.

Templeton, A. R. 1989. "The meaning of species and speciation." In *Speciation and its consequences*, D. Otte and J. Endler, eds., 3–27. Sunderland, MA: Sinauer Associates.

Tepper, C., L. Squiers, C. Hay, D. Gorbach, D. Friend, B. Black, B. Greenstein, and K. Strychar. 2012. "Cryptic species: A mismatch between genetics and morphology in *Millepora*." *Marine Science.* 2(5): 57–65.

Thomas, G. 1956. "The species conflict." In *The species concept in palaeontology*, P. C. Sylvester-Bradley, ed., 17–32. London: Systematics Association.

Trewavas, E. 1973. "What Tate Regan said in 1925." *Systematic Zoology.* 22: 92–93.

Trontelj, P., and C. Fišer. 2009. "Perspectives: Cryptic species diversity should not be trivialised." *Systematics and Biodiversity.* 7(1): 1–3.

Trueman, A. E. 1924. "The species concept in palaeontology." *Geological Magazine.* 61: 355–360.

Turner, A. 1986. "Species, speciation, and human evolution." *Human Evolution.* 1: 419–430.

Upchurch, P., P. M. Barrett, and P. Dodson. 2004. "Sauropoda." In *The Dinosauria*, D. B. Weishampel, P. Dodson, and H. Osmolska, eds., 259–322. Berkeley: University of California Press.

Van Bocxlaer, B., and R. Schultheiß. 2010. "Comparison of morphometric techniques for shapes with few homologous landmarks based on machine-learning approaches to biological discrimination." *Paleobiology.* 36(3): 497–515.

Vianey-Liaud, M., H. Gomes-Rodrigues, and J. Michaux. 2011. "L'espèce en paléontologie: De l'utilisation du binôme linnéen chez les rongeurs fossiles (Mammalia, Rodentia)." *Comptes Rendus Palevol.* 10: 117–131.

Vokes, H. E., and Vokes, E. H. 1992. "Neogene paleontology in the northern Dominican Republic. 12. The genus *Spondylus* (Bivalvia: Spondylidae)." *Bulletins of American Paleontology.* 102: 5–13.

Vrba, E. S. 1980. "Evolution, species, and fossils: How does life evolve?" *South African Journal of Science.* 76: 61–84.

Waller, T. 1969. "The evolution of the *Argopecten gibbus* stock (Mollusca, Bivalvia), with emphasis on the Tertiary and Quaternary species of eastern North America." *Paleontological Society Memoir.* 3: 1–125.

Waller, T. R. 2011. "Neogene paleontology of the northern Dominican Republic. 24. Propeamussiidae and Pectinidae (Mollusca: Bivalvia: Pectinoidea) of the Cibao Valley." *Bulletins of American Paleontology.* 2011 (381): 1–205.

Webber, A. J., and B. R. Hunda. 2007. "Quantitatively comparing morphological trends to environment in the fossil record (Cincinnatian Series; Upper Ordovician)." *Evolution.* 61(6): 1455–1465.

Webster, M. 2009. "Ontogeny, systematics, and evolution of the effaced Early Cambrian trilobites *Peachella* and *Eopeachella* new genus (Olenelloidea)." *Journal of Paleontology.* 83(2): 197–218.

Webster, M. 2011. "The structure of cranidial shape variation in three early ptychoparioid trilobite species from the Dyeran-Delamaran (traditional "lower-middle" Cambrian) boundary interval of Nevada, USA." *Journal of Paleontology.* 85(2): 179–225.

Westrop, S. R., and J. M. Adrain. 2007. "*Bartonaspis* new genus, a trilobite species complex from the base of the Upper Cambrian Sunwaptan Stage in North America." *Canadian Journal of Earth Sciences.* 44(7): 987–1003.

Wheeler, Q. D. 2007. "Invertebrate systematics or spineless taxonomy?" *Zootaxa.* 1668: 11–18.

Wheeler, Q. D., and R. Meier, eds. 2000. *Species concepts and phylogenetic theory: A debate.* New York: Columbia University Press.

White, T. 2003. "Early hominids—Diversity or distortion?" *Science.* 299: 1994–1997.

White, T. 2013. "Paleoanthropology: Five's a crowd in our family tree." *Current Biology.* 23(3): R112–R115.

Wiens, J. J. 2004a. "What is speciation and how should we study it?" *American Naturalist.* 163(6): 914–923.

Wiens, J. J. 2004b. "Speciation and ecology revisited: Phylogenetic niche conservatism and the origin of species." *Evolution.* 58(1): 193–197.

Wiens, J. J. 2007. "Species delimitation: New approaches for discovering diversity."
    Systematic Biology. 56(6): 875–878.

Wiens, J. J., and M. R. Servedio. 2000. "Species delimitation in systematics: Infer-
    ring diagnostic differences between species." *Proceedings of the Royal Society
    of London B*. 267: 631–636.

Wiley, E. O. 1978. "The evolutionary species concept reconsidered." *Systematic
    Zoology*. 27: 17–26.

Wiley, E. O. 1981. *Phylogenetics*. New York: John Wiley and Sons.

Wiley, E. O. 2002. "On species and speciation with reference to the fishes." *Fish
    and Fisheries*. 3: 1–10.

Wiley, E. O., and B. S. Lieberman. 2011. *Phylogenetics: The theory of phylogenetic
    systematics*. 2nd ed. Hoboken, NJ: Wiley-Blackwell.

Wiley, E. O., and R. L. Mayden. 2000a. "The evolutionary species concept." In
    Species concepts and phylogenetic systematics: A debate, Q. D. Wheeler and
    R. Meier, eds., 70–89. New York: Columbia University Press.

Wiley, E. O., and R. L. Mayden. 2000b. "A critique from the evolutionary spe-
    cies concept perspective." In *Species concepts and phylogenetic systematics: A
    debate*, Q. D. Wheeler and R. Meier, eds., 146–158. New York: Columbia Uni-
    versity Press.

Wiley, E. O., and R. L. Mayden. 2000c. "A defense of the evolutionary species
    concept." In *Species concepts and phylogenetic systematics: A debate*, Q. D.
    Wheeler and R. Meier, eds., 198–208. New York: Columbia University Press.

Wilke, T., and A. Falniowski. 2001. "The genus *Adriohydrobia* (Hydrobiidae: Gas-
    tropoda): Polytypic species or polymorphic populations?" *Journal of Zoologi-
    cal Systematics and Evolutionary Research*. 39(4): 227–234.

Wilkins, J. S. 2006. "A list of 26 species concepts." http://scienceblogs.com/evolving
    thoughts/2006/10/01/a-list-of-26-species-concepts/

Wilkins, J. S. 2009. *Species: A history of the idea*. Berkeley: University of California
    Press.

Williams, G. C. 1992. *Natural selection: Domains, levels, and applications*. Oxford:
    Oxford University Press.

Wilmott, A. J. 1949. "Intraspecific categories of variation." In *British flowering
    plants and modern systematic methods*, A. J. Wilmott, ed., 28–45. London: Tay-
    lor and Francis.

Wood, B. 1992. "Early hominid species and speciation." *Journal of Human Evolu-
    tion*. 22: 351–365.

Wood, B. 1993. "Early *Homo*: How many species?" In *Species, species concepts,
    and primate evolution*, W. H. Kimbel and L. Martin, eds., 485–522. New York:
    Plenum Press.

Wood, B., and N. Lonergan. 2008. "The hominin fossil record: Taxa, grades, and
    clades." *Journal of Anatomy*. 212: 354–376.

Wood, R. M., and R. L. Mayden. 2002. "Speciation and anagenesis in the genus

Cyprinella of Mexico (Teleostei: Cyprinidae): A case study of model III allopatric speciation." *Reviews in Fish Biology and Fisheries.* 12: 253–271.

Yates, A. M. 2003. "A new species of the primitive dinosaur *Thecodontosaurus* (Saurischia: Sauropodomorpha) and its implications for the systematics of early dinosaurs." *Journal of Systematic Palaeontology.* 1(1): 1–42.

Zapata, F., and I. Jiménez. 2012. "Species delimitation: Inferring gaps in morphology across geography." *Systematic Biology.* 61(2): 179–194.

The Stages of Speciation: A Stepwise Framework for Analysis of Speciation in the Fossil Record

Warren D. Allmon and Scott D. Sampson

Neither neobiologists nor paleobiologists can usually "see" speciation in action; both are left to infer what might have happened from a variety of indirect evidence.—Erwin and Anstey 1995, 11

Introduction

To the degree that "species" as identified among living organisms can be recognized in the fossil record (Allmon, chapter 3, this volume), other important methodological and substantive questions about species and speciation in the fossil record remain. To what degree can fossil species be studied as modern species are? What, if anything, can we learn about species and speciation from the fossil record that we could not learn from studying them only in the Recent? These questions can collectively be called the "fossil species study problem" (Allmon 2013, chapter 3 of this volume). As illustrated by the papers in this volume (as well as many others), numerous authors have attempted to address this problem. Yet, as also illustrated by these contributions, little in the way of standardization of method exists for such studies. Natural history is about both particulars and relative frequency (Gould 1991), and a lack of standard methods makes relative frequency very difficult to judge. Not all instances of speciation will be the same, of course. Yet to the degree that commonalities are present across at least biparental animals, a common approach or framework has potential to facilitate comparison and estimation of general patterns.

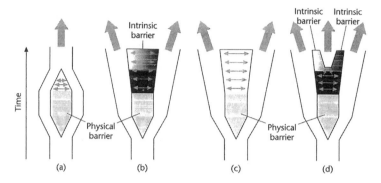

Figure 4.1 Outcomes of contact between previously isolated populations. "Possible outcomes of contact between populations that were previously isolated by a physical barrier. Broad arrows represent evolutionary lineages (species). Narrow horizontal arrows indicate gene flow between otherwise independently evolving lineages. The region between diverging lineages depicts the evolution of intrinsic isolating barriers (white, no intrinsic barrier; black, weak intrinsic barrier; gray, strong intrinsic barrier). (a) The populations merge and evolve as a single species; (b) reinforcement completes the formation of reproductive isolation initiated in allopatry; (c) the populations retain genetic identity, but form hybrid swarms in zones of contact; (d) the recombination of differentiated genomes results in a new, 'hybrid' species." (Wood and Rieseberg 2002, 416)

Species are evolutionarily distinct lineages (Allmon, chapter 3, this volume; Miller, chapter 2, this volume), and their formation—speciation—is the origin of such lineages (Simpson 1951; Turelli et al. 2001; Gavrilets 2004b; Wiens 2004a; Wiley and Lieberman 2011). Although it can appear abrupt on geological timescales, at ecological timescales speciation (by which we here mean the multiplication of species, not just their transformation) is usually not an "event," but rather the outcome of a more or less continuous process—a sequence of changes in populations over some length of time (e.g., Grant and Grant 2006; Nosil 2008; Butlin et al. 2009; Rieppel 2009; Seehausen 2009), from thousands to millions of years (McCune 1997; Coyne and Orr 2004; Allmon and Smith 2011, 252; Etienne and Rosindell 2012; Norris and Hull 2012; Rosenblum et al. 2012). During this process, the level of "reproductive isolation can vary from none, to weak, to intermediate, to strong, to complete" (Nosil 2008, 27) (fig. 4.1). Speciation is generally "considered complete when reproductive isolation is strong or near complete, or when genotypic clusters become largely non-overlapping (i.e., a discontinuity has developed)" (Nosil 2008, 27). A multitude of different factors, extrinsic and intrinsic to organisms, can affect this process along the way, but all instances of species formation must proceed along a generally similar path, from a single evolutionary entity to multiple ones.

Speciation is usually recognizable only in retrospect, or by comparing different species or populations that appear to be at different points in the process. Identification of equivalent points, or stages, however, is challenging. Because the processes of speciation are continuous, carving speciation up into separate phases or stages will always be somewhat arbitrary. Anyone seeking to do so faces additional difficulty from the fact that we are almost never able to observe the process of speciation from start to finish. Further complications result from speciation varying in different taxa and different environmental conditions.

Speciation can be studied at a variety of temporal scales, from small changes over a few generations to major changes over millions of years. At the smallest scale (ca. 10^1–10^3 years), many studies of speciation in living organisms focus on the acquisition of differences that function as "reproductive barriers" or "reproductive isolating mechanisms" between populations (e.g., Coyne and Orr 2004), or on the role of selection or adaptation (e.g., Schluter 2000; Nosil 2012). At slightly larger scales (ca. 10^3–10^5 years), other studies focus on the geography of speciation (e.g., allopatry vs. sympatry). Most neontological studies of speciation are at these temporal levels. At larger scales (ca. >10^5 years), investigations typically focus on a small number of individual species-level lineages or speciation events over thousands to millions of years, and attempt to determine specifics surrounding the pattern and process of speciation. Many paleontological studies of speciation (table 4.1) occur at these scales. At still larger scales, speciation (often called, agnostically, "origination") is examined via patterns of taxonomic diversity over millions to tens of millions of years (e.g., Sepkoski 1998; Lieberman 2001a,b). Much recent paleontological attention has been devoted to this largest scale, perhaps (as suggested by Alroy 2009, 301) at the expense of species-level studies.

A major challenge for evolutionary biologists at all of these scales is to determine which of many possible factors have had the greatest impact on speciation within a given clade, and whether or not general principles apply to the origin of species across taxa. Although investigators commonly invoke one or two specific causal mechanisms as the key factor influencing speciation in any particular case, much more rarely is it made explicit exactly how this influence has occurred, or how other factors might have been involved. More generally, theoretical insights into the nature of speciation have been obscured by the frequent failure to correlate a given mechanism

TABLE 4.1 **Some examples of lineage-level studies of speciation in various taxonomic groups of animals in the fossil record.**

Taxon	Reference
Corals	Budd (2010); Budd and Pandolfi (chapter 7, this volume)
Bivalves	Waller (1969); Hallam (1982); Stanley (1986); Stanley and Yang (1987); Johnson (1993) Roopnarine (1995); Crampton and Gale (2005)
Gastropods (terrestrial and freshwater)	Gould (1969); Williamson (1981); Reif (1983b); Chiba (2007); Glaubrecht (2011)
Gastropods (marine)	Fisher et al. (1964); Rodda and Fisher (1964); Schindel (1982); Michaux (1987); McKinney and Allmon (1995); Geary (1990a,b, 1995); Nehm and Geary (1994); Allmon and Smith (2011)
Ammonoids	Yacobucci (1999, chapter 8 of this volume)
Brachiopods	Ager (1983); Johnson (1975)
Bryozoans	Jackson and Cheetham (1990, 1994); Anstey and Pachut (2004)
Trilobites	Eldredge (1971, 1974); Abe and Lieberman (2009, 2012); Adrain and Westrop (2005, 2006); Hopkins and Webster (2009)
Crustaceans (other than ostracodes)	Tasch (1979); Yamaguchi (1980); Rode and Lieberman (2005)
Ostracodes	Cronin (1985, 1987); Kamiya (1992)
Echinoids	Smith (1984); McNamara (1987)
Conodonts	Dzik (1999)
Fishes	McCune (1996, 1997, 2004); Smith (1987); Bell and Haglund (1982)
Birds	Stewart (2002)
Mammals	Martin (1993); Flynn et al. (1995); Gingerich (1976); Masters (1993); Vrba (1987); Wilson (1969)

with a particular component of the speciation process. For example, to what degree does environmental change lead to speciation because of alteration of patterns of gene flow or because of new selection regimes (e.g., Thorp et al. 2008; Surget-Groba et al. 2012)? This problem is especially important in paleontology, where actual events and processes cannot be directly observed. If we are to identify and analyze speciation on both ecological (microevolutionary) and geological (macroevolutionary) timescales, it may be helpful to divide speciation into its components so that we can try to identify exactly where and when different changes occur.

In short, one thing that seems to be missing from the speciation discussion is a consistent theoretical framework that would enable us to dissect the process—to compare what happens within and between clades, and thus construct and pursue a research agenda that would elucidate the causes of speciation, especially over deep timescales that only the fossil record can reveal.

Stages of Speciation

A four-stage framework

Although the processes of speciation among animals vary, at least four elements appear to be common to all (figs. 4.2, 4.3): (1) isolated populations (sometimes called "daughter" populations or "incipient species") must form, becoming separate from parent populations; (2) these populations must persist in isolation, neither going extinct nor merging with the parental population by interbreeding should the opportunity arise; (3) these populations must diverge (become differentiated) genetically from the parent such that they will maintain their separate evolutionary status and not merge with the parent; and (4) these new species must stabilize or expand their population sufficiently to survive long enough to play a separate evolutionary role. Speciation requires all four of these elements or stages; it cannot be said to occur if any one of them fails. Notwithstanding difficulties

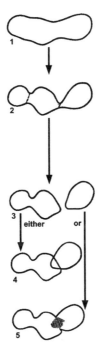

Figure 4.2 "Stages of speciation."
From Mayr (1942, 160).

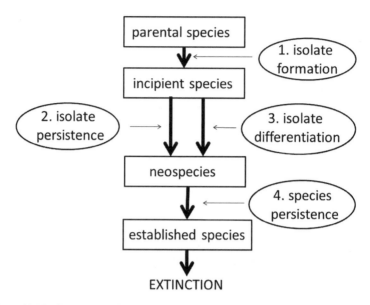

Figure 4.3 The Stages of Speciation (SOS) framework.
Rectangles represent the status of populations, from parental species to extinction of a descendant (daughter) species. Numbered ovals indicate the four stages of the speciation process. Downward-pointing arrows represent the relative timing of each of the four stages. Stages 2 (isolate persistence) and 3 (isolate differentiation) can occur simultaneously; persistence, however, is required for differentiation to proceed, whereas the reverse is not true.

in identification, these stages are not entirely arbitrary or merely heuristic. They are biologically distinct phenomena, even though they may frequently involve common mechanisms and/or overlap in time.

In this chapter, we expand upon earlier studies (Allmon 1992, 1994, 2001, 2003, 2009; McKinney and Allmon 1995; Allmon et al. 1998; Sampson 1999; Allmon and Ross 2001; Allmon and Smith 2011; Allmon and Martin 2014) to more formally propose the stages of speciation" (SOS) framework for studying speciation in fossil animals. We use the word *framework* (sensu Ravitch and Riggan 2012) because we regard the approach advocated here as a theoretical scaffold upon which to reconstruct and compare species originations, both pattern and process. We argue that this four-stage framework permits greater resolution of the impact of the causal factors involved, and so can serve as a standard method for studying speciation, especially in the fossil record, offering a useful tool for identifying the effects of extrinsic and intrinsic factors acting at various times during the speciation process.

Definitions

When referring to living species, we use the biological species concept: species are groups of actually or potentially interbreeding natural populations that are reproductively isolated from other such groups (Mayr 1963; Coyne and Orr 2004). For fossil species, we use a modification of the general lineage concept (which is approximately equivalent to the evolutionary species concept), as discussed in Allmon (chapter 3, this volume): *fossil species are separately evolving lineages recognized as groups of morphologically distinct populations within which variation is of the magnitude displayed by closely related, or presumably analogous, living species and their local populations, and between which the differences are of the kind and degree expected to result from reproductive isolation of populations in such related or analogous species.*

In this paper, *isolate* and *isolated population* refer to a population or group of populations prevented from free gene exchange with other such populations of the same species, and having the potential to become a new species (cf. Mayr 1963, 366; Stanley 1979, 195; Taylor 1990, 429); this usage is equivalent to "demes" of Damuth (1985). A *metapopulation* is an ensemble of such populations connected by occasionally dispersing individuals (Hanski and Gilpin 1991). Both populations and metapopulations have finite lifetimes, that is, expected times to extinction (Hanski and Gilpin 1991). The processes of population formation, persistence, and differentiation occur on what Hanski and Gilpin call the "metapopulation scale . . . at which individuals infrequently move from one place (population) to another, typically across habitat types which are not suitable for their feeding and breeding activities, and often with substantial risk of failing to locate another suitable habitat patch in which to settle" (1991, 7).

All isolates are *incipient species*; that is, populations that have the potential to become established species (Dobzhansky 1940; Mayr 1942). We refer to newly formed species (i.e., those incipient species that have achieved effective reproductive isolation) as *neospecies* (Carson 1976). The point at which a population should be called a neospecies is arbitrary; there is "a sliding scale" for assigning species status (fig. 4.1). "As reproductive barriers become stronger, taxa become more 'species-like', and when reproductive isolation is complete we consider taxa to be 'good species'" (Coyne and Orr 2004, 31, 34; the term "good species" can also refer to reproductively isolated species that are also morphologically distinct; see Allmon, chapter 3, this volume).

A persistent neospecies is referred to here as an *established species*. Although designating a species as "established" is somewhat arbitrary, this distinction is an important and very real one because it indicates that a species has expanded sufficiently in population size and geographic range to have a reasonable probability of entering the fossil record (see below).

Stage 1: Isolate formation

For a population to have the potential to become a distinct evolutionary lineage, it must somehow separate from other populations; that is, gene flow must cease. The exact nature of this separation varies. Mayr (1942, 1963) maintained that, at least among sexually reproducing organisms, cessation of gene flow is almost always associated with geographic separation (see discussion in Coyne 1994; Nosil 2008), and geographic isolation came to be widely accepted as the most common mode of speciation in animals (Coyne 1994; Coyne and Orr 2004; Mallet 2008a; Nosil 2008). Considerable evidence, however, has since accumulated detailing exceptions to this description; that is, genetic isolation apparently can, under certain circumstances, occur without large-scale significant geographic isolation (i.e., "sympatric speciation"; Via 2001; Bolnick and Fitzpatrick 2007; Fitzpatrick et al. 2009; Bird et al. 2012), and significant divergence is possible in the presence of gene flow (e.g., Turelli et al. 2001; Nosil 2008, 2012; Feder et al. 2012). Even without geographic isolation, however, gene flow still must cease (or be dramatically reduced) for populations to have a chance of becoming separate species. The SOS framework therefore applies equally to all of the various "models" of speciation (table 4.2).

Geographic isolation can occur in many ways (e.g., Mayr 1942, 1963; Wiley 1981; Futuyma 2002; Wiley and Lieberman 2011). However it occurs, it has long been thought that many more daughter populations form than go on to become established descendant species. Mayr (1963, 367), for example, admitted that "we know, as yet, little about the frequency of genuine isolates in various groups of animals." Yet he claimed that "most species bud off peripheral isolates at regular intervals . . . [and] nearly all of them either reestablish contact with the parental species or else die out" (Mayr 1963, 554), and he estimated that "peripheral isolates are produced 50 or 100 or 500 times as frequently as new species" (1963, 513). He noted that four families of birds for which he tabulated data have many more isolates than subspecies, and many more weakly than strongly differentiated isolates. Comparing the list of species that seem to have originated in a single postulated Pleistocene forest refuge in Amazonia with the list

TABLE 4.2 **Classification of models of speciation.**

	ecological speciation		nonecological speciation		
geographic mode	by-product	competitive	ecological persistence	drift, etc.	sexual selection
Allopatric – **Vicariance 1** (parent ≅ daughter; bifurcation, dumbbell allopatry)					
Allopatric – **Vicariance 2** (parent >> daughter; = peripatry)					
Allopatric – Dispersal (parent >> daughter; = peripatry)					
Parapatric					
Sympatric					

(left margin label: ← continuum →)

of species that could potentially have been isolated in that refuge, Mayr (1969, 16) found the number of new, derived species to be relatively small, which he explained in part by the extinction of a large proportion of the isolated incipient species. Stanley (1979, 175) similarly stated that "some unknown percentage of the myriads of isolates issuing from any species must technically attain the status of a distinct species without actually being recognized as such because they are snuffed out before expanding to become fully established." Stanley (1978) labeled these ephemeral isolates "aborted species."

Direct field studies of the fates of populations in the context of speciation are few. Such studies are more numerous, however, in the conservation and population biology literature (e.g., Berger 1990; Anstett et al. 1997; Morris and Doak 2002), and, to a lesser extent, in ecology. For example, some ecological studies of marine gastropods suggest that species are broken into many small short-lived local populations that are repeatedly reestablished by recolonization (Spight 1974; Quinn et al. 1989). Similarly, Cain and Cook (1989) followed 8 replicate populations of the land snail *Cepaea* in enclosures over almost 20 years; 7 of the populations became extinct and the 8th almost did. Studies of the fates of local populations within metapopulations also suggest high rates of population extinction

TABLE 4.3 **Examples of factors that may influence speciation at each of the stages of the SOS framework, sorted by whether they are studyable in living or fossil species.**

Stage	Mechanism/evidence— neontological	Mechanism/evidence— paleontological
Isolate formation	• Environmental disturbance • Dispersal ability	• Environmental disturbance • Dispersal ability
Isolate persistence	• Permissive environment • Adaptation	• Permissive environment • Adaptation
Isolate differentiation	• Environment • Selection • Genetics • Behavior	• Environment
Persistence of species	• Environment • Abundance • Competition	• Environment • Abundance

(e.g., Ebenhard 1991; Harrison 1991, and references therein). All of these studies, as well as modeling (e.g., Ludwig 1996; Middleton and Nisbet 1997; Frank and Wissel 2002; Morris and Doak 2002), suggest that the great majority of populations have relatively short durations (compared to the species as a whole). This implies that frequently only a small number of populations persist long enough to become neospecies.

If we examine the processes by which these populations are produced, it is clear that a variety of causal factors may be involved (table 4.3), including: extrinsic processes that divide parental populations by vicariance events (such as habitat disturbance); availability of habitat patches beyond the geographic range of the parental species; and intrinsic features of organisms and populations that increase the likelihood of geographic isolation (e.g. dispersal ability, population structure, and resource specialization).

Extrinsic factors potentially increasing the rate of isolate formation, such as environmental disturbance or shifts in predation intensity, can often be identified based on geological or paleontological evidence (e.g., Cracraft 1985; Stanley 1986; Cronin 1985, 1987; Cronin and Ikeya 1990; Allmon and Smith 2011). For example, Abe and Lieberman (2012) argued that an evolutionary radiation of Devonian trilobites was driven more by geographic isolation (largely the result of sea-level rise and fall) rather than competition. Allmon et al. (1998) proposed an intermediate disturbance hypothesis that suggested maximal speciation would occur at moderate levels of environmental perturbation; too little would produce few isolates, whereas too much would cause too many isolates to become

extinct. Intrinsic factors affecting isolate formation, such as dispersal ability, may also be recognizable in fossils (e.g., Hansen 1978, 1982; Jablonski
and Lutz 1983; Taylor 1989).

Stage 2: Isolate persistence

If, as discussed above, many more isolated populations form than go on
to become neospecies, then the "supply" of isolates is not a limiting factor
in controlling the occurrence of speciation. It is then to the other stages of
the speciation process that we must look.

Once formed, isolated populations face four possible fates (fig. 4.2; table 4.3; Mayr 1969, 13): (1) disappearance by merging with the parent population or other daughter populations; (2) extinction by the death of all of
their component individuals; (3) persistence in isolation without achieving
reproductive isolation; or (4) persistence and differentiation to become
neospecies. Different factors have different effects on the probabilities of
each of these outcomes.

Isolate persistence involves at least two aspects: persistence of the population and persistence of isolation. Isolation can be maintained by geography (distance or barriers) and/or by ecological and environmental differences, such as limited or biased dispersal or selection against migrants due
to local adaptation (see Bradburd et al. 2013). The persistence of the condition of isolation may be gradational with the events or processes that led
originally to the formation of the isolated population, or may involve new
conditions. Persistence of isolates may be more common, for example, in
"novel environments" that lack predators or competitors (Schluter 2000,
79). This has been called the ecological persistence model of speciation,
in contrast to ecological speciation (Schluter 2000, 189); see below.

To the extent that isolated populations are vulnerable to extinction by
the death of all of their constituent individuals, factors conducive to the
survival of those individuals will lead to higher probability of isolate persistence (cf. Stanley 1979; Glazier 1987; Allmon 1992). Such factors might
include those intrinsic to organisms, such as adaptation, as well as extrinsic factors such as the nature of the environment. Higher abundance and
larger geographic ranges (Blackburn and Gaston 1997; Gaston and Blackburn 2000; Gaston 2003) are widely thought to reduce the probability of
extinction of neospecies (Hutchinson 1959; MacArthur and Wilson 1967;
Jackson 1974; Leigh 1981; Diamond 1984; Stanley 1986; Lande 1998; Cardillo et al. 2005; Payne and Finnegan 2007; see "Stage 4," below), and this

likely pertains as well to isolated populations that are not yet reproductively isolated (Allmon 1992). Isolated populations that are—for either extrinsic or intrinsic reasons—able to expand rapidly to a larger size, and concomitantly larger ranges, will have a greater chance of persistence than populations that lack this capacity (e.g., Ebenhard 1991).

Survival of individuals can be influenced by the "rigor" (Hutchinson 1959) or "permissiveness" (Vermeij 2002; Hardie and Hutchings 2010) of the environment—that is, how metabolically expensive or otherwise difficult is it for the average individual organism to survive and reproduce. All other factors being equal, isolates that occur in less rigorous, more permissive environments will show higher rates of persistence than isolates in more challenging environments (Stanley 1979; Allmon 1992; Vermeij 2012).

Populations of individuals that possess morphological, behavioral, or biochemical characteristics that enhance their relative fitness in their local environments ("aptations" sensu Gould and Vrba 1982) should be able to persist longer than populations of individuals that lack such features. Several authors (e.g., Stanley 1979; Glazier 1987) have stressed that, to the degree that daughter populations from the same parent differ in characteristics that render them differentially susceptible to extinction, sorting will take place at the level of the isolate. If so, any characteristic "which would increase survivorship is expected to be preferentially represented in species newly formed from those isolates" (Glazier 1987, 325). The result of such a process, Stanley (1979, 197) suggests, will be bias in the direction of speciation, since descendant species will have been drawn from only a subset of ancestral variation (see also Lloyd 1988, on "avatar selection"). Importantly, morphological novelties recognizable in fossils may be identified as adaptive, or at least as functional (e.g. Hickman 1988), with possible impacts on the rate of isolate persistence (e.g., Vermeij 1987).

Stage 3: Isolate differentiation

In the context of speciation, differentiation of isolated populations manifests in two ways: development of reproductive isolation between populations, and production of phenotypic differences between populations. These are not necessarily correlated, as indicated by the frequent occurrence of polymorphism and cryptic species (Allmon, chapter 3, this volume) (fig. 4.4).

The first of these, acquisition of reproductive isolation, involves some kind of reorganization and redefinition of the fertilization system rela-

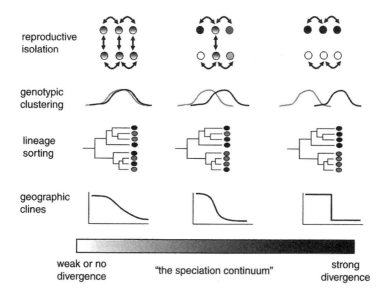

Figure 4.4 "Schematic illustration of the continuous nature of divergence during speciation." From Nosil 2012, 4.

tive to that of the parent population. Reproductive isolation is generally thought to arise as a by-product or side-consequence of divergence in other traits (Mayr 1942, 1963; Dobzhansky 1951; Simpson 1953; Schluter 2000, 70–71, 189ff.; Coyne and Orr 2004, 37), usually during isolation of populations. It may also, however, occur during secondary contact between populations, incipient species, or neospecies, via reinforcement and character displacement (Pfennig and Pfennig 2012). A variety of genes appear to be involved in directly contributing to the evolution of reproductive isolation ("speciation genes"; see Coyne and Orr 2004; Nosil and Schluter 2011).

The achievement of reproductive isolation as a by-product of divergent selection due to ecological differences between isolated populations has been labeled ecological speciation (Schluter 2000, 2001, 2009; Rundle and Nosil 2005; Jiggins 2008; Nosil 2012), adaptive speciation (Dieckmann et al. 2004), competitive speciation (Rosenzweig 1978; Polechová and Barton 2005), and gradient speciation (Moritz et al. 2000), and has become the focus of intensive neontological research (e.g., Baker 2005; Funk 2009; Sobel et al. 2009; Doebeli 2011; Weissing et al. 2011; Nosil 2012). Divergent selection also may be due to nonecological factors, such as sexual selection (Ritchie 2007). (Speciation that occurs by mechanisms other than selection related to ecological differences has been called noneco-logical or nonadaptive speciation [Schluter 2000, 2009; Rundell and Price 2009; Hoso 2012; Svensson 2012; see also Kopp 2010].)

Divergence depends not only on extrinsic factors but on the genome itself and on the supply of available and appropriate genetic variation. In-trinsic biases in the direction of variation, such as variation "along genetic lines of least resistance" (Schluter 2000, 215ff.) or related to developmental "constraints" such as heterochrony or homology of varying depth (Klin-genberg 1998; Smith 2003; McCune and Schimenti 2012), may strongly affect the direction, amount, and timing of population differentiation.

Finally, if differentiation or divergence does not proceed far enough, it may be reduced by speciation reversal (Wolfe 2003; Seehausen et al. 2008) or speciation collapse (Taylor et al. 2006). In other words, popula-tion differentiation, potentially even to a degree recognizable in the fossil record, can be reversed if a daughter population is subsumed into a parent population (or another daughter).

Stage 4: Neospecies persistence

The fourth stage of the speciation process is the persistence of newly formed species as independent evolutionary units in geological time. This stage is frequently marked by an increase in population size together with a concomitant expansion in geographic range, sometimes into sympatry with the parental species (Mayr 1942, 1963; Phillimore and Price 2009) (figs. 4.1, 4.2). This process may also involve further divergence beyond that present at the onset of reproductive isolation (e.g., by competitive dis-placement or reinforcement) (Pfennig and Pfennig 2012).

Just as most geographic isolates do not persist for long enough to achieve reproductive isolation, it is likely that only a small percentage of neospecies persist to become established species (Mayr 1963; Stanley 1979; Rosenblum et al. 2012). Thus, causal factors affecting the persistence of neospecies clearly grade into factors affecting persistence (or extinction) of established species (Maynard Smith 1989; Raup 1991). In practice, it may therefore be difficult or impossible to make a distinction between the persistence of population isolates and established neospecies. When we examine the history of any clade, particularly in the fossil record, we are usually observing only established neospecies (Eldredge and Gould 1972; Eldredge 1995; Kimbel 1991, 362–363; Liow and Stenseth 2007). Indeed, it is speciation that typically generates the kinds of morphological changes necessary for a species to persist and be recognized in the fossil record (Futuyma 1987, 1989; Kimbel 1991, 362).

In light of this continuity (isolate → neospecies → established species), it might be objected that this fourth stage should not be regarded as part of the speciation process, given that it occurs following the establishment of reproductive isolation. Yet a similar argument might be applied to stages 1 and 2. After all, it is likely that the most population isolates do not become distinct species. Another possible objection to the necessity or utility of stage 4 is that neospecies persistence should not be regarded as a step in the speciation process because there is no definitive end-point other than extinction. As discussed above, however, prior to becoming "established," the vast majority of newly formed species must first pass through a neospecies phase in which they expand population size and geographic range. In general, there is some minimum population size below which the probability of extinction becomes so great as to jeopardize the long-term survival of the species (e.g., Belovsky 1987; Calder 2000). As Phillimore and Price (2009, 240) note, analysis of the speciation process would not be complete without consideration of this stage, because without range expansions, "newly produced species would remain geographical replacements of one another, limiting both sympatric diversity and the total number of species that could be produced from a common ancestor."

This expansion phase, encompassed in stage 4, may be accompanied by an accumulation of adaptations, and these may increase the probability of survival in a new niche. As Darwin emphasized in his principle of divergence (1859), neospecies that do not diverge from their sister neospecies, or from the parental species, are unlikely to persist over geologic time spans because of resource competition. Some of these adaptations may be accumulated prior to reproductive isolation, others may occur afterward, while the neospecies exists in isolation. Still others may result post-contact, through such processes as competitive displacement.

The central point here is that, for the majority of animal species, any population (or subset of populations) must go through all four stages in order to form and survive over geologic timescales. All four stages are necessary for an animal species to appear in the fossil record, and all four stages probably usually encompass a relatively brief time span (10^3–10^6 years) relative to the typical duration of a fossil species. In this sense, speciation begins with the isolation of populations (though the bulk of isolates do not form distinct species) and ends with the transition from neospecies to established species (fig. 4.1).

Discussion

Previous proposals of speciation "stages"

Discussions of stages of speciation have been presented previously, with various degrees of explicitness (e.g., Allmon 1992; Lowry 2012; fig. 4.2). Several of these discussions have been in the context of a "continuum" of process instead of, or in addition to, consideration of discrete stages (e.g., Hendry et al. 2009; Merrill et al. 2011; fig. 4.4).

For example, Lowry (2012, 1) states that "while most biologists would agree that speciation occurs across a continuum over time . . . biologists must study speciation at various points along the continuum of the process. Repeatedly, those who have taken up this challenge have found that dividing speciation into stages is a useful framework for better understanding the entire process."

Wiens (2004a, 920) has proposed that a "new research program in allopatric speciation . . . should minimally address three general questions": (1) "what are the extrinsic ecological factors that cause geographic range splitting?"; (2) "what intrinsic organismal traits underlie these ecological factors?"; and (3) "what microevolutionary factors impede adaptive evolution in these limiting organismal traits [and thereby keep them from evolving/adapting] during the time frame of speciation?" To answer these three questions, Wiens proposes four investigations: (1) "identify a single vicariance event based on the co-occurrence of multiple sister species in the same area sharing similar habitat preferences and a similar geographic disjunction"; (2) "identify the nature and timing of the environmental change that created this disjunction of habitats based on concordance between geological evidence and levels of molecular divergence between sister species"; (3) "focus on a particular pair of sister species and identify the specific ecological factors that underlie the inability to cross the geographic barrier; (4) "determine what population genetic factors may limit adaptation in these traits to ecological conditions at the geographic barrier."

Similarly, Butlin et al. (2009, 7) suggest that "it would be helpful to understand the contributions of different mechanisms to speciation rates and their variation. Unfortunately, as Coyne and Orr (2004) have emphasized in their thorough review of the literature, this is a question about which we know rather little." Nosil and Harmon (2009, 128) note that, while divergence of populations during speciation is continuous, different degrees of divergence can be thought of "as arbitrary 'stages' of the continuous process of speciation . . . , with greater divergence equating to greater progress toward the completion of speciation." Rundle and Nosil (2005, 337) explicitly divide ecological speciation into "three necessary components: an ecological source of divergent selection, a form of reproductive isolation, and a genetic mechanism to link them."

Phillimore and Price (2009, 240–241) propose "three steps that limit the rate at which new species form": (1) "gene flow between populations must be restricted" (i.e., "populations become geographically isolated"); (2) "populations diverge in various traits that generate reproductive isolation"; and (3) "populations must expand ranges." Miller (2001, 7) similarly writes that "the establishment of a new species probably involves something like the following chain of events . . . 1) appearance of a new innovation . . . 2) isolation or separation to conserve or protect the innovation . . . 3) significant morphologic differentiation . . . When the sequence includes these steps, most paleobiologists would agree that a new species-lineage has been established."

In discussing ecological speciation, Foote (2012) refers to "a three-stage process that encompasses and links ecological divergence, phenotypic divergence, and reproductive isolation: [1] ecological contrast or an ecological gradient promotes divergent natural selection; [2] this causes adaptive divergence of phenotypic traits between individuals in the different ecological contexts; [3] when one or several of these adaptations are also associated with reproductive compatibility, divergence in these traits results in reproductive isolation either through assortative mating or low hybrid fitness" (2012, 447–448). Other recent discussions of stages in the speciation process include Streelman and Danley (2003), Tautz (2004), Wolf et al. (2008), and Liow (2010).

A number of other neontological studies have used or otherwise acknowledged the utility of previous versions of the SOS framework presented here (Allmon 1992; e.g., Chown 1997; Barraclough et al. 1998; Funk 1998; Dynesius and Jansson, 2000; Gavrilets et al. 2000; Schluter 2000, 79; Barraclough and Savolainen 2001; Jansson and Dynesius 2002; Gavrilets 2004a,b; Levin 2005; Ricklefs and Bermingham 2007; Rabosky 2015). Dynesius and Jansson (2014) explicitly "restructured and generalized" the framework of Allmon (1992), "distinguishing three controls of speciation rate: splitting, persistence and duration." Their *splitting control* "resembles

Allmon's 'isolate formation,' but indicates the rate at which within-species lineages are initiated by whatever means, also when the differentiation process is the initiator (e.g. ecological speciation)." Their *persistence control* "includes the level of persistence of the population(s) constituting a lineage, as well as of the processes and states that keep them apart." Their *duration control* "is the time period needed for speciation to be completed" which determines which lineages are what Dynesius and Jansson call "full-fledged species" (referred to herein as "established species" and which are "within-species lineages" ("incipient species" herein) (Dynesius and Jansson 2014, 923). It therefore combines what we refer to here as stages 2 and 4.

The preceding summary suggests that an approach similar to the SOS framework discussed here has broad utility in studies of speciation. The particular approach we describe here, however, is likely to be more effective as a tool for investigating species originations than previous proposals because it is both more explicit and more encompassing. It is more explicit in that it links specific processes with specific stages, enabling especially paleobiologists to explore where exactly in the speciation process a given factor has been most impactful. The SOS framework is more encompassing in the way it recognizes and focuses attention not solely on the fate of incipient species, but also on that of neospecies, including the factors that constrain or accentuate the probability of formation and persistence of these neospecies.

General applications of the SOS framework

The SOS framework does not require that all four stages be temporally or causally disjunct. Indeed, the stages may be concurrent and/or closely connected (figs. 4.3, 4.5; table 4.2). For example, isolate formation, persistence, and differentiation can all be affected by natural selection (see below), and stages 2 and 3 (persistence of geographic isolates and acquisition of reproductive isolation, respectively) will inevitably overlap at least partly in time. The key point is that the stages represent distinct biological phenomena that generate distinct evolutionary results. All are necessary for the formation of new established species, and the likelihood of each can vary independently in any particular instance. A given process or phenomenon (e.g., environmental perturbation) may affect the formation and persistence of isolated populations, but unless those populations persist, differentiate sufficiently, and subsequently endure as newly formed species, they will have no impact on macroevolutionary patterns.

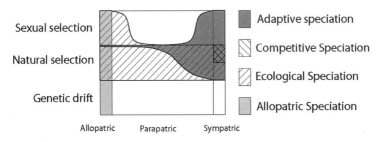

Figure 4.5 "Notions of speciation."
From Dieckmann et al. 2004, 8.

Butlin et al. (2009, 2–3) argue that "many speciation mechanisms have been shown to be plausible, but there remains considerable uncertainty over which of these scenarios are responsible for most of the species on Earth." The four-stage SOS framework presented here is a useful tool for addressing this uncertainty insofar as it compels us to focus on one stage or process of speciation at a time, and to ask how a given factor affects each particular aspect of the overall process. Using the framework, speciation analysis begins with assessment of the causal mechanisms that may have affected each stage. This information can then be applied in a comparative phylogenetic context.

The framework should be especially useful when judging which factors may have been most important in either limiting or promoting species formation (table 4.3). For example, although intrinsic or extrinsic factors may result in production of numerous isolated populations (stage 1), the probability of speciation may remain unaltered if factors conspire to limit the persistence of isolates (stage 2; e.g., Jablonski and Flessa 1986; Allmon 1992) and/or reproductive isolation (stage 3; e.g., Coyne and Orr 2004). Alternatively, various causal factors (environmental change, sexual selection) may result in increased probabilities of populations successfully completing stages 1 to 3, whereas other factors (e.g., natural selection) might restrict the likelihood of neospecies persistence (Phillimore and Price 2009). This issue has occasionally been phrased as whether speciation is inherently "easy" or "hard" (e.g., Gavrilets 2005; Mallet 2008b). In other words, is speciation the default, occurring almost invariably in the absence of suppression? Or does speciation more frequently require promotion or driving by unusual circumstances?

For example, it has been suggested that rate of diversification can be reduced by effects of density dependence (e.g., McPeek 2008; Price 2008;

Rabosky and Lovette 2008; Weir and Mursleen 2012); that is, by a decrease in isolate persistence. Similar patterns, however, could also result from decrease in rate of isolate formation, due for example to abiotic environmental change (e.g., Esselstyn et al. 2009; Abe and Lieberman 2012), which could be largely stochastic (e.g., Phillimore and Price 2008; Etienne and Haegeman 2011). Similarly, instances of increased diversification (radiation) have been attributed to a number of factors, including key innovations (Heard and Hauser 1995; Schluter 2000, 181), creation or invasion of empty or otherwise permissive environments (e.g., Grant and Grant 1995; Vermeij 2002), changes in dispersal ability (Hansen 1978, 1980; Palumbi 1994), and occurrence of abundant instances of vicariance caused by environmental disturbance or fragmentation (e.g., Cracraft 1985, 1992; Lieberman 2012) (see next section). The macroevolutionary implications are dramatically different for each of these scenarios (Stanley 1986; Allmon 1992, 1994; Johnson et al. 1995). Being more explicit about links between causal factors and stages are likely to result in more explicit hypotheses about the process of speciation.

Each of the 4 stages, then, can be regarded as a "portal of probability." Metaphorically, the chances of speciation occurring during a given interval are directly related to the size of each portal, which is in turn controlled by various causal forces. Populations may pass through one portal in great numbers only to be denied passage through the next. The "larger" the portal, the more daughter populations or neospecies pass through to the next stage, and the greater the probability that species diversity will increase. For example, the probability of speciation may increase if environmental perturbations partition a given habitat, affecting formation and/ or persistence of isolates (stages 1 and 2). Conversely, the probability of isolated populations forming and persisting may remain relatively stable over an interval of time, but the evolution of a particular character— connected to reproduction (for example, secondary sexual characteristics modified under the influence of sexual selection) or to ecology—may increase the likelihood of reproductive isolation (stage 3).

The SOS framework also may offer a way to assess the overall role of natural selection in speciation, a topic that has been important and controversial within evolutionary biology since 1859 (e.g., Darwin 1859; Mayr 1963; Gould 1990, 2002; Via 2001; Allmon 2013). With the recent rise in popularity of theories of ecological speciation, natural selection has moved closer to the center of much discussion of processes of speciation (e.g., Butlin et al. 2009 and references above). Indeed, Sobel et al. (2009,

295) have argued that the term "ecological speciation" is unnecessary, since "natural selection is a ubiquitous part of speciation." Yet (as others have argued: e.g., Haegeman and Etienne 2009; Kopp 2010; Etienne and Haegeman 2011), this is not necessarily true. Application of the SOS framework may contribute to addressing this issue by sharpening focus on the role of selection at different points in the speciation process. Most recent attention has focused on stage 3, during which advocates of ecological speciation argue that ecological differences create divergent selection pressures on populations.

Yet selection can also play a major role at each of the other three stages (Schluter 2000; Butlin et al. 2009). Selection may contribute to either or both of the first two stages (formation and persistence of isolates) by eliminating individual organisms, for example by predation or abiotic environmental change (e.g., Stanley 1986; Allmon 2001). Disruptive selection or intraspecific character displacement can also contribute to isolate formation via polymorphisms that lead to physical or ecological separation (Pfennig and Pfennig 2012). Population persistence can be enhanced by adaptation. Stage 4, persistence of neospecies, depends strongly on the ability to individuals and populations to survive, often alongside and in competition with closely related, perhaps "ecologically redundant" species (Rosenzweig 1995; Pfennig and Pfennig 2012). Thus, once reproductive isolation has been acquired, morphological divergence via natural selection may well determine whether a species will persist (Schluter 1994, 2000; Phillimore and Price 2009).

Evolutionary radiations

The SOS framework has potential to elucidate the processes behind evolutionary (or adaptive) radiations, which remains an area of intensive research (Erwin 1992; Heard and Hauser 1995; Hunter 1998; Schluter 2000; Seehausen 2006; Rabosky and Lovette 2008; Gavrilets and Losos 2009; Hodges and Derieg 2009; Yoder et al. 2010) Frequently it is unclear how a proposed key innovation has resulted in elevated rates of speciation (Lauder and Liem 1989; Hunter 1998). The 4-stage perspective can help by encouraging investigators to be explicit as to how a particular feature may have affected speciation rates. Key innovations may have a variety of effects depending on which stage(s) of speciation they act upon (Allmon 1992). For example, the syrinx of passerine birds may have facilitated the acquisition of reproductive isolation (stage 3) by permitting a greater potential range of vocalizations, thereby contributing to the high diversity of this clade (Raikow 1986). Conversely, evolutionary radiations are sometimes characterized by substantial morphological innovation without associated increases in species diversity (novelty events of Erwin 1992). Such radiations might conceivably be due largely to one or more adaptations acting on stages 2 or 4. That is, causal factors driving divergence could elevate the rate of population or neospecies persistence without associated changes at the other stages.

Evolutionary radiations are typically thought to involve elevated rates of species formation (Stanley 1979; Schluter 2000; Rundell and Price 2009; Lieberman 2012). Whereas such radiations may entail increased likelihoods at only one stage of speciation, or at one more than another, the SOS framework also suggests an additional hypothesis. Assuming a relatively constant rate of extinction for successful species, evolutionary radiations may frequently result from elevated likelihoods at not one but two or more stages. In effect, multiple causal factors may open "portals of probability" at several stages, with the resulting wave of isolated populations and neospecies culminating in accelerated or cascading rates of established species formation. This cascade hypothesis resembles the multistep key innovation concept proposed by Heard and Hauser (1995).

Various examples from the literature can be cited as potential support for such a hypothesis. One of the most remarkable and debated evolutionary radiation is that of haplochromine cichlid fishes in the East African Great Lakes (see, e.g., Dorit 1990; Kocher 2004; Seehausen 2006). Both extrinsic factors (e.g., changing lake levels) and intrinsic factors (e.g., mouthbrooding; jaw flexibility) have been cited as increasing the likelihood of population isolate formation and persistence (stages 1 and 2). Sexual selection may also have accelerated the divergence and genetic isolation of populations (Dominey 1984), largely through modifications of species-specific coloration (stage 3). The derived pharyngeal jaw apparatus may have permitted a large degree of ecological flexibility in the face of competition from other cichlid species (Liem 1974), as well as the diversity of head morphologies related to exploiting food resources, both possibly resulting in higher levels of neospecies persistence (stage 4). Seehausen (2006) posits an alternative stage 4 hypothesis, arguing that the speciation rate in African cichlids declined as niche space filled up. Specifically, early phases of local radiations were characterized by higher rates of speciation than were later phases. None of the above hypotheses is mutually exclusive, and it may that the radiation of Great Lakes cichlids resulted from increased "success" at all four stages. It would be useful to test the cascade hypothesis against other evolutionary radiations.

Application to the fossil record

Hunt and Rabosky (2014, 435) posed the provocative conundrum of "a near-perfect paleontological demonstration of speciational trait change in a single lineage, in which we are able to unambiguously rule out immigration and other potential biases as the cause of the pattern." Such a pattern, they argue, could result from peripatric speciation (sensu Mayr 1982 and Eldredge and Gould 1972), but it could also be produced by "speciation attributable to divergent natural selection" (i.e., ecological speciation), or by "postspeciation character displacement in ecological or reproductive traits." Given the limitations of the fossil record, determining which of these causal scenarios is more likely is difficult or impossible. The only way forward is to attack speciation in the fossil record with a general causal taxonomy that allows available evidence to be grouped into categories. The SOS framework provides such a framework.

The SOS framework may be useful for studying speciation in the fossil record within two broad arenas. First, at the level of individual clades, it focuses our attention on searching for particular patterns. For example, how did a given suite of species pass through each of the 4 stages? Is there evidence for critical influence at one or more stages? How complete, or incomplete, is the record, and what other factors may have played a part in generating the observed patterns? Second, the SOS framework may help to broaden our understanding of the speciation process from a more general cross-taxon perspective. For example, compiling cases from the literature into formal categories (i.e., which stage or stages were most important) might allow us to estimate relative frequency of different causes, and to generate hypotheses of causes through time. Do we find the same general patterns, say, among vertebrates in marine and terrestrial realms? Or in Paleozoic versus Mesozoic gastropods? Tackling such large-scale questions could result in considerable advances to macroevolutionary theory as it applies to species originations.

Paleontology has access to time, space, morphology, and environment. Fossils can therefore, at least sometimes, allow analysis of the likelihood of different stages in the SOS framework (table 4.3). Examples might include isolate formation by extrinsic versus intrinsic factors (e.g., vagility of adults, larval dispersal ability), or isolate persistence in response to varying environmental rigor, adaptation, geographic range, and abundance. Although it is unlikely that the fossil record will contain abundant records of incipient species, due to their usually small size and short temporal duration (Eldredge and Gould 1972), inferences can be drawn based on a variety of observations at larger temporal and spatial scales. For example, we can examine episodes of environmental disturbance (e.g., Allmon et al. 1998; Allmon 2001, 2003) and hypothesize whether it is more likely that diversification was a result of vicariance, establishment of isolates by dispersal, or maintenance of isolates in refugial conditions. Changes in sea level on shallow shelves is a frequent source of environmental disturbance and cause of potential isolate formation (stage 1) that can be studied (Allmon and Smith 2011; Lieberman 2012). Sea level change is also of course a significant source of sedimentary hiatuses, but this juxtaposition of potential evolutionary cause and stratigraphic incompleteness (dubbed the common cause hypothesis: e.g., Peters and Foote 2002; Peters 2006; Allmon and Smith 2011; Butler et al. 2011; Patzkowski and Holland 2012) has recently become a focus of research in its own right, and the subject may not be as intractable as previously thought.

Changes in climate are frequently accessible in the geologic record, and also frequently proposed as causes or triggers for speciation, even if the exact mechanism is left unclear. In the context of the SOS framework, such suggestions likely focus on formation of isolates (e.g., Clarke et al. 1992; Clarke and Gaston 2006; Hua and Wiens 2013). The possible isolation of populations in refugia as a result of frequent climate changes in the Quaternary is a particular case of such causation that has attracted considerable attention (e.g., Coope 2004; Hewitt 2000, 2004; Lister 2004; Zink et al. 2004; Barnosky 2005), although not yet a consensus on what factors may have been most important when and why.

Depth gradients in marine environments are also tractable in the geological record and have been highlighted as potential causes for speciation (e.g., Cisne et al. 1982; White 1988; Carlon and Budd 2002; Crame 2002; Ingram 2011). Most of these suggestions have focused on the divergence of populations (stage 3), in either allopatry or parapatry, due to differential selection associated with differing environmental conditions along such gradients.

Numerous paleontological studies have discussed processes, phenomena, or conditions—both intrinsic and extrinsic to organisms—that may have led to increased speciation due to increased formation of isolates (stage 1), even if the mechanism(s) have not usually been made explicit

(Fleming 1962; Johnson 1975; Vrba 1980, 1985; Hallam 1982, 1988; Ager 1983; Valentine and Jablonski 1983; Cronin 1985, 1987; Stanley 1986, 1990; Cronin and Ikeya 1990; McCune 2004; Abe and Lieberman 2012).

Fossilizable morphology sometimes suggests aspects of ecology and life history that may have important effects on speciation. The most frequently cited example is dispersal capability in marine invertebrates, which can affect the likelihood of isolate formation (stage 1: e.g., Hansen 1978, 1980; Jablonski 1986; Palumbi 1994; Wani 2011). In addition, isolate differentiation (stage 3) via a niche shift can sometimes be detected in the fossil record. Nehm and Geary (1994), for example, documented a change in apparent ecological niche in Neogene marginellid gastropods that they suggested led ultimately to speciation. Schluter (2000, 38), however, cautions that this approach is probably not frequently applicable because the niche of a fossil species cannot be reconstructed for all ancestors, which is an overarching challenge for all such fossil studies.

Morphology also occasionally provides insight into sexual dimorphism and thus also to potential sources of sexual selection and/or changes in mate recognition systems. The most cited case is the evolution of African antelopes, which provides an example of an evolutionary radiation that may have been driven by multiple causal factors operating at several of the stages discussed here. Vrba (1980, 1983) employed comparisons of the more diverse and specialized Alcelaphinae (wildebeest and kin; about 27 fossil and extant species) with the less diverse, less specialized Aepycerotinae (impala; 1–3 fossil and extant species) to formulate the resource-use hypothesis. She postulated that alcelaphines have been more likely to fragment into isolated populations during periods of environmental perturbations due to associated changes in the distribution of key resources. Therefore, environmental perturbations are thought to have acted in concert with resource specialization to increase the rate of formation (and perhaps the persistence) of geographical isolates (stages 1 and 2). Sister taxa among African antelopes are, however, typically distinguished not on the basis of viability-related characters, but rather on reproductive characters, namely horns. Therefore, the differentiation of horn morphologies (stage 3) has likely been important in reproductive isolation of species, and the processes involved (e.g., sexual selection, mate recognition selection) may well have contributed to the elevated rates of speciation in alcelaphines (Sampson 1999).

The SOS framework also allows for analysis of the connections between trophic resources, abundance, and speciation in the fossil record,

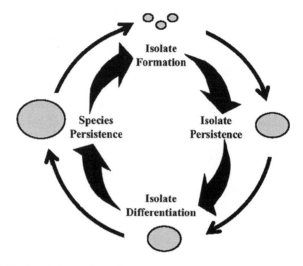

Figure 4.6 The "speciation cycle" and expected population sizes at each different points in the cycle.
From Allmon and Martin (2014).

via a model dubbed the speciation cycle (Allmon 2009; Allmon and Martin 2014; fig. 4.6). This model, which is an expansion of the taxon cycle of Wilson (1961) and the speciation cycle of Grant and Grant (1997, 2008), predicts the most likely population sizes during different stages of the speciation process. According to this model, cyclic series of changes are most likely (but see Pigot et al. 2012). Initially, a relatively abundant parental population gives rise to one or more smaller daughter populations—either by vicariance or dispersal. The daughter population(s) either persist(s) in isolation, merge(s) with the parental population, or become(s) extinct. If an isolated population persists long enough, it will become sufficiently differentiated that reproductive isolation is maintained if/when contact with the parent occurs. Differentiation may be more rapid when population size is small (e.g., Reiss 2013), but population size may increase following differentiation, during longer intervals of persistence. Once it becomes an established species, expanding its range and population size, it may give rise to its own daughter populations, and the cycle begins anew.

The speciation cycle integrates the sometimes conflicting relationships between trophic resources and species diversity noted in the literature (see Allmon and Martin 2014). Widespread (and therefore usually abundant) parental species are more likely to form a larger number of isolated daughter populations (usually at their margins), and so should give rise to more daughter species. These more narrowly distributed (and usually less abundant) populations and species may change more quickly, perhaps under the influence of genetic drift, but are more susceptible to extinction. For an initially small population, probability of speciation should therefore increase with increasing abundance, but only up to a point, after which it declines rapidly. An intermediate rate of isolate formation, not too low or too high, may thus confer the greatest probability of neospecies persistence (stage 4: Lawton 1993; Chown 1997; Allmon 2009). Testing the speciation cycle in the fossil record is difficult because of the challenges of estimating absolute abundance of individuals in fossils. Although some approaches may nevertheless be possible, at present it seems that their preliminary application produces ambiguous results (see discussion in Allmon and Martin 2014).

Conclusions

This chapter argues that the analysis of animal speciation in the fossil record would benefit from an explicit 4-stage framework, here termed the stages of speciation (SOS) framework. The stages consist of: (1) formation of population isolates; (2) persistence of population isolates; (3) differentiation of isolates; and (4) persistence of neospecies. Allopatric speciation in biparental animals involves all 4 stages. Although they may overlap temporally, and the onset and completion of each may be arbitrarily defined, all 4 stages pertain to real biological events, the rates of which can be accelerated or constrained by distinct suites of causal factors.

The SOS framework is more than a mere refinement of previous methods for analyzing speciation. Although the various components of this perspective are familiar to all evolutionary biologists, a coherent agenda incorporating them into analyses of speciation has been lacking, with significant consequences for our understanding of evolutionary processes. The SOS framework encourages investigators to examine the influence of multiple causal forces acting at distinct phases of the speciation process, leading to more specific questions about species origins. Is speciation in a given clade the result of elevated rates at one stage or multiple stages? Which causal factors have restricted the rate of speciation in a given clade relative to more species-rich sister taxa? Does one stage generally dominate others in controlling rates of speciation across taxa? How do causal factors interact in a given ecological context at each stage?

One corollary of the SOS framework that pertains most directly to evolutionary radiations is that each stage may constitute a "portal of probability," enhancing or inhibiting the odds of populations/neospecies passing through to the subsequent stage. This insight, in turn, leads to the cascade hypothesis, which postulates that evolutionary radiations may frequently result from the opening of multiple portals that collectively accelerate the rate of species originations.

In sum, application of the SOS framework will likely result in novel insights about the process of speciation, from specific origination events to large-scale radiations. Once key parameters are more clearly defined, it may be possible to produce mathematical models relating various causal factors and stages to the production of diversity (see McKinney and Allmon, 1995, for some early attempts). Given that all 4 speciation stages may be directly associated with ecological factors, and occur on ecological timescales, adoption of this methodology has potential to facilitate the integration of ecology and macroevolution. Finally, the SOS framework provides an explicit framework for paleontologists to use to compare equivalent parts of the speciation process, and may thereby foster greater synthesis of paleontological insights on the causes and effects of speciation.

Acknowledgments

We are grateful to John Hunter, Ron Heinrich, Mike Bell, and several anonymous reviewers for comments on much earlier versions of this manuscript; to Dana Geary, Bruce Lieberman, Chuck Mitchell, and Peg Yacobucci for comments on more recent drafts; and to Andrew Matthiessen for assistance with references. This work has been supported in part by NSF grants EAR-0719642 and EAR-1053517 to WDA.

References

Abe, F. R., and B. S. Lieberman, 2009. "The nature of evolutionary radiations: A case study involving Devonian trilobites." *Evolutionary Biology*. 36:225–234.

Abe, F. R., and B. S. Lieberman, 2012. "Quantifying morphological change during an evolutionary radiation of Devonian trilobites." *Paleobiology*. 38(2): 292–307.

Adams, C. E., and F. A. Huntingford, 2004. "Incipient speciation driven by phenotypic plasticity? Evidence from sympatric populations of Arctic char." *Biological Journal of the Linnean Society*. 81: 611–618.

Adrain, J. M., and S. R. Westrop, 2005. "Late Cambrian ptychaspidid trilobites from western Utah: Implications for trilobite systematics and biostratigraphy." *Geological Magazine.* 142(4): 377–398.

Adrain, J. M., and S. R. Westrop, 2006. "New earliest Ordovician trilobite genus *Millardicurus*: The oldest known hystricurid." *Journal of Paleontology.* 80(4): 650–671.

Ager, D. V., 1983. "Allopatric speciation—An example from the Mesozoic Brachiopoda." *Palaeontology.* 26(3): 555–565.

Allmon, W. D., 1992. "A causal analysis of stages in allopatric speciation." *Oxford Surveys in Evolutionary Biology.* 8: 219–257.

Allmon, W. D., 1994 "Taxic evolutionary paleoecology and the ecological context of macroevolutionary change." *Evolutionary Ecology.* 8: 95–112.

Allmon, W. D., 2001. "Nutrients, temperature, disturbance, and evolution: A model for the Late Cenozoic marine record of the Western Atlantic." *Palaeogeography, Palaeoclimatology, Palaeoecology.* 166: 9–26.

Allmon, W. D., 2003. "Boundaries, turnover, and the causes of evolutionary change: A perspective from the Cenozoic." In *From greenhouse to icehouse: The marine Eocene-Oligocene transition.* D. Prothero, L. Ivany, E. Nesbitt, eds., 511–521. New York: Columbia University Press.

Allmon, W. D., 2009. "Speciation and shifting baselines: Prospects for reciprocal illumination between evolutionary paleobiology and conservation biology." In *Conservation paleobiology: Using the past to manage for the future.* In G. P. Dietl and K. Flessa, eds., 245–273. Paleontological Society Papers, 15.

Allmon, W. D., 2013. "Species, speciation, and paleontology up to the Modern Synthesis: Persistent themes and unanswered questions." *Palaeontology.* 56(4): 1199–1223.

Allmon, W. D., and R. E. Martin, 2014. "Seafood through time revisited: The Phanerozoic increase in marine trophic resources and its macroevolutionary consequences." *Paleobiology.* 40(2):256–287.

Allmon, W. D., and R. M. Ross, 2001. "Nutrients and evolution in the marine realm." In *Evolutionary paleoecology: The ecological context of macroevolutionary change.* W. D. Allmon and D. J. Bottjer, eds., 105–148. New York: Columbia University Press.

Allmon, W. D., and U. E. Smith, 2011. "What, if anything, can we learn from the fossil record about speciation in marine gastropods? Biological and geological considerations." *American Malacological Bulletin.* 29(1): 247–276.

Allmon, W. D., P. J.Morris, and M. L.McKinney, 1998. "An intermediate disturbance hypothesis of maximal speciation." In *Biodiversity dynamics: Turnover of populations, taxa, and communities.* M. L. McKinney and J. A. Drake, eds., 349–376. New York: Columbia University Press.

Alroy, J., 2002. "How many named species are valid?" *Proceedings of the National Academy of Sciences of the USA.* 99(6): 3706–3711.

Alroy, J., 2009. "Speciation and extinction in the fossil record of North American mammals." In *Speciation and patterns of diversity*. R. Butlin, J. Bridle, and D. Schluter, eds., 301–323. New York: Cambridge University Press.

Andersson, M., 1994. *Sexual selection*. Princeton, NJ: Princeton University Press.

Andersson, M., and Y. Iwasa, 1996. "Sexual selection." *Trends in Ecology and Evolution*. 11: 53–58.

Anstett, M.-C., M. Hossaert-McKey, and D. McKey, 1997. "Modeling the persistence of small populations of strongly independent species: Figs and fig wasps." *Conservation Biology*. 11(1): 204–213.

Anstey, R. L., and J. F. Pachut, 2004. "Cladistic and phenetic recognition of species in the Ordovician bryozoan genus *Peronopora*." *Journal of Paleontology*. 78(4):651–674.

Baker, J. M., 2005. "Adaptive speciation: The role of natural selection in mechanisms of geographic and non-geographic speciation." *Studies in History and Philosophy of Biological and Biomedical Sciences*. 36: 303–326.

Bambach, R. K. 1993. "Seafood through time: Changes in biomass, energetics, and productivity in the marine ecosystem." *Paleobiology*. 19:372–397.

Barnosky, A. D., 2005. "Effects of Quaternary climatic change on speciation in mammals." *Journal of Mammalian Evolution*. 12(1/2): 247–264.

Barraclough, T. G., and V. Savolainen, 2001. "Evolutionary rates and species diversity in flowering plants." *Evolution*. 55(4): 677–683.

Barraclough, T. G., A. P. Vogler, and P. H. Harvey, 1998. "Revealing the factors that promote speciation." *Philosophical Transactions of the Royal Society of London B*. 353: 241–249.

Bell, M. A., and T. R. Haglund, 1982. "Fine-scale temporal variation of the Miocene stickleback *Gasterosteus doryssus*." *Paleobiology*. 8(3): 282–292.

Belovsky, G. E. 1987. "Extinction models and mammalian persistence." In *Viable populations for conservation*. M. E. Soule, ed., 35–57. Cambridge, UK: Cambridge University Press.

Benton, M. J., and P. N. Pearson, 2001. "Speciation in the fossil record." *Trends in Ecology and Evolution*. 16(7): 405–411.

Berger, J., 1990. "Persistence of different-sized populations: An empirical assessment of rapid extinctions in bighorn sheep." *Conservation Biology*. 4(1): 91–98.

Bird, C. E., I. Fernandez-Silva, D. J. Skillings, and R. J. Toonen, 2012. "Sympatric speciation in the post "Modern Synthesis" era of evolutionary biology." *Evolutionary Biology*. 39: 158–180.

Blackburn, T. M., and K. J. Gaston, 1997. "The relationship between geographic area and the latitudinal gradient in species richnesss in New World birds." *Evolutionary Ecology*. 11(2): 198–204.

Bolnick, D. I., and B. M. Fitzpatrick, 2007. "Sympatric speciation: Models and empirical evidence." *Annual Review of Ecology, Evolution, and Systematics*. 38: 459–487.

Bradburd, G. S., P. L. Ralph, and G. M. Coop, 2013. "Disentangling the effects of geographic and ecological isolation on genetic differentiation." *Evolution.* 67(11): 3258–3273.

Budd, A. F. 2010. "Tracing the long-term evolution of a species complex: Examples from the *Montastrea "annularis"* complex." *Palaeoworld.* 19: 348–356.

Butler, R. J., R. B. J. Benson, M. T. Carrano, P. D. Mannion, and P. Upchurch, 2011. "Sea level, dinosaur diversity, and sampling biases: investigating the 'common cause' hypothesis in the terrestrial realm." *Proceedings of the Royal Society B.* 278: 1165–1170.

Butlin, R. K., J. Galindo, and J. W. Grahame, 2008. "Sympatric, parapatric, or allopatric: The most important way to classify speciation?" *Philosophical Transactions of the Royal Society B.* 363: 2997–3007.

Butlin, R., J. Bridle, and D. Schluter, eds., 2009. *Speciation and patterns of diversity.* Cambridge: Cambridge University Press.

Cain, A. J., and L. M. Cook, 1989. "Persistence and extinction in some *Cepaea* populations." *Biological Journal of the Linnean Society.* 38: 183–190.

Calder, W. A. 2000. "Diversity and convergence: Scaling for conservation." In *Scaling in biology.* J. H. Brown and G. B. West, eds., 297–323. Oxford, UK: Oxford University Press.

Cardillo, M., G. M. Mace, K. E. Jones, J. Bielby, O. R. P. Bininda-Emonds, W. Sechrest, C. D. L. Orme, and A. Purvis, 2005. "Multiple causes of high extinction risk in large mammal species." *Science.* 309: 1239–1241.

Carlon, D. B., and A. F. Budd, 2002. "Incipient speciation across a depth gradient in a scleractinian coral?" *Evolution.* 56(11): 2227–2242.

Carson, H. L., 1976. "Inference of the time of origin of some *Drosophila* species." *Nature.* 259: 395–396.

Chiba, S., 2007. "Taxonomic revision of the fossil land snail species of the genus *Mandarina* in the Ogasawara Islands." *Paleontological Research.* 11(4): 317–329.

Chown, S. L., 1997. "Speciation and rarity: separating cause from consequence." In *The biology of rarity: Causes and consequences of rare-common differences.* W. E. Kunin and K. J. Gaston, eds., 91–109. London: Chapman & Hall.

Cisne, J. L., G. O. Chandlee, B. L. Rabe, and J. A. Cohen, 1982. "Clinal variation, episodic evolution, and possible parapatric speciation: The trilobite *Flexicalymene senaria* along an Ordovician depth gradient." *Lethaia.* 15(4): 325–341.

Clarke, A., and J. A. Crame, 2010. "Evolutionary dynamics at high latitudes: Speciation and extinction in polar marine faunas." *Philosophical Transactions of the Royal Society B.* 365: 3655–3666.

Clarke, A., and K. J. Gaston, 2006. "Climate, energy, and diversity." *Proceedings of the Royal Society B.* 273: 2257–2266.

Clarke, A., J. A. Crame, J-O. Stromberg, and P. F. Barker, 1992. "The Southern Ocean benthic fauna and climate change: A historical perspective." *Philosophical Transactions of the Royal Society of London B.* 338: 299–309.

Coope, G. R. 2004. "Several million years of stability among insect species because of, or in spite of, Ice Age climatic instability?" *Philosophical Transactions of the Royal Society B.* 359: 209–214.

Coyne, J. A., 1994. "Ernst Mayr and the origin of species." *Evolution.* 48: 19–30.

Coyne, J. A., and H. A. Orr, 2004. *Speciation.* Sunderland, MA: Sinauer Associates.

Cracraft, J., 1985. "Biological diversification and its causes." *Annals of the Missouri Botanical Garden.* 72: 794–822.

Cracraft, J., 1992. "Explaining patterns of biological diversity: Integrating causation at different spatial and temporal scales." In *Systematics, ecology, and the biodiversity crisis.* N. Eldredge, ed., 59–76. New York: Columbia University Press.

Crame, J. A., 2002. "Evolution of taxonomic diversity gradients in the marine realm: A comparison of Late Jurassic and Recent bivalve faunas." *Paleobiology.* 28(2): 184–207.

Crampton, J. S., and A. S. Gale, 2005. "A plastic boomerang: Speciation and intraspecific evolution in the Cretaceous bivalve *Actinoceramus*." *Paleobiology.* 31(4): 559–577.

Cronin, T. M., 1985. "Speciation and stasis in marine Ostracoda: Climatic modulation of evolution." *Science.* 227: 60–63.

Cronin, T. M., 1987. "Speciation and cyclic climatic change." In *Climate: History, periodicity, and predictability.* M. R. Rampino, J. E. Sanders, W. S. Newman, and L. K. Königsson, eds., 333–342. New York: Van Nostrand Reinhold Co.

Cronin, T. M., and N. Ikeya, 1990. "Tectonic events and climatic change: Opportunities for speciation in Cenozoic marine Ostracoda." In *Causes of evolution: A paleontological perspective.* R. M. Ross and W. D. Allmon, eds., 210–248. Chicago: University of Chicago Press.

Damuth, J., 1985. "Selection among 'species': A formulation in terms of natural functional units." *Evolution.* 39: 1132–1146.

Darwin, C. R., 1859. *On the origin of species.* London: John Murray.

Diamond, J. M., 1984. "'Normal' extinctions of isolated populations." In *Extinctions.* M. Nitecki, ed., 191–246. Chicago: University of Chicago Press.

Dieckmann, U., J. A. J. Metz, M. Doebeli, and D. Tautz, 2004. "Introduction." In *Adaptive speciation.* U. Dieckmann, J. A. J. Metz, M. Doebeli, and D. Tautz, eds., 1–16. Cambridge, UK: Cambridge University Press.

Dobzhansky, T., 1940. "Speciation as a stage in evolutionary divergence." *American Naturalist.* 74: 302–321.

Dobzhansky, T., 1951. *Genetics and the origin of species.* 3rd ed. New York: Columbia University Press.

Doebeli, M., 2011. *Adaptive diversification.* Princeton, NJ: Princeton University Press.

Dominey, W. J., 1984. "Effects of sexual selection and life history on speciation: Species flocks in African cichlids and Hawaiian *Drosophila.*" In *Evolution of fish species flocks.* A. A. Echelle and I. Kornfield, eds., 231–249. Orono: University of Maine Press.

Dorit, R. L., 1990. "The correlates of high diversity in Lake Victoria haplochromine cichlids: A neontological perspective." In *Causes of evolution: A paleontological perspective.* R. M. Ross and W. D. Allmon, eds., 322–353. Chicago: University of Chicago Press.

Dynesius, M., and R. Jansson, 2000. "Evolutionary consequences of changes in species' geographical distributions driven by Milankovitch climate oscillations." *Proceedings of the National Academy of Sciences of the USA.* 97(16): 9115–9120.

Dynesius, M., and R. Jansson, 2014. "Persistence of within-species lineages: A neglected control of speciation rates." *Evolution.* 68(4): 923–934.

Dzik, J., 1999. "Relationship between rates of speciation and phyletic evolution: Stratophenetic data on pelagic conodont chordates and benthic ostracodes." *Geobios.* 32(2): 205–221.

Ebenhard, T. 1991. "Colonization in metapopulations: A review of theory and observations." *Biological Journal of the Linnean Society.* 42: 105–121.

Eldredge, N., 1971. "The allopatric model and phylogeny in Paleozoic invertebrates." *Evolution.* 25: 156–167.

Eldredge, N., 1974. "Stability, diversity, and speciation in Paleozoic epeiric seas." *Journal of Paleontology.* 48(3): 541–548.

Eldredge, N., 1989. *Macroevolutionary dynamics: Species, niches, and adaptive peaks.* New York: McGraw Hill.

Eldredge, N., 1995. *Reinventing Darwin.* New York: John Wiley and Sons.

Eldredge, N., and S. J. Gould, 1972. "Punctuated equilibria: An alternative to phyletic gradualism." In *Models in paleobiology.* T. J. M. Schopf, ed., 82–115. San Francisco: Freeman, Cooper & Co.

Erwin, D. H., 1992. "A preliminary classification of evolutionary radiations." *Historical Biology.* 6: 133–147.

Erwin, D. H., and R. L. Anstey, 1995. Speciation in the fossil record. In *New approaches to speciation in the fossil record.* D. Erwin and R. Anstey, eds., 11–38. New York: Columbia University Press.

Esselstyn, J. A., R. M. Timm, and R. M. Brown, 2009. "Do geological or climatic processes drive speciation in dynamic archipelagos? The tempo and mode of diversification in southeast Asian shrews." *Evolution.* 63(10): 2595–2610.

Etienne, R. S., and B. Haegeman, 2011. "The neutral theory of biodiversity with random fission speciation." *Theoretical Ecology.* 4: 87–109.

Etienne, R. S., and J. Rosindell, 2011. "Prolonging the past counteracts the pull of the present: Protracted speciation can explain observed slowdowns in diversification." *Systematic Biology.* 61(2): 204–213.

Feder, J. L., S. P. Egan, and P. Nosil, 2012. "The genomics of speciation-with-gene-flow." *Trends in Genetics.* 28(7): 342–350.

Fisher, W. L., P. U. Rodda, and J. W. Dietrich, 1964. "Evolution of *Athleta petrosa* stock (Eocene, Gastropoda) of Texas." *Bureau of Economic Geology, University of Texas, Publication* 6413.

Fitzpatrick, B. M., J. A. Fordyce, and S. Gavrilets, 2009. "Pattern, process, and geographic modes of speciation." *Journal of Evolutionary Biology.* 22(11): 2342–2347.

Fleming, C. A., 1962. "Paleontological evidence for speciation preceded by geographic isolation." In *The evolution of living organisms.* G. W. Leper, ed., 225–241. Melbourne: Melbourne University Press.

Flynn, L. J., J. C. Barry, M. E. Morgan, D. Pilbeam, L. L. Jacobs, and E. H. Lindsay, 1995. "Neogene Siwalik mammalian lineages: Species longevities, rates of change, and modes of speciation." *Palaeogeography, Palaeoecology, Palaeoclimatology.* 115: 249–264.

Foote, A. D., 2012. "Investigating ecological speciation in non-model organisms: A case study of killer whale ecotypes." *Evolutionary Ecology Research.* 14: 447–465.

Frank, K., and C. Wissel, 2002. "A formula for the mean lifetime of metapopulations in heterogeneous landscapes." *American Naturalist.* 159(5):530–552.

Funk, D. J., 1998. "Isolating a role for natural selection in speciation: Host adaptation and sexual isolation in *Neochlamisus bebbianae* leaf beetles." *Evolution.* 52(6): 1744–1759.

Funk, D. J., 2009. "Investigating ecological speciation." In *Speciation and patterns of diversity.* R. Butlin, J. Bridle, and D. Schluter, eds., 195–218. Cambridge, UK: Cambridge University Press.

Futuyma, D. J., 1987. "On the role of species in anagenesis." *American Naturalist.* 130: 465–473.

Futuyma, D. J., 1989. "Speciational trends and the role of species in macroevolution." *American Naturalist.* 134(2): 318–321.

Futuyma, D. J., 2002. Speciation. In *Encyclopedia of evolution.* M. Pagel, ed., vol. 2, 1072–1078. New York: Oxford University Press.

Gaston, K. J. 2003. *The structure and dynamics of geographic ranges.* Oxford, UK: Oxford University Press.

Gaston, K. J., and T. M. Blackburn, eds. 2000. *Patterns and process in macroecology.* Oxford, UK: Blackwell Science.

Gavrilets, S., 2003. "Models of speciation: What have we learned in 40 years?" *Evolution.* 57(10): 2197–2213.

Gavrilets, S., 2004a. "Speciation in metapopulations." In *Ecology, genetics, and evolution of metapopulations.* I. Hanski and O. E. Gaggiotti, eds., 275–303. Amsterdam: Elsevier Academic Press.

Gavrilets, S., 2004b. *Fitness landscapes and the origin of species.* Princeton, NJ: Princeton University Press.

Gavrilets, S., 2005. "'Adaptive speciation'—It is not that easy: A reply to Doebeli et al." *Evolution.* 59(3): 696–699.

Gavrilets, S., and J. B. Losos, 2009. "Adaptive radiation: Contrasting theory with data." *Science.* 323: 732–737.

Gavrilets, S., R. Acton, and J. Gravner, 2000. "Dynamics of speciation and diversification in a metapopulation." *Evolution.* 54(5): 1493–1501.

Geary, D. H., 1990a. "Patterns of evolutionary tempo and mode in the radiation of *Melanopsis* (Gastropoda; Melanopsidae)." *Paleobiology.* 16: 492–511.

Geary, D. H., 1990b. "Exploring the roles of intrinsic and extrinsic factors in the evolutionary radiation of *Melanopsis.*" In *Causes of evolution: A paleontological perspective.* R. M. Ross and W. D. Allmon, eds., 305–321. Chicago: University of Chicago Press.

Geary, D. H., 1995. "The importance of gradual change in species-level transitions." In *New approaches to speciation in the fossil record.* D. H. Erwin and R. L. Anstey, eds., 67–86. New York: Columbia University Press.

Gingerich, P. D., 1976. "Paleontology and phylogeny: Patterns of evolution at the species level in early Tertiary mammals." *American Journal of Science.* 276(1): 1–28.

Glaubrecht, M. 2011. "Towards solving Darwin's 'mystery': Speciation and radiation in lacustrine and riverine freshwater gastropods." *American Malacological Bulletin.* 29(1/2):187–216.

Glazier, D. S., 1987. "Toward a predictive theory of speciation: The ecology of isolate selection." *Journal of Theoretical Biology.* 126: 323–333.

Gould, S. J., 1969. "An evolutionary microcosm: Pleistocene and Recent history of the land snail P. (*Poecilozonites*) in Bermuda." *Bulletin of the Museum of Comparative Zoology.* 138: 407–532.

Gould, S. J., 1990. "Speciation and sorting as the source of evolutionary trends; or 'things are seldom what they seem.'" In *Evolutionary trends.* K. McNamara, ed., 3–27. London: Belhaven Press.

Gould, S. J., 1991. "Exaptation: A crucial tool for an evolutionary psychology." *Journal of Social Issues.* 47(3): 43–65.

Gould, S. J., 2002. *The structure of evolutionary theory.* Cambridge, MA: Harvard University Press.

Gould, S. J., and E. S. Vrba, 1982. "Exaptation—a missing term in the science of form." *Paleobiology.* 8(1): 4–15.

Gourbière, S., and J. Mallet, 2009. "Are species real? The shape of the species boundary with exponential failure, reinforcement, and the 'missing snowball.'" *Evolution.* 64(1): 1–24.

Grant, P. R., and B. R. Grant, 1995. "The founding of a new population of Darwin's finches." *Evolution.* 49: 229–240.

Grant, P. R., and B. R. Grant, 1997. "Genetics and the origin of bird species." *Proceedings of the National Academy of Sciences of the USA.* 94:7768–7775.

Grant, P. R., and B. R. Grant, 2006. "Species before speciation is complete." *Annals of the Missouri Botanical Garden.* 93: 94–102.

Grant, P. R., and B. R. Grant, 2008. *How and why species multiply: The radiation of Darwin's finches.* Princeton, NJ: Princeton University Press.

Haegeman, B., and R. S. Etienne, 2009. "Neutral models with generalised speciation." *Bulletin of Mathematical Biology.* 71(6): 1507–1519.

Hallam, A., 1982. "Patterns of speciation in Jurassic *Gryphaea.*" *Paleobiology.* 8(4): 354–366.

Hallam, A., 1988. "The contribution of palaeontology to systematics and evolution." In *Prospects in systematics.* D. L. Hawksworth, ed., 128–147. Oxford, UK: Systematics Association and Clarendon Press.

Hanski, I., and M. Gilpin, 1991. "Metapopulation dynamics: Brief history and conceptual domain." *Biological Journal of the Linnean Society.* 42: 3–16.

Hansen, T. A., 1978. "Larval dispersal and species longevity in Lower Tertiary gastropods." *Science.* 199: 885–887.

Hansen, T. A., 1980. "Influence of larval dispersal and geographic distribution on species longevity in neogastropods." *Paleobiology.* 6: 193–207.

Hardie, D. C., and J. A. Hutchings, 2010. "Evolutionary ecology at the extremes of species' ranges." *Environmental Reviews.* 18: 1–20.

Harrison, S., 1991. "Local extinction in a metapopulation context: An empirical evaluation." *Biological Journal of the Linnean Society.* 42: 73–88.

Heard, S. B., and D. L. Hauser, 1995. "Key evolutionary innovations and their ecological mechanisms." *Historical Biology.* 10:151–173.

Hendry, A. P., D. I. Bolnick, D. Berner, and C. L. Peichel, 2009. "Along the speciation continuum in sticklebacks." *Journal of Fish Biology.* 75(8): 2000–2036.

Hewitt, G. M. 2000. "The genetic legacy of the Quaternary ice ages." *Nature.* 405: 907–913.

Hewitt, G. M. 2004. "Genetic consequences of climatic oscillations in the Quaternary." *Philosophical Transactions of the Royal Society B.* 359: 183–195.

Hickman, C. S., 1988. "Analysis of form and function in fossils." *American Zoologist.* 28: 775–793.

Hopkins, M. P., and M. Webster, 2009. "Ontogeny and geographic variation of a new species of the corynexochine trilobite *Zacanthopsis* (Dyeran, Cambrian)." *Journal of Paleontology.* 83(4): 524–547.

Hoso, M., 2012. Non-adaptive speciation of snails by left-right reversal is facilitated on oceanic islands." *Contributions to Zoology.* 81(2): 79–85.

Hua, X., and J. J. Wiens, 2013. "How does climate influence speciation?" *American Naturalist.* 182(1): 1–12.

Hunt, G., and D. L. Rabosky, 2014. "Phenotypic evolution in fossil species: Pattern and process." *Annual Review of Earth and Planetary Sciences.* 42: 421–441.

Hunter, J. P., 1998. "Key innovations and the ecology of macroevolution." *Trends in Ecology and Evolution.* 13(1): 31–36.

Hutchinson, G. E. 1959. "Homage to Santa Rosalia: Why are there so many kinds of animals?" *American Naturalist.* 93: 145–159.

Ingram, T., 2011. "Speciation along a depth gradient in a marine adaptive radiation." *Proceedings of the Royal Society B*. 278: 613–618.

Iwasa, Y., and A. Pomiankowski, 1995. "Continual change in mate preference." *Nature*. 377: 420–422.

Jablonski, D., 1986. "Larval ecology and macroevolution in marine invertebrates." *Bulletin of Marine Science*. 39: 565–587.

Jablonski, D. and K. W. Flessa, 1986. "The taxonomic structure of shallow-water marine faunas: Implications for Phanerozoic extinctions." *Malacologia*. 27: 43–66.

Jablonski, D., and R. A. Lutz, 1983. "Larval ecology of marine benthic invertebrates: Paleobiological implications." *Biological Reviews*. 58: 21–89.

Jablonski, D., and K. Roy, 2002. "Geographic range and speciation in fossil and living molluscs." *Proceedings of the Royal Society B*. 270: 401–406.

Jackson, J. B. C. 1974. "Biogeographic consequences of eurytopy and stenotopy among marine bivalves and their evolutionary significance." *American Naturalist*. 108: 541–560.

Jackson, J. B. C., and A. H. Cheetham, 1990. "Evolutionary significance of morphospecies: A test with cheilostome Bryozoa." *Science*. 248:579–583.

Jackson, J. B. C., and A. H. Cheetham, 1994. "Phylogeny reconstruction and the tempo of speciation in cheilostome Bryozoa." *Paleobiology*. 20: 407–423.

Jackson, J. B. C., and A. H. Cheetham, 1999. "Tempo and mode of speciation in the sea." *Trends in Ecology and Evolution*. 14(2): 72–77.

Jansson, R., and M. Dynesius, 2002. "The fate of clades in a world of recurrent climatic change: Milankovitch oscillations and evolution." *Annual Review of Ecology and Systematics*. 33: 741–777.

Jiggins, C. D., 2008. "Ecological speciation in mimetic butter-flies." *BioScience*. 58: 541–548.

Johnson, A. L. A., 1981. "Detection of ecophenotypic variation in fossils and its application to a Jurassic scallop." *Lethaia*. 14: 277–285.

Johnson, A. L. A. 1993. "Punctuated equilibria versus phyletic gradualism in European Jurassic *Gryphaea* evolution." *Proceedings of the Geologists' Association*. 104: 209–222.

Johnson, J. G. 1975. "Allopatric speciation in fossil brachiopods." *Journal of Paleontology*. 49(4): 646–661.

Johnson, K. G., A. F. Budd, and T. A. Stemann, 1995. "Extinction selectivity and ecology of Neogene Caribbean reef corals." *Paleobiology*. 21 (1): 52–73.

Kamiya, T., 1992. "Heterochronic dimorphism of *Loxoconcha uranouchiensis* (Ostracoda) and its implication for speciation." *Paleobiology*. 18(2): 221–236.

Kimbel, W. H., 1991. "Species, species concepts and hominid evolution." *Journal of Human Evolution*. 20: 355–371.

Klingenberg, C. P. 1998. "Heterochrony and allometry: The analysis of evolutionary change in ontogeny." *Biological Reviews*. 73(1): 79–123.

Kocher, T. D., 2004. "Adaptive evolution and explosive speciation: The cichlid fish model." *Nature Reviews Genetics.* 5: 288–298.

Kopp, M., 2010. "Speciation and the neutral theory of biodiversity." *Bioessays.* 32: 564–570.

Lande, R., 1981. "Models of speciation by sexual selection on polygenic traits." *Proceedings of the National Academy of Science of the USA.* 78:721–725.

Lande, R., 1982. "Rapid origin of sexual isolation and character divergence in a cline." *Evolution.* 36: 213–223.

Lande, R. 1998. "Anthropogenic, ecological, and genetic factors in extinction." In *Conservation in a changing world.* G. M. Mace, A. Balmford, and J. R. Ginsberg, eds., 29–52. Cambridge, UK: Cambridge University Press.

Langerhans, R. B., and R. Riesch, 2013. "Speciation by selection: A framework for understanding ecology's role in speciation." *Current Zoology.* 59(1): 31–52.

Lauder, G. V., and K. F. Liem, 1989. "The role of historical factors in the evolution of complex organismal functions." In *Complex organismal functions: Integration and evolution in vertebrates.* D. B. Wake and G. Roth, eds., 63–78. New York: John Wiley and Sons.

Lawton, J. H. 1993. "Range, population abundance, and conservation." *Trends in Ecology and Evolution.* 8: 409–413.

Leigh, E. G., 1981. "The average lifetime of a population in a varying environment." *Journal of Theoretical Biology.* 90:213–239.

Levin, D. A. 2005. "Isolate selection and ecological speciation." *Systematic Botany.* 30: 233–241.

Lieberman, B. S. 2001a. "A probabilistic analysis of rates of speciation during the Cambrian radiation." *Proceedings of the Royal Society B.* 268: 1707–1714.

Lieberman, B. S. 2001b. "Analyzing speciation rates in macroevolutionary studies." In *Fossils, phylogeny, and form: An analytical approach.* J. Adrain, G. D. Edgecombe, and B. S. Lieberman, eds., 340–358. New York: Plenum Press/Kluwer Academic Publishers.

Lieberman, B. S. 2012. "Adaptive radiations in the context of macroevolutionary theory: A paleontological perspective." *Evolutionary Biology.* 39: 181–191.

Liem, K. F., 1974. "Evolutionary strategies and morphological innovations: Cichlid pharyngeal jaws." *Systematic Zoology.* 22: 425–441.

Liow, L.-H. 2010. "Speciation and the fossil record." In *Encyclopedia of Life Sciences.* Chichester, UK: John Wiley & Sons, Ltd. DOI: 10.1002/9780470015902. a0001666.pub2

Liow, L.-H., and N. C. Stenseth, 2007. "The rise and fall of species: Implications for macroevolutionary and macroecological studies." *Proceedings of the Royal Society B.* 274: 2745–2752.

Lister, A. M. 2004. "The impact of Quaternary Ice Ages on mammalian evolution." *Philosophical Transactions of the Royal Society B.* 359: 221–241.

Lloyd, E. A., 1988. *The structure and confirmation of evolutionary theory.* New York: Greenwood Press.

Lowry, D. B., 2012. "Ecotypes and the controversy over stages in the formation of new species." *Biological Journal of the Linnean Society*. 106(2): 241–257.

Ludwig, D., 1996. "The distribution of population survival times." *American Naturalist*. 147(4): 506–526.

MacArthur, R. H., and E. O. Wilson, 1967. *The theory of island biogeography*. Princeton, NJ: Princeton University Press.

Mallet, J., 2008a. "Mayr's view of Darwin: Was Darwin wrong about speciation?" *Biological Journal of the Linnean Society*. 95:3–16.

Mallet, J. 2008b. "Hybridization, ecological races, and the nature of species: Empirical evidence for the ease of speciation." *Philosophical Transactions of the Royal Society B*. 363: 2971–2986.

Mallet, J., A. Meyer, P. Nosil, and J. L. Feder, 2009. "Space, sympatry, and speciation." *Journal of Evolutionary Biology*. 22(11): 2332–2341.

Martin, R. A., 1993. "Patterns of variation and speciation in Quaternary rodents." In *Morphological change in Quaternary mammals of North America*. R. A. Martin and A. D. Barnosky, eds., 226–280. Cambridge: Cambridge University Press.

Masters, J. C., 1993. "Primates and paradigms: Problems with the identification of genetic species." In *Species, species concepts, and primate evolution*. W. H. Kimbel and L. B. Martin, eds. 43–64. New York: Plenum Press.

Maynard Smith, J., 1989. "The causes of extinction." *Philosophical Transactions of the Royal Society of London B*. 325: 241–252

Mayr, E., 1942. *Systematics and the origin of species*. New York: Columbia University Press.

Mayr, E., 1963. *Animal species and evolution*. Cambridge, MA: Harvard University Press.

Mayr, E., 1969. "Bird speciation in the tropics." *Biological Journal of the Linnean Society*. 1:1–17.

Mayr, E., 1982. "Processes of speciation in animals." In *Mechanisms of speciation*. C. Barigozzi, ed., 1–20. New York: Alan R. Liss.

McCune, A. R. 1996. "Biogeographic and stratigraphic evidence for rapid speciation in semionotid fishes." *Paleobiology*. 22(1): 34–48.

McCune, A. R., 1997. "How fast is speciation? Molecular, geological, and phylogenetic evidence from adaptive radiations of fishes." In *Molecular evolution and adaptive radiation*. T. J. Givnish and K. J. Sytsma, eds., 587–610. Cambridge, UK: Cambridge University Press.

McCune, A. R., 2004. "Diversity and speciation of semionotid fishes in Mesozoic rift lakes." In *Adaptive speciation*. U. Dieckmann, J. A. J. Metz, M. Doebeli, and D. Tautz, eds., 362–379. Cambridge, UK: Cambridge University Press.

McCune, A. R., and J. C. Schimenti, 2012. "Using genetic networks and homology to understand the evolution of phenotypic traits." *Current Genomics*. 13(1):74–84.

McKinney, M. L., and W. D. Allmon, 1995. "Metapopulations and disturbance: From patch dynamics to biodiversity dynamics." In *New approaches to speciation in the fossil record*. D. Erwin and R. Anstey, eds., 123–183. New York: Columbia University Press.

McNamara, K. J., 1987. "Taxonomy, evolution, and functional morphology of southern Australian Tertiary hemiasterid echinoids." *Palaeontology*. 30(2): 319–352.

Mcpeek, M. A., 2008. "The ecological dynamics of clade diversification and community assembly." *American Naturalist*. 172: E270–E284.

Merrill, R. M., Z. Gompert, L. M. Dembeck, M. R. Kronforst, W. O. McMillan, and C. D. Jiggins, 2011. "Mate preference across the speciation continuum in a clade of mimetic butterflies." *Evolution*. 65(5): 1489–1500.

Michaux, B. 1987. "An analysis of allozymic characters of four species of New Zealand *Amalda* (Gastropoda: Olividae: Ancillinae)." *New Zealand Journal of Zoology*. 14: 359–366.

Middleton, D. A. J., and R. M. Nisbet, 1997. "Population persistence time: Estimates, models, and mechanisms." *Ecological Applications*. 7(1):107–117.

Miller, W., III, 2001. "The structure of species, outcomes of speciation, and the 'species problem': Ideas for paleobiology." *Palaeogeography, Palaeoclimatology, Palaeoecology*. 176: 1–10.

Miller, W., III, 2006. "What every paleontologist should know about species: New concepts and questions." *Neues Jahrbuch fur Geologie und Paläontologie Monatenschiefen*. 2006(9): 557–576.

Morris, W. F., and D. F. Doak, 2002. *Quantitative conservation biology*. Sunderland, MA: Sinauer Associates.

Moritz, C., J. L. Patton, C. J. Schneider, and T. B. Smith, 2000. "Diversification of rainforest faunas: An integrated molecular approach." *Annual Review of Ecology and Systematics*. 31: 533–563.

Nehm, R., and D. H. Geary, 1994. "A gradual morphologic transition during a rapid speciation event in marginellid gastropods (Neogene: Dominican Republic)." *Journal of Paleontology*, 68(4): 787–795.

Norris, R. D., and P. M. Hull, 2012. "The temporal dimension of marine speciation." *Evolutionary Ecology*. 26(2): 393–415.

Nosil, P., 2008. "Ernst Mayr and the integration of geographic and ecological factors in speciation." *Biological Journal of the Linnean Society*. 95: 26–46.

Nosil, P., 2012. *Ecological speciation*. Oxford, UK: Oxford University Press.

Nosil, P., and B. J. Crespi, 2006. "Experimental evidence that predation promotes divergence in adaptive radiation." *Proceedings of the National Academy of Sciences of the USA*. 103: 9090–9095.

Nosil, P., and L. J. Harmon, 2009. "Niche dimensionality and ecological speciation." In *Speciation and patterns of diversity*. R. Butlin, J. Bridle, and D. Schluter, eds., 127–154. Cambridge: Cambridge University Press.

Nosil, P., and J. L. Feder, 2012. "Genomic divergence during speciation: Causes and consequences." *Philosophical Transactions of the Royal Society B.* 367: 332–342.

Nosil, P., and D. Schluter, 2011. "The genes underlying the process of speciation." *Trends in Ecology and Evolution.* 26(4): 160–167.

Nosil, P., L. J. Harmon, and O. Seehausen, 2009. "Ecological explanations for (incomplete) speciation." *Trends in Ecology and Evolution.* 24(3): 145–156.

Palumbi, S. R., 1994. "Reproductive isolation, genetic divergence, and speciation in the sea." *Annual Review of Ecology and Systematics.* 25: 547–572.

Paterson, H. E. H., 1985. "The recognition concept of species." In *Species and speciation.* Transvaal Museum Monograph 4, E. S. Vrba, ed., pp. 21–29. Pretoria, South Africa: Transvaal Museum Press.

Patzkowski, M. E., and S. M. Holland, 2012. *Stratigraphic paleobiology: Understanding the distribution of fossil taxa in time and space.* Chicago: University of Chicago Press.

Payne, J. L., and S. Finnegan, 2007. "The effect of geographic range on extinction risk during background and mass extinction." *Proceedings of the National Academy of Sciences of the USA.* 104:10506–10511.

Peters, S. E., 2006. "Genus extinction, origination, and the durations of sedimentary hiatuses." *Paleobiology.* 32: 387–407.

Peters, S. E., and M. Foote, 2002. "Determinants of extinction in the fossil record." *Nature.* 416: 420–424.

Pfennig, D. W., and K. S. Pfennig, 2012. *Evolution's wedge: Competition and the origins of diversity.* Berkeley: University of California Press.

Pfennig, D. W., M. A. Wund, E. C. Snell-Rood, T. Cruickshank, C. D. Schlichting, and A. P. Moczek, 2010. "Phenotypic plasticity's impact on diversification and speciation." *Trends in Ecology and Evolution.* 25: 459–467.

Phillimore, A. B., and T. D. Price, 2008. "Density-dependent cladogenesis in birds." *PLoS Biology.* 6: 483–489.

Phillimore, A. B., and T. D. Price, 2009. "Ecological influences on the temporal pattern of speciation." In *Speciation and patterns of diversity.* R. Butlin, J. Bridle, and D. Schluter, eds., 240–256. Cambridge, UK: Cambridge University Press.

Pigot, A. L., I. P. Owens, and C. D. L. Orme, 2012. "Speciation and extinction drive the appearance of directional range size evolution in phylogenies and the fossil record." *PLoS Biol.* 10(2):p.e1001260.

Polechová, J., and N. H. Barton, 2005. "Speciation through competition: A critical review." *Evolution.* 59(6): 1194–1210.

Price, T., 2008. *Speciation in birds.* Greenwood Village, CO: Roberts and Company Publishers.

Quinn, J. G., C. L. Wolin, and M. L. Judge, 1989. "An experimental analysis of patch size, habitat subdivision, and extinction in a marine intertidal snail." *Conservation Biology.* 3: 242–251.

Rabosky, D. L., 2015. "Reproductive isolation and the causes of speciation rate variation in nature." *Biological Journal of the Linnean Society.* Published online 19 Dec. 2015.

Rabosky, D. L., and I. J. Lovette, 2008. "Explosive evolutionary radiations: Decreasing speciation or increasing extinction through time?" *Evolution.* 62: 1866–1875.

Raikow, R. J., 1986. "Why are there so many kinds of passerine birds?" *Systematic Zoology.* 35: 255–259.

Raup, D. M., 1991. *Extinction: Bad genes or bad luck.* New York: W. W. Norton.

Raup, D. M., and S. M. Stanley, 1978. *Principles of paleontology.* 2nd ed. San Francisco: W. H. Freeman.

Ravitch, S. M., and J. M. Riggan, 2012. *Reason and rigor: How conceptual frameworks guide research.* Thousand Oaks, CA: Sage Publications.

Reif, W.-E., 1983a. "Hilgendorf's (1863) dissertation on the Steinheim planorbids (Gastropoda; Miocene): The development of a phylogenetic research program for paleontology." *Paläontologische Zeitschrift.* 57:7–20.

Reif, W.-E., 1983b. "The Steinheim snails (Miocene; Schwäbische Alb) from a neodarwinian point of view: A discussion." *Paläontologische Zeitschrift.* 57: 21–26.

Reiss, J. O., 2013. "Does selection intensity increase when populations decrease? Absolute fitness, relative fitness, and the opportunity for selection." *Evolutionary Ecology.* 27: 477–488.

Ricklefs, R. E., and E. Bermingham, 2007. "The causes of evolutionary radiations in archipelagoes: Passerine birds in the Lesser Antilles." *American Naturalist.* 169(3): 285–297.

Rieppel, O., 2009. "Species as a process." *Acta Biotheoretica.* 57: 33–49.

Ritchie, M. G., 2007. "Sexual selection and speciation." *Annual Review of Ecology, Evolution, and Systematics.* 38: 79–102.

Rodda, P. U., and W. L. Fisher, 1964. "Evolutionary features of *Athleta* (Eocene, Gastropoda) from the Gulf Coastal Plain." *Evolution.* 18(2): 235–244.

Rode, A. L., and B. S. Lieberman, 2005. "Integrating biogeography and evolution using phylogenetics and PaleoGIS: A case study involving Devonian crustaceans." *Journal of Paleontology.* 79:267–276.

Roopnarine, P. D. 1995. "A re-evaluation of evolutionary stasis between the bivalve species *Chione erosa* and *Chione cancellata* (Bivalvia: Veneridae)." *Journal of Paleontology.* 69(2): 280–287.

Rosenberg, A. 1994, *Instrumental biology or the disunity of science.* Chicago: University of Chicago Press.

Rosenblum, E. B., B. A. J. Sarver, J. W. Brown, S. Des Roches, K. M. Hardwick, T. D. Hether, J. M. Eastman, M. W. Pennell, and L. J. Harmon, 2012. "Goldilocks meets Santa Rosalia: An ephemeral speciation model explains patterns of diversification across time scales." *Evolutionary Biology.* 39(2): 255–261.

Rosenzweig, M. L., 1978. "Competitive speciation." *Biological Journal of the Linnean Society.* 10(3): 275–289.

Rosenzweig, M. L. 1995. *Species diversity in space and time*. New York: Cambridge University Press.

Rundell, R. J., and T. D. Price, 2009. "Adaptive radiation, nonadaptive radiation, ecological speciation, and nonecological speciation." *Trends in Ecology and Evolution*. 24(7): 394–399.

Rundle, H. D., and P. Nosil, 2005. "Ecological speciation." *Ecology Letters*. 8: 336–352.

Ryan, M. J., 1990. "Signals, species, and sexual selection." *American Scientist*. 78: 46–52.

Sampson, S. D., 1999. "Sex and destiny: The role of mating signals in speciation and macroevolution." *Historical Biology*. 13: 173–197.

Schindel, D. E., 1982. "Punctuations in the Pennsylvanian evolutionary history of *Glabrocingulum* (Mollusca: Archaeogastropoda)." *Geological Society of America Bulletin*. 93: 400–408.

Schluter, D., 1994. "Experimental evidence that competition promotes divergence in adaptive radiation." *Science*. 266: 798–801

Schluter, D.. 2000. *The ecology of adaptive radiation*. New York: Oxford University Press.

Schluter, D., 2001. "Ecology and the origin of species." *Trends in Ecology and Evolution*. 16(7): 372–380.

Schluter, D., 2009. "Evidence for ecological speciation and its alternative." *Science*. 323: 737–741.

Schwartz, J. H., 2009. "Reflections on systematics and phylogenetic reconstruction." *Acta Biotheoretica*. 57: 295–305.

Seehausen, O., 2006. "African cichlid fish: A model system in adaptive radiation research." *Proceedings of the Royal Society B*. 273: 1997–98.

Seehausen, O., 2007. "Evolution and ecological theory—Chance, historical contingency, and ecological determinism jointly determine the rate of adaptive radiation." *Heredity*. 99: 361–363.

Seehausen, O., 2009. "The sequence of events along a 'speciation transect' in the Lake Victoria cichlid fish *Pundamilia.*" In R. Butlin, D. Schluter, D., J. R. Bridle, eds., 155–176. *Speciation and Ecology*. Cambridge, UK: Cambridge University Press.

Seehausen, O., G. Takimoto, D. Roy, and J. Jokela, 2008. "Speciation reversal and biodiversity dynamics with hybridization in changing environments." *Molecular Ecology*. 17: 30–44.

Sepkoski, J. J., Jr., 1998. "Rates of speciation in the fossil record." *Philosophical Transactions of the Royal Society B*. 353: 315–326.

Simpson, G. G., 1951. "The species concept." *Evolution*. 5: 285–298.

Simpson, G. G., 1953. *Major features of evolution*. New York: Columbia University Press.

Simpson, G. G., 1961. *Principles of animal taxonomy*. New York: Columbia University Press.

Smith, A. B., 1984. *Echinoid palaeobiology*. London: Allen & Unwin.

Smith, G. R., 1987. "Fish speciation in a western North American Pliocene rift lake." *Palaios*. 2: 436–445.

Smith, K. K., 2003. "Time's arrow: Heterochrony and the evolution of development." *International Journal of Developmental Biology*. 47(7/8): 613–622.

Sobel, J. M., G. F. Chen, L. R. Watt, and D. W. Schemske, 2010. "The biology of speciation." *Evolution*. 64(2): 295–315.

Spight, T. M., 1974. "Sizes of populations of a marine snail." *Ecology*. 55: 712–729.

Stanley, S. M., 1978. "Chronospecies' longevities, the origin of genera, and the punctuational model of evolution." *Paleobiology*. 4:26–40.

Stanley, S. M., 1979. *Macroevolution: Pattern and process*. San Francisco: W. H. Freeman.

Stanley, S. M., 1986. "Population size, extinction, and speciation: The fission effect in Neogene Bivalvia." *Paleobiology*. 12: 89–110.

Stanley, S. M., 1990. "The general correlation between rate of speciation and rate of extinction: Fortuitous causal linkages." In *Causes of evolution: A paleontological perspective*. R. M. Ross and W. D. Allmon, eds., 103–127. Chicago: University of Chicago Press.

Stanley, S. M., and X. Yang, 1987. "Approximate evolutionary stasis for bivalve morphology over millions of years: A multivariate, multilineage study." *Paleobiology*. 13(2): 113–139.

Stewart, J. R. 2002. "The evidence for the timing of speciation of modern continental birds and the taxonomic ambiguity of the Quaternary fossil record." In *Proceedings of the 5th Symposium of the Society of Avian Paleontology and Evolution*, Beijing, 1–4 June 2000. Z. Zhou and F. Zhang, eds., 259–280. Beijing: Science Press.

Streelman, J. T., and P. D. Danley, 2003. "The stages of vertebrate evolutionary radiation." *Trends in Ecology and Evolution*. 18: 126–131.

Surget-Groba, Y., H. Johansson, and R. S. Thorpe, 2012. "Synergy between allopatry and ecology in population differentiation and speciation." *International Journal of Ecology*. 2012: Article ID 273413, 2012. doi:10.1155/2012/273413

Svensson, E. I., 2012. "Non-ecological speciation, niche conservatism, and thermal adaptation: How are they connected?" *Organisms, Diversity, and Evolution*. 12(3): 229–240.

Tasch, P., 1979. "Crustacean branchiopod distribution and speciation in Mesozoic lakes of the Southern continents." In *Terrestrial Biology III*, Antarctic Research Series, 30:65–74. Washington, DC: American Geophysical Union.

Tautz, D., 2004. "Phylogeography and patterns of incipient speciation." In *Adaptive speciation*. U. Dieckmann, J. A. J. Metz, M. Doebeli, and D. Tautz, eds., 305–321. Cambridge, UK: Cambridge University Press.

Taylor, A. D., 1990, "Metapopulations, dispersal, and predator-prey dynamics: An overview." *Ecology*. 71: 429–33.

Taylor, E. B., J. W. Boughman, M. Groenenboom, M. Sniatynski, D. Schluter, and J. L. Gow, 2006. "Speciation in reverse: Morphological and genetic evidence of the collapse of a threespined stickleback (*Gasterosteus aculeatus*) species pair." *Molecular Ecology.* 15: 343–355.

Thorpe, R. S., Y. Surget-Groba, and H. Johansson, 2008. "The relative importance of ecology and geographic isolation for speciation in anoles." *Philosophical Transactions of the Royal Society B.* 363: 3071–3081.

Turelli, M., N. H. Barton, and J. A. Coyne, 2001. "Theory and speciation." *Trends in Ecology and Evolution.* 16(7): 330–343.

Valentine, J. W. 1971. "Resource supply and species diversity patterns." *Lethaia.* 4:51–61.

Valentine, J. W., and D. Jablonski, 1983. "Speciation in the shallow sea: General patterns and biogeographic controls." In *Evolution, time, and space: The emergence of the biosphere.* R. W. Sims, J. H. Price, and P. E. S. Whalley, eds., 201–226. Systematics Association, special volume 23. London: Academic Press.

Vermeij, G. J., 1987. *Evolution and escalation: An ecological history of life.* Princeton, NJ: Princeton University Press.

Vermeij, G. J., 2002. "The geography of evolutionary opportunity: Hypothesis and two cases in gastropods." *Integrative and Comparative Biology.* 42(5): 935–940.

Vermeij, G. J., 2012. "Crucibles of creativity: The geographic origins of tropical molluscan innovations." *Evolutionary Ecology.* 26: 357–373.

Via, S., 2001. "Sympatric speciation in animals: The ugly duckling grows up." *Trends in Ecology and Evolution.* 16: 381–390.

Via, S., 2009. "Natural selection in action during speciation." *Proceedings of the National Academy of Sciences of the USA.* 106 (suppl.1): 9939–9946.

Vrba, E. S., 1980. "Evolution, species, and fossils: How does life evolve?" *South African Journal of Science.* 76:61–84.

Vrba, E. S., 1983. "Macroevolutionary trends: New perspectives on the roles of adaptation and incidental effect." *Science.* 221: 387–389.

Vrba, E. S., 1985. "African Bovidae: Evolutionary events since the Miocene." *South African Journal of Science.* 81: 264–266.

Vrba, E. S., 1987. "Ecology in relation to speciation rates: Some case histories of Miocene-Recent mammal clades." *Evolutionary Ecology.* 1(4): 283–300.

Vrba, E. S., 1992. "Mammals as a key to evolutionary theory." *Journal of Mammalogy.* 73: 1–28.

Vrba, E. S., 1995. "Species as habitat specific, complex systems." In *Speciation and the recognition concept: Theory and application.* D. M. Lambert and H. G. Spencer, eds., 3–44. Baltimore: Johns Hopkins University Press.

Waller, T., 1969. "The evolution of the *Argopecten gibbus* stock (Mollusca, Bivalvia), with emphasis on the Tertiary and Quaternary species of eastern North America." *Paleontological Society Memoir* 3, *Journal of Paleontology*, vol. 43, no. 5, supp., 1–125.

Wani, R., 2011. "Sympatric speciation drove the macroevolution of fossil cephalopods." *Geology*. 39(11): 1079–1082.

Weir, J. T., and S. Mursleen, 2012. "Diversity-dependent cladogenesis and trait evolution in the adaptive radiation of the auks (Aves: Alcidae)." *Evolution*. 67(2): 403–416.

Weissing, F. J., P. Edelaar, and G. S. van Doorn, 2011. "Adaptive speciation theory: A conceptual review." *Behavioral Ecology and Sociobiology*. 65(3): 461–480.

West-Eberhard, M. J., 1983. "Sexual selection, social competition, and speciation." *Quarterly Review of Biology*. 58: 155–183.

White, B. N. 1988. "Oceanic anoxic events and allopatric speciation in the deep sea." *Biological Oceanography*. 5(4): 243–259.

Wiens, J. J., 2004a. "What is speciation and how should we study it?" *American Naturalist*. 163(6): 914–923.

Wiens, J. J., 2004b. "Speciation and ecology revisited: Phylogenetic niche conservatism and the origin of species." *Evolution*. 58(1): 193–197.

Wiley, E. O., 1981. *Phylogenetics*. New York: John Wiley and Sons.

Wiley, E. O., and B. S. Lieberman, 2011. *Phylogenetics*. 2nd ed. New York: J. Wiley & Sons.

Williamson, P. G. 1981. "Palaeontological documentation of speciation in Cenozoic molluscs from Turkana Basin." *Nature*. 293:437–443.

Wilson, E. O., 1961. "The nature of the taxon cycle in the Melanesian ant fauna." *American Naturalist*. 95:169–193.

Wilson, W., 1969. "Problems in the speciation of American fossil bison." In *Post-Pleistocene man and his environment on the Northern Plains*. Proceedings of the First Annual Paleo-environmental Workshop. 178–200. Calgary: Student's Press, University of Calgary.

Wolf, J. B. W., C. Harrod, S. Brunner, S. Salazar, F. Trillmich, and D. Tautz, 2008. "Tracing early stages of species differentiation: Ecological, morphological, and genetic divergence of Galapagos sea lion populations." *BMC Evolutionary Biology*. 8:150. Doi: 10.1186/1471-2148-8-150

Wolfe, K., 2003. "Speciation reversal." *Nature*. 422: 25–26.

Wood, R. M., and R. L. Mayden, 2002. "Speciation and anagenesis in the genus *Cyprinella* of Mexico (Teleostei: Cyprinidae): A case study of Model III allopatric speciation." *Reviews in Fish Biology and Fisheries*. 12: 253–271.

Wood, T. E., and L. H. Rieseberg, 2002. "Speciation: Introduction." In *Encyclopedia of Life Sciences*. 415–422. New York: John Wiley & Sons.

Yacobucci, M. M., 1999. "Plasticity of developmental timing as the underlying cause of high speciation rates in ammonoids: An example from the Cenomanian Western Interior Seaway of North America." In *Advancing Research in Living and Fossil Cephalopods*. F. Olóriz and F. J. Rodríguez-Tovar, eds., 59–76. New York: Plenum Press.

Yamaguchi, T. 1980. "A new species belonging to the *Balanus amphitrite* Darwin group (Cirripedia, Balanomorpha) from the late Pleistocene of Japan: An example of peripheral speciation." *Journal of Paleontology*. 54(5): 1084–1101.

Yoder, J. B., E. Clancey, S. Des Roches, J. M. Eastman, L. Gentry, W. Godsoe, T. J. Hagey, D. Jochimsen, B. P. Oswald, J. Robertson, and B. A. J. Sarver, 2010. "Ecological opportunity and the origin of adaptive radiations." *Journal of Evolutionary Biology*. 23(8): 1581–1596.

Zink, R. M., J. Klicka, and B. R. Barber, 2004. "The tempo of avian diversification during the Quaternary." *Philosophical Transactions of the Royal Society B*. 359:215–220.

Morphology and Molecules: An Integrated Comparison of Phenotypic and Genetic Rates of Evolution

Steven J. Hageman

Introduction

The ability to successfully apply species and species concepts to real world problems such as biostratigraphy and conservation biology makes a strong argument for the existence of species (i.e., a level of organization that exists in the phenotype and genotype), whether or not we fully understand the processes involved in their origin. What we do know about the origin of biological species is drawn from four general perspectives: (1) Biological: variation within and among the broader phenotypes of closely related living organisms (including behavior, skeletal and nonskeletal morphology, physiology, and biochemistry, all generally limited to ecological timescales) and cross-breeding hybridization experiments; (2) Molecular: genetic data consisting of coded nucleotide sequences (DNA) ranging from short spans of a few dozen base pairs of uncertain placement or function to well-documented genes to entire genomes; (3) Paleontological: documentation of patterns and rates of morphologic change in closely related lineages through geologic time (primarily skeletal phenotypes); and (4) Theoretical considerations: some informed by the other three areas and others pure and independent.

In the past 150 years, an enormous amount of data has been generated from these four perspectives toward understanding the processes and mechanisms of speciation. These include: (1) hundreds of empirical studies of rates of morphologic change (e.g., Cheetham, 1986; Lazarus, 1986; Stanley and Yang, 1987; Geary, 1990; Budd and Klaus, 2008; Groves and Reisdorph, 2009; Pachut and Anstey 2009; Geary et al., 2010; Miller, chapter 2, this volume); (2) thousands of empirical studies of speciation in modern organisms based on variation of the phenotype (e.g., Belk, 1989; McPeek et al., 2010; Oros et al., 2010; Arkhipkina et al., 2012); and (3) molecular (genetic) data, requiring the organized efforts of federal governments and their consortia to house and store our current data set (e.g., National Center for Biotechnology Information, US National Institutes of Health). Finally, numerous theoretical ideas (e.g., Allmon and Ross, 1990; Levinton, 2001; Roopnarine, 2003; Coyne and Orr, 2004; Estes and Arnold, 2007; Gingerich, 1983, 2009; Hunt, 2010; Allmon, chapter 3, this volume; Miller, chapter 2, this volume) have been proposed to interpret and explain these data.

Because of these efforts we do understand a great deal about how evolution works and the resulting diversity of life. However, I would argue that in the grand scheme, a gaping hole still exists in the middle ground between these perspectives that is Darwin's original curiosity: "The origin of species." How do the genes, combined with the whole phenotype and its environment, shift within populations to generate a new species, which in many cases can be identified by a portion of its phenotype (e.g., skeleton) and act in ways that appear to be disconnected across different scales of time?

Attempts at integrating phenotypic data with genetic data have largely concentrated on using data sets that combine molecular and morphologic data in a single cladistics analysis (e.g., Mishler, 1994; Giribet et al., 2000) or superimposing the results of independent molecular phylogenies with stratophenetic character states (e.g., Lydeard et al., 1995; Wahlberg and Nylin, 2003; Decellea et al., 2012; Jagadeeshan and O'Dea, 2012). Typically, these studies and have concentrated on rates of cladogenesis rather than documenting change within clades (Nee, 2004; Quental and Marshall, 2010).

The goal of this project is to integrate morphologic (skeletal phenotype) and molecular (genetic) data into a single analysis and to provide a model that can eventually be extended to methods that estimate rates of genotypic change within and among species-level clades based on calibrated rates of change through fossil phenotypes.

General Properties of an Integrated Model

A key to moving forward with our understanding of the process of speciation will be to study extant taxa that have an excellent fossil record, that is, to simultaneously perform molecular and morphologic studies on a single suite of modern individuals and then trace their lineage through geologic time using both molecular phylogenies and stratophenetics.

Two simple models for morphologic change (fig. 5.1.1) and molecular change (fig. 5.1.2) through geologic time can be scaled and integrated into a single model (figs. 5.1.3 and 5.2.2). Figure 5.1.1 depicts a hypothetical situation of complete morphologic stasis for two closely related species A and B. The y-axis represents geologic time, a numerical chronology beginning with the present at the top (0.0 years), developed from all available and relevant sources (e.g., biostratigraphy, radiometric dating, chemostratigraphy). The x-axis represents any morphologic feature that can be adequately preserved in the fossil record so as to be quantified and compared among specimens. In this model, morphology could represent a simple linear measurement such as "length of nose," or a value derived from the ordination of a multivariate suite of characteristics such as a principal component score for observations from dozens of features for a specimen (e.g., Budd and Pandolfi, chapter 7, this volume), or any of the metrics resulting from morphometric methods that represent relationships among landmarks or describe shapes of organisms or their parts.

Figure 5.1.2 is a model of molecular change for a gene segment used in phylogenetic analysis, e.g., COI or 16s mRNA) versus time based on an empirical mutation rate (Freeland, 2006) for two extant sister species. The molecular distance (thick line A_0B_0) represents the percentage of nucleotide base pairs in common for the gene.

The molecular distance (genetic difference) between any two organisms, whether of the same population or even different phyla, will reflect the relative amount of time since the two had a common ancestor (Freeland, 2006; Beebee and Rowe, 2008). Beginning at the time that these two taxa shared a common ancestor (X in fig. 5.1.2), they will accumulate random mutations at a relatively constant rate; thus, the amount accumulated by either one is one-half the total distance (AY and BY, fig. 5.1.2). The mutation rate can be calibrated, especially for more closely related groups, using well-constrained divergence times based on fossils (Donoghue and Benton, 2007). Virtually any sequence or combination of partial sequences can be used to define the molecular distance between taxa, so long as the sequence is directly comparable among the taxa involved (Freeland, 2006; Beebee and Rowe, 2008).

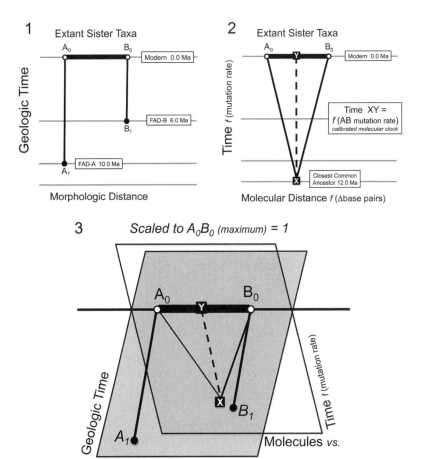

Figure 5.1 (1) Morphologic variation (phenotypic) versus geologic time, showing complete morphologic stasis of two sister species. The morphologic distance (thick line A_0B_0) is constant through time and the first appearance datum for species A is 10.0 Ma (specimen A_1) and 6.0 Ma for species B (specimen B_1). Specimen A_0 is same species as A_1, the subscripts indicate position in time. (2) Molecular change (genetic) versus time as a function of mutation rate, showing gradual divergence at a constant rate for two sister species A and B. The molecular distance (thick line A_0B_0) increases through time (0 at inferred time of last shared ancestor "X" 12.0 Ma. Specimens A_0 and B_0 are the same individuals as those characterized morphologically in 5.1.1. (3) Mathematical spaces 1 and 2 overlap when specimens A_0 and B_0 are scaled to the same metric (maximum distance = 1.0), i.e., same two specimens for species A and B in each mathematical space.

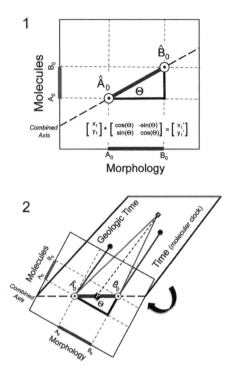

Figure 5.2 Model for an integrated analysis of morphology and molecules.
(1) Model for an integrated analysis of morphology and molecules. Each axis is scaled to the maximum value observed in the study (or another defined standard). A new, combined axis can be generated by rotating observations through the angle theta, where θ = arcsine of (Δ molecules/Δ morphology). These axes can be scaled because the line A_0B_0 for morphology is measured on the same specimens as the line A_0B_0 for molecules, thus the line between $\hat{A}_0$ and $\hat{B}_0$ is derived from the same specimens for both axes. (2) Because the combined axis for $\hat{A}_0$ and $\hat{B}_0$ represents both morphology and molecules (scaled), the geologic time axis of fig. 5.1.1 can be equilibrated with the molecular clock time axis of fig. 5.1.2. The result is an integrated model of morphology, molecules, and time (morpho-molecular-temporal space).

Derivation of an Integrated Model

Comparing figures 5.1.1 and 5.1.2, we can see that distance A_0B_0 for morphology and distance A_0B_0 for molecules are intimately related in that they are calculated on the exact same two specimens representing two different species. Thus scaled values for morphology and molecules A_0B_0 can be superimposed as an intersection of the two mathematical spaces (fig. 5.1.3). Values for morphologic A_0B_0 and molecular A_0B_0 can be represented for the two specimens on a scatter plot (fig. 5.2.1). These two dimensions (morphology and molecules) can be scaled to unity, such that morphologic $A_0B_0 = 1.0$, molecular $A_0B_0 = 1.0$, and the integrated distance $\hat{A}_0$ and $\hat{B}_0$ can be calculated.

When only two specimens are used in the analysis the distance $\hat{A}_0$ to $\hat{B}_0$ will equal $\sqrt{2}$ and the angle theta (θ) will equal 45° (x and y each scaled to 1.0). However, when multiple specimens are employed in an analysis, if x is scaled to 1.0 = maximum morphologic distance and y is scaled to 1.0 = maximum molecular distance (not required to be from the same specimen pair), then all other pairwise comparisons of integrated data for specimens $\hat{A}_0$ to $\hat{B}_0$ can be compared to this standardized distance based on maximum A_0B_0 for morphology and molecules of the group of specimens under consideration. In practice, A_0B_0 of morphology and molecules can be scaled to any defined standard.

Once a new combined axis is created for morphology and molecules scaled to unity, the third dimension (time) can be projected behind the plane (fig. 5.2.2), which represents both morphology vs. geologic time (fig. 5.1.1) and molecules vs. calibrated molecular clock time (fig. 5.1.2). Because points $\hat{A}_0$ and $\hat{B}_0$ are identical (i.e., exact same specimens), we have a reference point on which all planes can be tied (5.1.3) allowing for the integrated model of figure 5.2.2, which can viewed from any perspective of the three dimensions (morpho-molecular-temporal space).

A limiting requirement of the methodology proposed here is that direct comparisons of taxa can only be made within a unique set of morphologic and molecular variables, i.e., modification of the list of characters, either morphologic or molecular, results in a new mathematical space that is not directly comparable to others. Therefore, selection of morphologic features (Ciampaglio et al., 2001) and molecular data (Freeland, 2006; Beebee and Rowe, 2008) should be given careful consideration at the outset, so as to maximize the utility of a given study.

Interpretation of Integrated Data

On the plane of scaled morphology and molecules (fig. 5.3), the maximum (100%) is defined by the maximum value of the absolute difference in the suite of specimens under study. For example, in figure 5.3, if the maximum difference between any pairwise comparison of taxa for the character "nose-length" was 2.0 cm, the morphologic axis would range from 0% = 0.0 cm

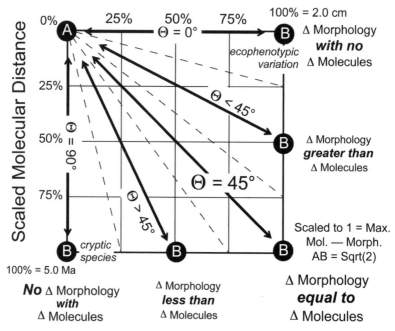

Figure 5.3 Model showing the relationship of angle theta (fig. 5.2.1) to relative rates of change in morphology and molecules in a system scaled by the maximum observed value on each axis.

Molecular distance of 0 = no divergence time between specimens = same population. Morphologic distance of 0 = no phenotypic variation between specimens, based only on the characteristics used in the analysis.

to 100% = 2.0 cm. The units of the scaled molecular distance will be based on the method used and number of nucleotide base pairs involved. This value (index of shared molecular code) can, however, be converted to a time interval using a calibrated molecular clock. Thus, for a study where the maximum difference in number of shared nucleotide base pairs at corresponding sites results in a maximum calculated molecular clock age of 5.0 Ma, the scaled molecular axis would range from 0% = 0.0 years (present day) to 100% = 5.0 Ma.

Scaling is a function of the angle theta (θ) through which each axis must be rotated in order to create a new combined axis through the two points (equation in fig. 5.2.1). Theta (θ) = arctangent of (Molecular A_0B_0 ÷ Morphologic A_0B_0). On plots of morphology vs. molecules, the angle theta

(θ from the rotational angle in fig. 5.2.1) will be $45°$ if there is a 1:1 relationship (fig. 5.3). All distances between pairwise specimen comparisons are scaled to this standard, the maximum observed morphologic and molecular distances (potentially from two different specimen pairs). The ratio of morphologic change to molecular change will be greater than, less than, or equal to the standard of maximum observed ($\hat{A}_0$ to $\hat{B}_0 = 1.0$ and $\theta = 45°$).

Theta will *decrease* as the magnitude of molecular variation decreases relative to morphologic variation to the point where $\theta = 0°$ with no difference in molecules (all morphologic variation is environmentally controlled = ecophenotypic (i.e., no time for any accumulated molecular differentiation = same population, fig. 5.3). Theta will *increase* as the magnitude of molecular variation increases relative to morphologic variation to the point where $\theta = 90°$ with no difference in morphology (morphologically cryptic species, fig. 5.3).

Comparing differential rates of change using integrated data

Several hypothetical scenarios are presented in figure 5.4 in order to demonstrate how patterns can be interpreted. The mathematical space is defined on ten hypothetical species (A, B, C, D, F, G, H, X, Y, Z) and data are scaled to 100% morphologic change = 2.0 cm (A–G) and 100% molecular change = 5.0 Ma (A–Z). All distances in are plotted relative to species A (fig. 5.4), though each pairwise distance could be plotted in this space.

In figure 5.4.1, species F, G, and H have diverged from species A at rates where morphologic change exceeds molecular change (for variables that define this space). Species F and G have evolved for the same amount of time from species A (scaled molecular distance = 25% = 1.25 Ma for both), but species G has shown a greater amount of morphologic change. In comparison, species F and H show the same degree of morphologic change relative to species A (scaled morphologic distance = 75% = 1.50 cm of net change), but species H has diverged from A for twice as much time as F has from A (25% vs. 50% = 1.25 Ma vs. 2.50 Ma).

Figure 5.4.2 provides a scenario where four species (A, B, C, D) belong to one putative genus (lightly shaded circles) and three species (X, Y, Z) belong to a second putative genus (darkly shaded circles). This example shows that species B, C, and D have evolved at the same rate ($\theta < 45°$, morphologic change faster than molecular change) and that their times since divergence from species A can be ordered (B, D, C). Species X,

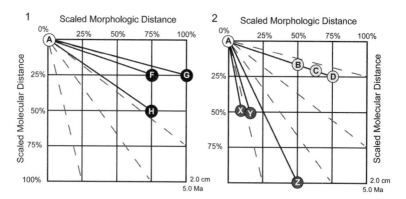

Figure 5.4 Hypothetical examples of different rates of morphologic and molecular evolution, relative to the standardized distance (maximum Δ morphology = AC = 1.0; maximum Δ molecules = AZ = 1.0).
(1) Three species (F, G, H), each compared to species A. Species F and G diverged from species A at the same time (25% of 5.0 Ma = 1.25 Ma), but species G had a greater amount of morphologic change than species F (100% vs. 75% of 2.0 = 2.0 cm and 1.5m change). In comparison, species H and F had equal morphologic change from species A (75% of 2.0 = 1.5 cm), but species H diverged from species A twice as long ago as species F did from A (50% vs. 25% of 5.0 Ma, 2.5 Ma and 1.25 Ma). (2. Three species of the same genus (B, C, D) showing successive divergence times from species A, but all at the same rate (morphology more rapid than molecules as compared to the standard). Two species of a different genus, species X and Y, diverged from species A at about the same time, with comparable rates of change (molecules changing faster than morphology relative to standard). Species Z is an outlier, inviting further investigation of its relationships to both species X and Y and it relative position to species A.

Y, and Z show faster relative molecular change (θ > 45°). Species X and Y diverged from species A twice as long ago as did species B from A (50% = 2.50 Ma vs. 25% = 1.25 Ma). Species Z diverged from species A over three times as long ago as species B and its morphology has changed more rapidly than X and Y relative to the degree of its molecular change (fig. 5.4.2).

Evaluating hypothetical timing of first occurrence and common ancestors

By plotting expected molecular distance through geologic time using a calibrated molecular clock, scaled to the best chronologic data available for stratophenetics, one of three results could be expected (fig. 5.5).

If one species is a true sister taxon of the other (e.g., species B is a direct descendant of species A, fig. 5.5.1), the molecular estimate of the closest common ancestor X should be coincident with the first appearance of the descendant species. The reason that these dates might not align exactly, even when the phylogenetic assumption is correct, include inaccuracies/imprecision in molecular clock calibration and or in chronostratigraphic resolution. Also, lack of alignment can occur when the first appearance datum does not adequately reflect the true origin of the species due to preservation or sampling bias. Statistical methods related to maximum likelihoods (i.e., error bars) can most likely be developed in order to determine the degree to which disagreement of molecular clock and chronostratigraphic dates are acceptable.

If the two taxa are not true sister species, but are part of a relatively close clade originating from other species not included in the plot, the calibrated molecular date will be older than the stratophenetic date of either. That is, figure 5.5.2 shows that based on molecular distance, species B shared a common ancestor with species A much earlier than the first occurrence of species B. Again, methods to determine whether this can be accounted for by an incomplete fossil record of species B, or inaccuracies of dating with molecular clock/chronostratigraphy, will have to be developed. But plots such as this can direct those inquiries.

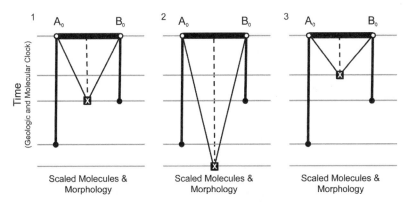

Figure 5.5 Models for three possible outcomes of comparing predicted times of youngest common ancestor for species pairs (assuming valid calibrated molecular clock and chronostratigraphy).
(1) Predicted time from molecules matches observed first appearance (or predicted time of origin), i.e., one could likely be the direct descendent of the other. (2) Predicted time from molecules significantly exceeds the observed first appearance (or predicted time of origin), which would indicate that either the observed stratigraphic ranges do not adequately represent the true ranges of the taxa, or both taxa share third, unidentified, common ancestor. (3) Predicted time from molecules is significantly less than the observed first appearance (or predicted time of origin), which suggests that error exists in identification of some specimens within the study.

In a similar scenario, if the timing of a closest common ancestor is significantly younger than the first occurrence of either taxa (fig. 5.5.3), this should raise questions about the validity of the assumption that these are true sister species (again accounting for acceptable variation due to molecular and chronostratigraphic dating).

Selection of Molecules, Morphology, and Distance Metrics

Molecular characters

Very few genes have been tied directly to morphologic variation in the phenotype (Gompel et al., 2005). Genes and partial gene sequences are used in most phylogenetic applications, e.g., COI cytochrome oxidase is mitochondrial and codes for proteins with physiological functions (Freeland, 2006; Beebee and Rowe, 2008); 12s rRNA and 16s rRNA are also mitochondrial (plus prokaryotes) and code for parts of the ribosome. Commonly used nuclear genes (18S, 28S) also code for ribosomes (Freeland, 2006; Beebee and Rowe, 2008). In fact, many phylogenetic applications use noncoding parts of the genome such as internal transcribed spacer (ITS) between regions that code for ribosomes (Freeland, 2006; Beebee and Rowe, 2008). These random, nonfunctioning mutations collect in the ITS regions.

Thus, the molecules used in this and in most studies of evolution serve as accumulated mutations (molecular clocks originally proposed by Zuckerkandl and Pauling [1962, 1965]) or proxies for relative amount of time since two taxa shared a common ancestor. Molecular clocks can be calibrated using geochronologically established divergence times from the fossil record (Omland, 1997; Donoghue and Benton, 2007). Because rates of molecular change are not fixed absolutely even within clades (Ayala, 1997; Drummond et al., 2006), molecular clocks can be modeled as relaxed or variable across branches (Drummond et al., 2012; Jagadeeshan and O'Dea, 2012).

In addition to options for selecting a molecular distance metric, there are also several strategies for calibrating molecular clocks (Sanderson, 2002; Kumar, 2005; Dornburg et al., 2011), but similar to complexities with morphologic change through time (fig. 5.6), any of these could be incorporated into a more complex model than the one proposed here.

In a most simple calculation of molecular distance, two taxa may be compared based on how many nucleotide base pairs (bp) they share, e.g.,

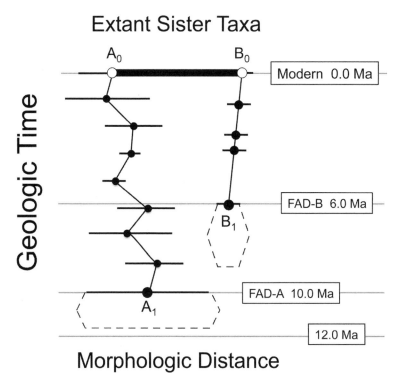

Extant Sister Taxa

A_0 B_0

Modern 0.0 Ma

Geologic Time

FAD-B 6.0 Ma

B_1

FAD-A 10.0 Ma

A_1

12.0 Ma

Morphologic Distance

Figure 5.6 A hypothetical model of morphologic variation (phenotype) versus geologic time. This model, which is more complex and realistic than fig. 5.1.1, shows both anagenesis in mean morphology for species B and reversals (zigzags) of mean phenotypes for species A. The amount of variation within samples (horizontal lines) is great for species A, but nearly constant for species B. The morphologic distance (thick line A_0B_0) is variable through time. The actual time and morphology at the origin of each species is approximated within the dashed borders.

two taxa sharing 95% of 650 bp would be much closer in molecular space than two taxa sharing 87% of the same sequence.

The method typically used to calculate uncorrected nucleotide distance between any two taxa (*p-distance*) (Nei and Kumar, 2000; Yang, 2006) is expressed as

$$\hat{p} = n_d/n, \text{where}$$

n_d = number of bp sites with different nucleotides between the two sequences, and

n = total number of nucleotide bp sites examined.
The variance of $\hat{p}$ (Nei and Kumar 2000; Yang, 2006) is

$$V(\hat{p}) = p(1-p)/n.$$

P-distance is usually an adequate index for closely related taxonomic comparisons, but it does not account for backwards or parallel substitutions (Yang, 2006) nor complexities that arise because of different processes of mutation (Nei and Kumar, 2000), i.e., substitution (exchange at one site), deletion (everything shifts), or inversion (two ends pivot around middle third) (Nei and Kumar, 2000). In addition, some of these processes operate with different likelihoods depending on the nucleotides involved, e.g., (T or C) vs. (A or G) (Nei and Kumar, 2000).

For large p (molecular distances), more complex models are required based on probabilities of nucleotide pairs, e.g., Kimura's Two-Parameter Method or Tamura and Nei's Method (Nei and Kumar, 2000). For this study, the Kimura 2-parameter model (K2P of Kimura, 1980) with a discrete approximation of the Γ distribution (K2P + Γ) was used to calculate molecular distance (data from Dick et al., 2003). This results in a p-distance expressed in units of percent difference (Nei and Kumar, 2000).

Morphologic characters

Virtually any kind of morphologic data can be used so long as they can be defined, collected, and treated uniformly across all specimens for study. Although the morphologic portion of the space will have greatest relevance when constructed as part of a hypothesis based in paleobiological theory, the methodology does not require this.

A suite of closely related specimens should be incorporated into a model (not just a single pair). Subsets of the total can be plotted individually, but once the space is defined by a set of specimens, morphologic characters and molecular description of a new model must be generated if any specimens are added.

Figure 5.6 represents some of the complexities of morphologic change and properties of empirically collected data. These complexities are intentionally omitted from the development of a model that integrates morphology and molecules; however, many of these issues have been studied in detail and in many cases practical solutions have been proposed that can be included directly into a more comprehensive integrated model. Complexities present in figure 5.6 include:

1. Patterns of change within species through time from stasis to anagenesis, random walk, or a combination (Bookstein, 1987; Gingerich, 1983, 2009; Cheetham and Jackson, 1995; Roopnarine et al., 1999; Hunt, 2010, 2012).

2. Variation within samples (range, confidence intervals, relative analysis of average only) (e.g., Hageman, 1994; Renaud et al., 2007; Pachut and Anstey, 2009), and changes in variance among samples through time, i.e., variable width of sample error bars (e.g., Hageman, 1994; Ricklefs, 2006).

3. Temporal resolution and correlation of samples (e.g., Sadler, 1981, 2004; Kidwell and Bosence, 1991; Kowalewski and Bambach, 2008).

4. Uncertainty of first occurrence vs. evolutionary origin of a taxon (Marshall, 1990, 1997; Weiss and Marshall, 1999; Hayek and Bura, 2001).

5. Amount of morphologic variation that is due to environmental effects rather than genetic control, i.e., estimating the partitioning of the amount of the total variance that is caused by heritable genetic factors vs. environmental/other factors (Falconer, 1981; Jackson and Cheetham, 1990; Cheetham et al., 1993, 1994; Hunter and Hughes, 1994; Riisgård and Goldson, 1997; Hageman et al., 1999, 2009, 2011).

Case Study of Integrated Molecular and Morphologic Data

In a series of three papers (Dick et al., 2003; Herrera-Cubilla et al., 2006, 2008), a research group analyzed molecular and morphologic data for the same suite of specimens assigned to nine species in two genera (*Cupuladria* and *Discoporella*) from a single family (Cupuladridae) of modern bryozoans. These data are well suited to demonstrating the integrated methodology proposed here. Jagadeeshan and O'Dea (2012) recently published a study using the same kinds of morphologic and molecular data, which refine some of the species-level taxonomy for Panamanian Cupuladridae. The nomenclature of Dick et al. (2003) and Herrera-Cubilla et al. (2006 and 2008) is used here because it applies to the particular specimens employed in this study. The data set and methods of Jagadeeshan and O'Dea (2012) are also very well suited to the methodology proposed here.

Although change associated with a single morphological character may be interpreted more intuitively, Cheetham (1987) cautioned against interpreting patterns of evolution based on single morphologic features. Suites of related characters can carry more paleobiological significance. Regardless of how many characters are incorporated into a single analysis, the selection of characters to represent the morphologic phenotype should be

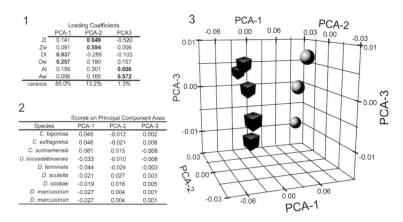

1

Loading Coefficients

	PCA-1	PCA-2	PCA3
Zl	0.141	**0.649**	-0.520
Zw	0.091	**0.594**	0.006
Ol	**0.937**	-0.268	-0.103
Ow	**0.257**	0.190	0.157
Al	0.159	0.301	**0.606**
Aw	0.056	0.165	**0.572**
variance	85.0%	13.2%	1.3%

2

Scores on Principal Component Axes

Species	PCA-1	PCA-2	PCA-3
C. biporosa	0.045	-0.012	0.002
C. exfragminis	0.046	-0.021	0.008
C. surinamensis	0.081	0.015	-0.008
D. bocasdeltoroensis	-0.033	-0.010	-0.008
D. terminata	-0.044	-0.024	-0.003
D. scutella	-0.021	0.027	0.003
D. cookae	-0.019	0.016	0.005
D. marcusorum	-0.027	0.004	0.001
D. marcusorum	-0.027	0.004	0.001

Figure 5.7 Principal component analysis of cupuladriid bryozoans (nine species, six characters). (1) Loading coefficients from principal components analysis of data from table 5.1, transformed (log10 (x+1)), with largest loadings in boldface. (2) PCA scores for each species on the first three axes. (3) Scatter plot of PCA scores. The loading coefficients for PCA-1 are dominated by opesia size, which differentiates between genera (spheres = *Cupuladria*, cubes = *Discoporella*). PCA-2 represents zooecia size and PCA-3 represents aperture size, both of which differentiate species within genera on the scatter plot. The first three axes account for 99.5% of total variance.

hypothesis driven. In order to identify a subset of characters with discriminatory power among cupuladriid species, data for over 25 characters available from Herrera-Cubilla et al. (2006, 2008) were analyzed using principal component analysis (PCA). A subset of six was chosen based on loading coefficients (fig. 5.7.1, table 5.1). These include the length and width of zooecia, opesia, and apertures (Zl, Zw, Ol, Ow, Al, Aw). These features are related to the shape and size of autozooecia (primary modules) in the bryozoan family Cupuladridae (characters defined and illustrated in Herrera-Cubilla et al., 2006, fig. 1.6, table 1).

Scores from PCA, nine species and six characters, were derived using PAST v. 2.15, with raw morphologic data (mm) transformed as log10 (1+x) prior to PCA analysis. For these data (table 5.1, fig. 5.7), the variables Ol and Ow (opesia length and width) dominate the loading coefficients of PCA-1 (fig. 5.7.1), whereas PCA-2 and PCA-3 are dominated by zooecia (Zl, Zw) and aperture (Al, Aw) size. To demonstrate the methodology, I use the score for each species on PCA-1 (fig. 5.7.2). From a plot of PCA scores (fig. 5.3), we can expect PCA-1 to differentiate species between the two genera, whereas PCA-2 and PCA-3 differentiate species within genera. These insights may also guide the selection of characters used to define the morphospace for study.

TABLE 5.1. **Raw data for six morphologic characters from nine species of cupuladriid bryozoans, from Herrera-Cubilla et al. (2006, table 1) and Herrera-Cubilla et al. (2008, table 5), Z = zooecia, O = opesia, A = aperture, l = length and w = width, all measurements in mm.**

Species	abbreviation	Zl	Zw	Ol	Ow	Al	Aw
Cupuladria 4 (Cupuladria biporosa)	Cbipor	0.447	0.317	0.321	0.180	0.214	0.161
Cupuladria 5 (Cupuladria exfragminis)	Cexfr	0.423	0.294	0.330	0.181	0.218	0.173
Cupuladria 6 (Cupuladria surinamensis)	Csuri	0.547	0.371	0.406	0.219	0.235	0.165
Discoporella 2 (Discoporella bocasdeltoroensis)	Dboca	0.440	0.290	0.120	0.120	0.170	0.140
Discoporella 8 (Discoporella terminata)	Dterm	0.390	0.270	0.100	0.120	0.150	0.140
Discoporella 7 (Discoporella scutella)	Dscut	0.500	0.370	0.120	0.160	0.210	0.180
Discoporella 3A (Discoporella cookae)	Dcook	0.480	0.340	0.130	0.150	0.220	0.170
Discoporella 3B (Discoporella marcusorum)	DmarB	0.450	0.320	0.120	0.140	0.200	0.150
Discoporella 3C (Discoporella marcusorum)	DmarC	0.450	0.320	0.120	0.140	0.200	0.150

The following is an outline of the protocol that can be used to produce the integrated model for the example case. Begin with a matrix of morphologic distances between every taxonomic pair (upper-right of table 5.2.1) calculated as the absolute value of $(x_i - x_n)$ from morphologic data (here, PCA-1 scores, fig. 5.7.2). Then calculate each pairwise difference as a percentage of the maximum observed morphologic difference (upper-right of table 5.2.2), e.g., $0.036 / 0.125 * 100 = 28.8\%$ (*C. biporosa* vs. *C. surinamensis*).

In a similar manner, use a pairwise matrix of molecular distances between every taxonomic pair (lower-left of table 5.2.1), here as absolute value (K2P + Γ) differences (from Dick et al., 2003, lower part of their table 3). Then calculate each pairwise difference as a percentage of the maximum observed molecular difference (lower-left of table 5.2.2), e.g., $17.27 / 25.8 * 100 = 66.9\%$ (*C. biporosa* vs. *C. surinamensis*).

Then, create a new table where x = % morphologic distance (upper-right of table 5.2.2) and y = % molecular distance (lower-left of table 5.2.2). Resulting scatter plots (fig. 5.8) will have x-axis values from 0% = 0.000 to 100% = 0.125 scores on PCA-1, dominated by the size of the opesia. The resulting y-axis has values from 0% = 0.00 to 100% = 25.8 (K2P + Γ distance), which can be converted to 14.03 Ma using an estimated

TABLE 5.2. (1) Similarity matrix for nine species assigned to two genera. Upper-right half of matrix is distances between paired taxa based on PCA-1 scores. The lower-left is molecular distances for 16s mRNA from Dick et al. (2003) for the same specimens. (2) Similarity matrix for the same specimens, expressed as a percentage of the maximum value observed in 1 (in bold). Percentages were used in scatter plots for fig. 5.9 (upper-right = morphology = x-axis; lower-left = molecules = y-axis).

1

Morphologic (PCA-Axis One) Distances →

	Cupuladria 4 (C. biporosa)	Cupuladria 5 (C. exfragminis)	Cupuladria 6 (C. surinamensis)	Discoporella 2 (D. bocasdeltoroensis)	Discoporella 8 (D. terminata)	Discoporella 7 (D. scutella)	Discoporella 3A (D. cookae)	Discoporella 3B (D. marcusorum)	Discoporella 3C (D. marcusorum)
Cupuladria 4 (C. biporosa)	**0.00** / 0.90	0.002	0.036	0.077	0.089	0.065	0.064	0.072	0.072
Cupuladria 5 (C. exfragminis)	5.7	**0.00** / 0.46	0.034	0.079	0.090	0.067	0.066	0.074	0.074
Cupuladria 6 (C. surinamensis)	17.3	20.2	**0.00** / 0.73	0.113	0.125	0.101	0.100	0.108	0.108
Discoporella 2 (D. bocasdeltoroensis)	25.0	25.8	19.5	**0.00** / 0.14	0.011	0.012	0.013	0.005	0.005
Discoporella 8 (D. terminata)	20.5	23.8	15.7	13.6	**0.00** / 0.14	0.023	0.025	0.017	0.017
Discoporella 7 (D. scutella)	22.7	24.3	18.3	16.9	13.3	**0.00** / 0.18	0.001	0.007	0.007
Discoporella 3A (D. cookae)	22.9	25.4	19.1	17.2	12.3	7.5	**0.00** / 0.37	0.008	0.008
Discoporella 3B (D. marcusorum)	22.8	24.9	19.8	19.3	13.6	8.9	3.2	**0.00** / 0.26	0.000
Discoporella 3C (D. marcusorum)	22.7	24.8	20.4	19.2	15.8	10.0	4.2	4.5	**0.00** / 1.15

← *16s Molecular (K2P + Γ) Distances (%)*

Distance matrix (page 2). Lower-left triangle values are molecular distances; upper-right triangle values are morphologic distances. The diagonal shows the 0.00 morphologic self-distance (staircase); the small values on the diagonal are the molecular within-group distances.

	Cupuladria 4 (C. biprosa)	Cupuladria 5 (C. exfragminis)	Cupuladria 6 (C. surinamensis)	Discoporella 2 (D. bocasdeltoroensis)	Discoporella 8 (D. terminata)	Discoporella 7 (D. scutella)	Discoporella 3A (D. cookae)	Discoporella 3B (D. marcusorum)	Discoporella 3C (D. marcusorum)
Cupuladria 4 (C. biprosa)	3.3 / 0.00	1.3	28.8	62.1	71.2	52.5	51.3	57.8	57.8
Cupuladria 5 (C. exfragminis)	22.1	1.8 / 0.00	27.4	63.4	72.6	53.8	52.7	59.1	59.1
Cupuladria 6 (C. surinamensis)	66.9	78.4	2.8 / 0.00	90.9	100.0	81.3	80.1	86.6	86.6
Discoporella 2 (D. bocasdeltoroensis)	97.0	100.0	75.7	0.5 / 0.00	9.1	9.6	10.8	4.3	4.3
Discoporella 8 (D. terminata)	79.3	92.3	60.8	52.7	0.5 / 0.00	18.7	19.9	13.4	13.4
Discoporella 7 (D. scutella)	88.0	94.0	70.9	65.7	51.6	0.7 / 0.00	1.2	5.3	5.3
Discoporella 3A (D. cookae)	88.9	98.6	74.0	66.7	47.8	29.1	1.4 / 0.00	6.5	6.5
Discoporella 3B (D. marcusorum)	88.3	96.7	76.9	74.6	52.8	34.3	12.3	1.0 / 0.00	0.0
Discoporella 3C (D. marcusorum)	88.1	96.3	79.2	74.6	61.2	38.6	16.3	17.3	4.5 / 0.00

% of maximum Morphologic (PCA-Axis One) Distance

% of Maximum 16s Molecular (K2P + Γ) Distance

2

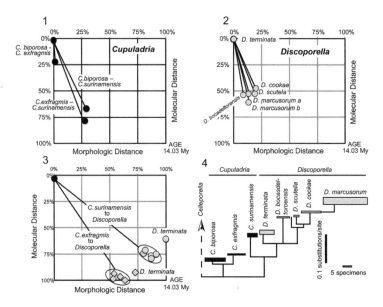

Figure 5.8 Single, scaled morpho-molecular space for nine cupuladriid species, with subsets of data plotted separately.
(1) Distances between paired species of the genus *Cupuladria* plotted in all combinations. *C. bipora* and *C. exfragmis* are the most closely related (diverged 22.1% of 14.03 Ma) and show very little morphologic change (1.3% of maximum PCA-1 difference for all nine species). *C. bipora* and *C. exfragmis* show about the same amount of morphologic divergence from *C. surinamensis*, but *C. exfragmis* apparently diverged earlier (78.4% of 14.03 Ma). (2) Distances between five species of the genus *Discoporella* each compared to *D. terminata*. See text for discussion. (3) Comparisons across genera. All *Discoporella* compared to *C. surinamensis* (maximum morphologic difference for PCA-1 scores) and *C. exfragmis* (maximum molecular difference). Species of *Discoporella* group in both comparisons, with the exception of *D. terminata*. (4) Phylogeny for the species represented, after Dick et al. (2003, fig. 4).

molecular clock for these specimens (25.8% bp difference maximum) ÷ (1.84% bp difference per million years mutation rate) based on the closing of the Isthmus of Panama (Dick et al., 2003).

A molecular phylogeny (16s) for the nine species from Dick et al. (2003, fig. 4) is summarized in fig. 5.8.4. Figure 5.8.1 shows the morphologic vs. molecular distance for the three species of the genus *Cupuladria* (scaled to maximum distances among all nine species). Results can be interpreted as follows. (1) Species pair *C. bipora* and *C. exfragmis* are nearly identical based on scores for PCA-1 (predominately opesia length and width) but have diverged in the 16s molecular composition over approximately 3.1 my (22.1% of 14.03 my). (2) Species *C. bipora* and *C. exfragmis* diverged from *C. surinamensis* at 9.39 Ma and 11.00 Ma (66.9% and 78.4%) respectively, but the amount of morphologic divergence is about the same (28.5% and 27.4% of the total for PCA-1).

Figure 5.8.2 shows the scaled morphologic vs. molecular distance for five species of the genus *Discoporella* all from a sixth species, *D. terminata* (other pairwise species combinations are not illustrated). Results can be interpreted as follows. 1) All five species diverged from *D. terminata* during a relatively short interval (6.01–8.59 Ma), and species *D. bocadeltorensis* and *D. marcusorum* b, diverged from *D. terminata* within 7.40–7.41 Ma (Fig. 5.8.2). This close timing of divergence does not necessarily mean that these species are sister taxa, only that molecularly they are equidistant from *D. terminata*. These species have had comparable rates of morphologic change (13.4% to 19.9% of the total variation on PCA-1) (Fig. 5.8.2).

Figure 5.8.3 shows the scaled morphologic vs. molecular distance for each species of the genus *Discoporella* to the cupuladriid species with maximum morphologic difference (*C. surinamensis*) distances plotted with circles (fig. 5.8.3) and the species with greatest molecular difference (*C. exfragmis*) distances plotted with diamonds (fig. 5.8.3). Because these are not the same species in this case, no species plots 100% simultaneously for both axes. Results can be interpreted as follows. (1) In both cases, five of the *Discoporella* species plot in a cluster (ovals in fig. 5.8.3). The molecular distance values near 100% for *C. exfragmis* and most *Discoporella* species suggest a divergence time for these genera at 14.03 Ma. The suggested time of divergence based on *C. surinamensis* and most *Discoporella* species of about 10.6 Ma (75.4% of 14.03 Ma) is comparable to the estimated divergence time of these two cupuladriid species *C. exfragmis* and *C. surinamensis* 11.0 Ma (78% of 14.03 Ma). (2) In both sets of comparisons, *D. terminata* is an odd specimen out (fig. 5.8.3). It is unclear why it plots separate from other *Discoporella* species, but one of the roles of this methodology is to screen for outliers and investigate potential explanations.

Results for relative rates of morphologic and molecular change in these nine cupuladriid species (fig. 5.8) demonstrate the viability of the methodology and interpretation of empirical results. Clearly, the scale of the morpho-molecular space one defines will affect the relative distances among taxa. In this case, for instance, defining the space based only on *Discoporella* specimens (as compared to one defined by *Discoporella* and *Cupuladria* combined) may provide different insights.

Discussion: The Disconnect between Morphology and Molecules

When the model for complete morphologic stasis of species is integrated with the model for a molecular clock (fig. 5.9), the concerns expressed about processes to explain punctuated equilibrium (Eldredge and Gould, 1972; Gould and Eldredge, 1993; Ruse, 2000) are evident in the inescapable expectation that as molecular difference is traced back in time (diagonal lines in fig. 5.9), the distance between two related taxa by definition must be shorter and shorter.

This disconnect between rates of morphologic evolution and molecular evolution results in all forms of confounding observations including differing amounts of molecular distance between taxa through time that otherwise show constant morphologic distance (cf. fig. 5.9, species A and B at times T_2, T_1, and T_0). In cases where two species are not distinguishable based on the morphologic characters analyzed (or potentially any morphologic character), the morphologic lines of the species in a comparable figure 5.9 would be centered and indistinguishable, while the morphologic distance would inevitably diverge, resulting in cryptic species (Knowlton, 1993; Bickford et al., 2006).

The simple answer to this apparent paradox is that molecules used in molecular phylogenies are not the molecules that code for morphology (Pagel et al., 2006; Zeh et al., 2009; Rebollo et al., 2010). An entire field of biology—evo-devo, evolution and development—has developed in the past few decades that addresses questions about development and the specific genes that control the phenotype (Hageman, 2003; Carroll, 2005, 2007, 2009; Prud'homme, 2006, 2007).

In theory, relatively small mutations in a short region of DNA (regulatory elements) act on a conserved region of DNA (e.g., homeotic regulatory genes) that codes for body plans and morphology (Stone and Wray, 2001; Hoekstra and Coyne, 2007). Thus, the notion of "hopeful monsters," i.e., large shifts in morphology in single generation/mutation, is not that far-fetched (Theissen, 2006; Rieseberg and Blackman, 2010). Theories from the field of evo-devo are intuitive and elegant in explaining observed paleontological patterns of macroevolution such as arthropod segment differentiation and specialization of appendages (Carroll et al., 2004; Carroll 2008, 2009) and the evolution of tetrapod limbs (Shubin, 2008). However, the genetic pathways through homeotic regulatory genes to specific phenotypes

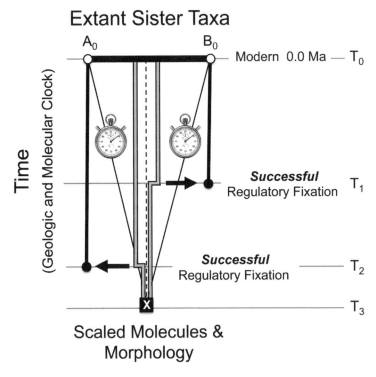

Figure 5.9 Integrated model of morphologic and molecular change through time. Line A_0B_0 represents scaled morphologic distance and molecular distance based on the maximum observed values in the suite of specimens being analyzed. Vertical axis represents geologic time calibrated to a molecular clock and chronostratigraphy. Straight vertical lines represent morphology of a species through time (complete stasis in this example). Diagonal lines represent molecular change (typical phylogenetic molecule), with the % base pair difference converging on zero at the time of the most recent common ancestor. Stepped thick lines represent base pair differences in regulatory elements that control the morphology in question. Accumulation of mutations in the molecules (thin diagonal lines) is steady and allows for a background clock. Accumulation of mutations in the regulatory elements (stepped thick lines) is irregular and can result in large, instant shifts in morphology.

can be identified only in model organisms thus far (Gompel et al., 2005; Davidson, 2010; Frankel et al., 2011; Wittkopp and Kalay, 2012).

Consider the model in figure 5.9. The diagonal lines represent the slow accumulation of mutations (all benign or entirely noncoding) in the genes typically used for phylogenetic analysis (e.g., COI or 16s). These accumulate at a relatively constant rate, as a predictable molecular clock (background stopwatch in fig. 5.9). The regulatory elements that are essential for coding morphology are represented by the thick gray lines in figure 5.9. Little is known about the actual mutation rate of these and other genes related to morphogenesis. They could mutate at exactly the same rate as those genes used for phylogenetic analysis or they could be faster or slower, but there is no reason to expect them to not mutate at some average rate. The difference, however, is that mutation of a regulatory element or other key gene involved in morphogenesis has a greater likelihood of being lethal to the individual. It would only be the rare event of a nonlethal mutation that is favorably selected for, is fixed, and eventually enters the broader gene pool that will appear in the fossil record as a novel morphology. Depending on the magnitude of the morphologic change, this could appear as a punctuated event or a small directional shift (fig. 5.9). The thick gray lines in fig. 5.9 depict successful shifts (fixation) of a hypothetical regulatory element or gene essential to morphogenesis. All of the other random mutations in these genes would be culled under negative selection, i.e., regulatory elements do not accumulate benign or noncoding sequences as do most genes applied in phylogenetics. Through time, the accumulated genetic distance based on regulatory elements would be minor (fig. 5.9, thick gray lines) and would probably not accumulate at a constant rate (though they may be mutating at a predictable rate in the background). We do not yet have data to document the specific regulatory elements (genetic code) that could provide a mechanism for these shifts.

Even if a "regulatory element" explanation for morphologic microevolution and speciation proves misdirected, this should not inhibit paleontologists from documenting the patterns of morpho-molecular change within and among closely related species in anticipation of the inevitable recognition of the systems that do control morphogenesis and its evolution (Carroll, 2009; Davidson, 2010). As paleontologists, we can begin to document the connections and disconnections between molecules (phylogenetic) and morphology and use them to identify/predict potential mutation and fixation of whichever molecules are responsible for morphogenesis. This knowledge can provide constraints on a framework for interpreting patterns of microevolution throughout geologic time.

Acknowledgments

I thank W. D. Allmon and M. M. Yacobucci for organizing the topical session on fossil species at the Minneapolis GSA meeting 2011. I also thank M. M. Yacobucci, J. Pachut, and G. Hunt for careful review and suggestions that improved this paper.

References

Arkhipkina, A. I., A. B. Vyacheslav, and D. Fuchs. 2012. Vestigial phragmocone in the gladius points to a deepwater origin of squid (Mollusca: Cephalopoda). *Deep Sea Research Part I: Oceanographic Research Papers* 61:109–122. doi:10.1016/j. dsr.2011.11.010

Allmon, W. D., and R. M. Ross. 1990. Specifying causal factors in evolution: The paleontological contribution. In *Causes of Evolution: A Paleontological Perspective*, edited by R. M. Ross, and W. D. Allmon, 1–17. Chicago: University of Chicago Press.

Ayala, F. J. 1997. Vagaries of the molecular clock. *Proceedings of the National Academy of Science USA* 94:7776–7783.

Beebee, T., and G. Rowe. 2008. *An Introduction to Molecular Ecology*, 2nd ed. New York: Oxford University Press.

Belk, D. 1989. Identification of species in the conchostracan genus *Eulimnadia* by egg shell morphology. *Journal of Crustacean Biology* 9:115–125. doi: 10.1163/193724089X00269

Bickford, D., D. J. Lohman, N. S. Sodhi, P. K. L. Ng, R. Meier, K. Winker, K. K. Ingram, and I. Das. 2006. Cryptic species as a window on diversity and conservation. *Trends in Ecology and Evolution* 22:148–155. doi:10.1016/j.tree.2006.11.004

Bookstein, F. L. 1987. Random walk and the existence of evolutionary rates. *Paleobiology* 13:446–464.

Budd, A. F., and J. S. Klaus. 2008. Early evolution of the *Montastraea "annularis"* species complex (Anthozoa: Scleractinia): Evidence from the Mio-Pliocene of the Dominican Republic. In *Evolutionary Stasis and Change in the Dominican Republic Neogene*. Topics in Geobiology, 30, edited by R. H. Nehm and A. F. Budd, 85–123. New York: Springer.

Carroll, S. B. 2005. *Endless Forms Most Beautiful: The New Science of Evo Devo and the Making of the Animal Kingdom*. New York: W. W. Norton & Company.

Carroll, S. B. 2007. *The Making of the Fittest: DNA and the Ultimate Forensic Record of Evolution*. New York: W. W. Norton & Company.

Carroll, S. B. 2008. Evo-devo and an expanding evolutionary synthesis: a genetic theory of morphological evolution. *Cell* 134:25–36. doi:10.1016/j.cell.2008. 06.030

Carroll, S. B. 2009. *Remarkable Creatures: Epic Adventures in the Search for the Origin of Species*. Boston: Houghton Mifflin Harcourt.

Carroll, S. B., J. K. Grenier, and S. D. Weatherbee. 2004. *From DNA to Diversity: Molecular Genetics and the Evolution of Animal Design*, 2nd ed. Oxford: Wiley-Blackwell.

Cheetham, A. H. 1986. Tempo of evolution in a Neogene bryozoan: Rates of morphologic change within and across species boundaries. *Paleobiology* 12:190–202.

Cheetham, A. H. 1987. Tempo of evolution in a Neogene bryozoan: Are trends in single morphologic characters misleading? *Paleobiology* 13:286–296.

Cheetham, A. H., and J. B. C. Jackson. 1995. Process from pattern: Tests for selection versus random change in punctuated bryozoan speciation. In *New Approaches to Speciation in the Fossil Record*, edited by D. H. Erwin and R. L.
Anstey, 184–207. New York: Columbia University Press.

Cheetham, A. H., J. B. C. Jackson, and L.-A. C. Hayek. 1993. Quantitative genetics of bryozoan phenotypic evolution. I. Rate tests for random change versus
selection in differentiation of living species. *Evolution* 47:1526–1538.

Cheetham, A. H., J. B. C. Jackson, and L.-A. C. Hayek. 1994. Quantitative genetics
of bryozoan phenotypic evolution. II. Analysis of selection and random change
in fossil species using reconstructed genetic parameters. *Evolution* 48:360–375.

Ciampaglio, C. N., M. Kemp, and D. W. McShea. 2001. Detecting changes in morphospace occupation patterns in the fossil record: Characterization and analysis of measures of disparity. *Paleobiology* 27:695–715.

Coyne, J. A. and H. A. Orr. 2004. *Speciation*. Sunderland, MA: Sinauer Associates,
Inc.

Davidson, E. H. 2010. *The Regulatory Genome: Gene Regulatory Networks in Development and Evolution*. New York: Academic Press.

Decellea, J., N. Suzukib, F. Mahéa, C. de Vargasa, and F. Nota. 2012. Molecular
phylogeny and morphological evolution of the *Acantharia* (Radiolaria). *Protist*
163:435–450. doi:10.1016/j.protis.2011.10.002

Dick, M. H., A. Herrera-Cubilla, and J. B. C. Jackson. 2003. Molecular phylogeny
and phylogeography of free-living Bryozoa (Cupuladriidae) from both sides
of the Isthmus of Panama. *Molecular Phylogenetics and Evolution* 27:355–371.
doi:10.1016/S1055-7903(03)00025-3

Donoghue, P. C. J., and M. J. Benton. 2007. Rocks and clocks: Calibrating the Tree
of Life using fossils and molecules. *Trends in Ecology & Evolution* 22:424–431.
doi:10.1016/j.tree.2007.05.005

Dornburg, A., J. M. Beaulieu, J. C. Oliver, and T. J. Near. 2011. Integrating fossil
preservation biases in the selection of calibrations for molecular divergence
time estimation. *Systematic Biology* 60:519–527. doi:10.1093/sysbio/syr019

Drummond, A. J., S. Y. W. Ho, M. J. Phillips, and A. Rambaut. 2006. Relaxed phylogenetics and dating with confidence. *PLoS Biol* 4:e88. doi:10.1371/journal.
pbio.0040088

Drummond, A. J., M. A. Suchard, M. A., X. Dong, and A. Rambaut. 2012. Bayesian phylogenetics with BEAUti and the BEAST 1.7. *Molecular Biology and
Evolution* 29:1969–973. doi:10.1093/molbev/mss075

Eldredge, N., and S. J. Gould. 1972. Punctuated equilibria: An alternative to phyletic gradualism. In *Models in Paleobiology*, edited by T. J. M. Schopf, 82–115.
San Francisco: Freeman Cooper.

Estes, S., and S. J. Arnold. 2007. Resolving the paradox of stasis: Models with stabilizing selection explain evolutionary divergence on all timescales. *American
Naturalist* 169:227–244.

Falconer, D. S. 1981. *Introduction to Quantitative Genetics*, 2nd ed. London: Longman Group Ltd.

Frankel, N., D. F. Erezyilmaz, A. P. McGregor, S. Wang, F. Payre, and D. L. Stern. 2011. Morphological evolution caused by many subtle-effect substitutions in regulatory DNA. *Nature* 474:598–603. doi:10.1038/nature10200

Freeland, J. 2006. *Molecular Ecology*. New York: John Wiley & Sons.

Geary, D. H. 1990. Patterns of evolutionary tempo and mode in the radiation of *Melanopsis* (Gastropoda; Melanopsidae). *Paleobiology* 16:492–511.

Geary, D. H., G. Hunt, I. Magyar, and H. Schreiber. 2010. The paradox of gradualism: Phyletic evolution in two lineages of lymnocardiid bivalves (Lake Pannon, central Europe). *Paleobiology* 36:592–614.

Giribet, G, D. L. Distel, M. Pol, W. Sterrer, and W. C. Wheeler. 2000. Triploblastic relationships with emphasis on the acoelomates and the position of Gnathostomulida, Cycliophora, Platheminthes, and Chaetognatha: A combined approach of 18S rDNA sequences and morphology. *Systematic Biology* 49:539–562. doi:10.1080/10635159950127385

Gingerich, P. D. 1983. Rates of evolution: Effects of time and temporal scaling. *Science* 222:159–161. doi:10.1126/science.222.4620.159-a

Gingerich, P. D. 2009. Rates of evolution. *Annual Review of Ecology, Evolution, and Systematics* 40:657–675. doi: 10.1146/annurev.ecolsys.39.110707.173457

Gompel, N., B. Prud'homme, P. Wittkopp, V. A. Kassner, and S. B. Carroll. 2005. Chance caught on the wing: Cis-regulatory evolution and the origin of pigment patterns in *Drosophila. Nature* 433: 481–487. doi: 10.1038/nature03235

Gould, S. J., and N. Eldredge. 1993. Punctuated equilibrium comes of age. *Nature* 366:223–227. doi:10.1038/366223a0

Groves, J. R., and S. Reisdorph. 2009. Multivariate morphometry and rates of morphologic evolution within the Pennsylvanian fusulinid *Beedeina* (Ardmore Basin, Oklahoma, USA). *Palaeoworld* 18:120–129. doi:10.1016/j.palwor.2008.12.001

Hageman, S. J. 1994. Microevolutionary implications of clinal variation in the Paleozoic bryozoan *Streblotrypa Lethaia* 27:209–222. doi: 10.1111/j.1502-3931.1994.tb01411.x

Hageman, S. J. 2003. Book Review. *From DNA to Diversity: Molecular Genetics and the Evolution of Animal Design*, by Carroll, Grenier, and Weatherbee (2001). *Journal of Paleontology* 77:598.

Hageman, S. J., M. Bayer, and C. D. Todd. 1999. Partitioning phenotypic variation: Genotypic, environmental, and residual components from bryozoan skeletal morphology. *Journal of Natural History* 33:1713–1735. doi:10.1080/002229399299815

Hageman, S. J., L. L. Needham, and C. D. Todd. 2009. Threshold effects of food concentration on the skeletal morphology of the bryozoan *Electra pilosa* (Linnaeus, 1767). *Lethaia* 42:438–451. doi:10.1111/j.1502-3931.2009.00164.x

Hageman, S. J., P. N. Wyse Jackson, A. R. Abernethy, and M. Steinthorsdottir. 2011. Calendar scale, environmental variation preserved in the skeletal phenotype of a fossil bryozoan (*Rhombopora blakei* n. sp.), from the Mississippian of Ireland. *Journal of Paleontology* 85:853–870.

Hayek, L.-A. C., and E. Bura. 2001. On the ends of the taxon range problem. In *Evolutionary Patterns: Growth, Form, and Tempo in the Fossil Record in Honor of Allan Cheetham*, edited by J. B. C. Jackson, S. Lidgard, and F. K. McKinney, 221–422. Chicago: University of Chicago Press.

Herrera-Cubilla, A., M. H. Dick, J. Sanner, and J. B. C. Jackson. 2006. Neogene Cupuladriidae of tropical America. I: Taxonomy of recent *Cupuladria* from opposite sides of the Isthmus of Panama. *Journal of Paleontology* 80:245–263.

Herrera-Cubilla, A., M. H. Dick, J. Sanner, and J. B. C. Jackson. 2008. Neogene Cupuladriidae of tropical America. II: Taxonomy of recent *Discoporella* from opposite sides of the Isthmus of Panama. *Journal of Paleontology* 82:279–298.

Hoekstra, H. E., and J. A. Coyne. 2007. The locus of evolution: Evo devo and the genetics of adaptation. *Evolution* 61:995–1026. doi: 10.1111/j.1558-5646.2007.00105.x

Hunt, G. 2010. Evolution in fossil lineages: Paleontology and the origin of species. *American Naturalist* 176:S61–S76. doi:10.1086/657057

Hunt, G. 2012. Measuring rates of phenotypic evolution and the inseparability of tempo and mode. *Paleobiology* 2012:351–373.

Hunter, E., and R. N. Hughes. 1994. The influence of temperature, food ration, and genotype on zooid size in *Celleporella hyalina* (L.). In *Biology and Palaeobiology of Bryozoans*, edited by P. J. Hayward, J. S. Ryland, and P. D. Taylor, 83–86, Fredensborg, Denmark: Olsen and Olsen.

Jagadeeshan, S., and A. O'Dea. 2012. Integrating fossils and molecules to study cupuladriid evolution in an emerging Isthmus. *Evolutionary Ecology* 26:337–355. doi:10.1007/s10682-011-9522-6

Jackson, J. B. C., and A. H. Cheetham. 1990. Evolutionary significance of morphospecies: A test with cheilostome Bryozoa. *Science* 248:579–583. doi: 10.1126/science.248.4955.579

Kidwell, S. M., and D. W. Bosence. 1991. Taphonomy and time-averaging of marine shelly faunas. In *Taphonomy: Releasing the data locked in the fossil record*, edited by P. A. Allison and D. E. G. Briggs, 115–209. New York: Plenum.

Kimura, M., 1980. A simple method for estimating evolutionary rate of base substitutions through comparative studies of nucleotide sequences. *Journal of Molecular Evolution* 16:111–120. doi:10.1007/BF01731581

Kowalewski, M., and R. K. Bambach. 2008. The limits of paleontological resolution. In *High-Resolution Approaches in Stratigraphic Paleontology*. Topics in Geobiology, vol. 21, edited by P. J. Harries, 1–48. Dordrecht: Springer Netherlands.

Kumar, S. 2005. Molecular clocks: Four decades of evolution. *Nature Reviews Genetics* 6:654–662. doi:10.1038/nrg1659

Knowlton, N. 1993. Sibling species in the sea. *Annual Review of Ecology and Systematics* 24:189–216. doi:10.1146/annurev.es.24.110193.001201

Lazarus, D. 1986. Tempo and mode of morphologic evolution near the origin of the radiolarian lineage *Pterocanium prismatium*. *Paleobiology* 12:175–189.

Levinton, J. S. 2001. *Genetics, Paleontology, and Macroevolution*, 2nd ed. Cambridge: Cambridge University Press.

Lydeard, C., M. C. Wooten, and A. Meyer. 1995. Molecules, morphology, and area cladograms: A cladistic and biogeographic analysis of *Gambusia* (Teleostei: Poeciliidae). *Systematic Biology* 44:221–236. doi:10.1093/sysbio/44.2.221

Marshall, C. R. 1990. Confidence intervals on stratigraphic ranges. *Paleobiology* 16:1–10.

Marshall, C. R. 1997. Confidence intervals on stratigraphic ranges with non-random distributions of fossil horizons. *Paleobiology* 23:165–173.

McPeek, M. A., L. B. Symes, D. M. Zong, and C. L. McPeek. 2010. Species recognition and patterns of population variation in the reproductive structures of a damselfly genus. *Evolution* 65:419–428. doi: 10.1111/j.1558-5646.2010.01138.x

Mishler, B. D. 1994. Cladistic analysis of molecular and morphological data. *American Journal of Physical Anthropology* 94:143–156. doi:10.1002/ajpa.133 0940111

Nee, S. 2004. Extinct meets extant: Simple models in paleontology and molecular phylogenetics. *Paleobiology* 30:172–178.

Nei, M., and S. Kumar. 2000. *Molecular Evolution and Phylogenetics*. New York: Oxford University Press.

Omland, K. E. 1997. Correlated rates of molecular and morphological evolution. *Evolution* 51:1381–1393.

Oros, M., T. Scholz, V. Hanzelová, and J. S. Mackiewicz. 2010. Scolex morphology of monozoic cestodes (Caryophyllidea) from the Palaearctic Region: A useful tool for species identification. *Folia Parasitologica* 57:37–46.

Pachut, J. F., and R. L. Anstey. 2009. Inferring evolutionary modes in a fossil lineage (Bryozoa: *Peronopora*) from the Middle and Late Ordovician. *Paleobiology* 35:209–230.

Pagel, M., C. Venditti, and A. Meade. 2006. Large punctuational contribution of speciation to evolutionary divergence at the molecular level. *Science* 314:119–121. doi: 10.1126/science.1129647

Prud'homme, B., N. Gompel, and S. B. Carroll. 2007. Emerging principles of regulatory evolution. *Proceedings of the National Academy of Science* 104:8605–8612. doi:10.1073/pnas.0700488104

Prud'homme, B., N. Gompel, A. Rokas, V. A. Kassner, T. M. Williams, S. D. Yeh, J. R. True, and S. B. Carroll. 2006. Repeated morphological evolution through cis-regulatory changes in a pleiotropic gene. *Nature* 440:1050–1053. doi:10.1038/nature04597

Quental, T. B., and C. R. Marshall. 2010. Diversity dynamics: Molecular phylogenies need the fossil record. *Trends in Ecology and Evolution* 25:434–441. doi: 10.1016/j.tree.2010.05.002

Renaud, S., J. C. Auffray, J. Michaux. 2007. Conserved phenotypic variation patterns, evolution along lines of least resistance, and departure due to selection in fossil rodents. *Evolution* 60:1701–1717. doi:10.1111/j.0014-3820.2006. tb00514.x

Rebollo, R., B. Horard, B. Hubert, and C. Vieira. 2010. Jumping genes and epigenetics: Towards new species. *Gene* 454:1–7. doi:10.1016/j.gene.2010.01.003

Rieseberg, L. H., and B. K. Blackman. 2010. Speciation genes in plants. *Annals of Botany* 106:439–455. doi:10.1093/aob/mcq126

Ricklefs, R. W. 2006. Time, species, and the generation of trait variance in clades. *Systematic Biology* 55:151–159. doi:10.1080/10635150500431205

Riisgård, H. U., and A. Goldson. 1997. Minimal scaling of the lophophore filter-pump in ectoprocts (Bryozoa) excludes physiological regulation of filtration rate to nutritional needs: Test of hypothesis. *Marine Ecology Progress Series* 156:109–120.

Roopnarine, P. D. 2003. Analysis of rates of morphologic evolution. *Annual Review of Ecology, Evolution, and Systematics* 34:605–632. doi:10.1146/annurev. ecolsys.34.011802.132407

Roopnarine, P. D, G. Byars, and P. Fitzgerald. 1999. Anagenetic evolution, stratophenetic patterns, and random walk models. *Paleobiology* 25:41–57.

Ruse, M. 2000. The theory of punctuated equilibria: Taking apart a scientific controversy. In *Scientific Controversies: Philosophical and Historical Perspectives*, edited by P. Machamer, M. Pera, and A. Baltas, 230–253. New York: Oxford University Press.

Sadler, P. M. 1981. Sediment accumulation rates and the completeness of stratigraphic sections. *Journal of Geology* 89:569–584.

Sadler, P. M. 2004. Quantitative biostratigraphy: Achieving finer resolution in global correlation. *Annual Reviews of Earth and Planetary Science* 32:187–213. doi: 10.1146/annurev.earth.32.101802.120428

Sanderson, M. J. 2002. Estimating absolute rates of molecular evolution and divergence times: A penalized likelihood approach. *Molecular Biology and Evolution* 19:101–109.

Shubin, N. 2008. *Your Inner Fish: A Journey into the 3.5-Billion-Year History of the Human Body*. New York: Pantheon Books.

Stanley, S. M., and X. Yang. 1987. Approximate evolutionary stasis for bivalve morphology over millions of years: A multivariate, multilineage study. *Paleobiology* 13:113–139.

Stone, J. R., and G. A. Wray. 2001. Rapid evolution of cis-regulatory sequences via local point mutations. *Molecular Biology and Evolution* 18:1764–1770.

Theissen, G. 2006. The proper place of hopeful monsters in evolutionary biology. *Theory in Biosciences* 124:349–369. doi:10.1016/j.thbio.2005.11.002

Wahlberg, N., and S. Nylin. 2003. Morphology versus molecules: Resolution of the positions of *Nymphalis*, *Polygonia*, and related genera (Lepidoptera: Nymphalidae). *Cladistics* 19:213–223. doi:10.1111/j.1096-0031.2003.tb00364.x

Wittkopp, P. J., and G. Kalay. 2012. Cis-regulatory elements: Molecular mechanisms and evolutionary processes underlying divergence. *Nature Reviews Genetics* 13:59–69. doi:10.1038/nrg3095

Weiss, R. E., and C. R. Marshall. 1999. The uncertainty in the true end point of a fossil's stratigraphic range when stratigraphic sections are sampled discretely. *Mathematical Geology* 31:435–453. doi: 10.1023/A:1007542725180

Yang, Z. 2006. *Computational Molecular Evolution*. New York: Oxford University Press.

Zeh, D. D., J. A. Zeh, and Y. C. Ishi. 2009. Transposable elements and an epigenetic basis for punctuated equilibria. *BioEssays* 31:715–726. doi: 10.1002/bies.200900026

Zuckerkandl, E., and L. Pauling. 1962. Molecular disease, evolution, and genetic heterogeneity. In *Horizons in Biochemistry*, edited by M. Kasha and B. Pullman, 189–225. New York: Academic Press.

Zuckerkandl, E., and L. Pauling. 1965. Evolutionary divergence and convergence in proteins. In *Evolving Genes and Proteins*, edited by V. Bryson and H. J. Vogel, 97–166. New York: Academic Press.

Fitting Ancestral Age-Dependent Speciation Models to Fossil Data

Lee Hsiang Liow and Torbjørn Ergon

Introduction

S peciation and its counterpart, extinction, are processes that create and remove species diversity globally. Both speciation and extinction rates have varied through time and across different clades, contributing to the differential diversity in varied habitats among communities and clades that we observe today, as well as in the fossil record. Extinction is a topic that is widely studied by paleontologists, not least because the fossil record gives us relatively direct information on the temporal distribution of extinct organisms. Speciation, on the other hand, is a lesser topic among paleontologists, in part because microevolutionary mechanisms of speciation are not readily studied using phenotypic information from fossils. Microevolutionary processes, such as isolating mechanisms, introgression, and genetic differentiation unlinked to morphological change, are more readily explored among incipient or newly formed species using molecular tools. Such young species are usually highly challenging to differentiate in the fossil record (Hunt 2010) since they are often rare and/or phenotypically similar to their ancestors (Liow et al. 2010). Whereas there is a fair amount of information on how (genus) extinction and origination rates vary through time (Alroy 2008; Foote 2000, 2003, 2005), much less is known about how speciation rates vary through time, in part because of the taxonomic resolution of the fossil record (Sepkoski 1998). On the other hand, biologists using phylogenetic trees of extant organisms have found support for different models of net diversification (i.e., the net result of speciation and extinction), ranging from diversity-dependence to niche-space filling processes (Purvis et al. 2011; Rabosky and Glor 2010; Ricklefs 2006). However, the extinction part of the diversification process is notoriously difficult to estimate with phylogenetic information based only on extant organisms (Quental and Marshall 2010) and is highly model dependent, not least for phylogenies that have deeper nodes. Similarly, while phylogenies of extant organisms give indications as to divergence times of species that have survived to the present, times of speciation are hard to estimate from only extant organisms. Additionally, as we will explain later, it is essential to account for extinctions of ancestral species when estimating rates of speciation.

Changing physical environments are commonly inferred to be drivers of diversification patterns in the fossil record (Benton 2009; Mayhew et al. 2008). However, intrinsic properties of a taxon, such as body size and life history, may also play a part in both survival and speciation/origination probabilities (Alroy 1998; Harnik 2011). In addition, the age of a taxon has been inferred as being a factor in survival (Doran et al. 2006; Finnegan et al. 2008), although age-independent extinction has also been inferred from empirical data (Van Valen 1973). While there is a long history of research on age-dependent extinction/survival using paleontological data (see references cited in Liow et al. 2011), age-dependent speciation is a newer topic (Agapow and Purvis 2002; Ezard et al. 2011a; Losos and Adler 1995; Purvis et al. 2011; Venditti et al. 2009). Correspondingly, there have only been a few mathematical treatments of age-dependent branching models (Crump and Mode 1968, 1969; Jones 2011a,b; Steel and McKenzie 2001).

Age-Dependent Speciation as a Theoretically Plausible Process

Speciation might be age-independent with respect to the age of the direct ancestor or parent species (in the rest of this chapter, we use the term "parent species" since "ancestral species" might refer to both direct and indirect ancestors, except when quoting previous literature where "ancestral species" is used). This may be the case when conditions under which speciation might likely occur are uniformly randomly distributed with respect to the age of the parent species. These conditions include the generation and maintenance of isolating barriers, for instance, in the form of sexual selection, vicariance events, or dispersal and subsequent geographic isolation (Coyne and Orr 2004). An analysis using more than a hundred molecular phylogenies of plants, animals, and fungi found support for age-independent branching events (Venditti et al. 2009).

However, one might also imagine scenarios under which speciation might be more likely earlier in the life of the parent species. Speciation may occur rapidly and repeatedly in the same lineage due to ecological opportunities arising because of the colonization of a previously un-invaded area or the evolution of key innovations in adaptive radiations (Schluter 2000). However, as ecological space is filled by newly evolved species, speciation rates are balanced by extinction rates, resulting in a temporally constant species richness, hence the classic pattern of diversity-dependent diversification (Phillimore and Price 2008; Rabosky and Glor 2010; Ricklefs et al. 2007). One way in which diversity-dependent diversification can result is from an "ease of speciation" at early ages of ancestral species earlier in clade history. This early ease of speciation might be due to the relative lack of interspecific competition as proposed in adaptive radiations, or the exposure of hidden genetic variation in a new environment followed by adaptation (Le Rouzic and Carlborg 2007; Pigliucci et al. 2006) and isolation among populations/individuals with different traits, or rapid geological processes such as repeated glaciations that fragment landscapes and isolate incipient species (Jones 2011b). While many authors have shown that diversity-dependent or early burst diversification can be used to describe their clades of interest (Phillimore and Price 2008; Rabosky and Glor 2010), the reference to "early" is early in the life of the clade and not in the life of the parent species. Note also that although the study by Venditti and colleagues (2009) argued for a constant rate of speciation, their data allow them only to explore sister taxa relationships, which include but are not exclusively parent-descendant relationships, the focus of the discussion here.

On the other hand, older, more "established" species might have a greater propensity to give rise to new species. For instance, Losos and Adler (1995) proposed a refractory period where genetic restructuring and stabilization occur in a new species during which this species is unlikely to give rise to yet another new species. Populations belonging to older species that have had time to both expand their geographic ranges (Losos and Adler 1995) and experience local extirpations (causing "holes" in their geographic range) may be more likely to be geographically isolated. Some of these populations might eventually become new species as a result of adaptation to the new local environment and isolation from their parent species.

Last but not least, we suggest here that the early and late speciation scenarios may combine within a given group to give rise to a "bathtub"-shaped speciation process (i.e., high both early and late in the life of the parent species). In this case, it may be "easier" to speciate earlier in life due to new opportunities as mentioned above, and also later in life due to the geographic isolation of far-dispersed populations. In the next section, we present an approach for describing age-dependent speciation, using hazard rate functions, clarifying similar ideas that had been presented in earlier studies (Ezard et al. 2011a; Jones 2011b), while also laying down assumptions underlying this approach.

Modeling age-dependent speciation

Instantaneous age-dependent speciation rates may be described by a continuous hazard function $h(x)$ (e.g., Kalbfleisch and Prentice 1980). In this model, species may give rise to any number of new species (including none) within its lifetime, and the number of speciation events from age a to age b, *given* that the parent species is extant at age b, is Poisson distributed with expectation

$$\textbf{(6.1)} \qquad \lambda_{a,b} = \int_a^b h(x)dx.$$

The speciation hazard function $h(x)$ here describes the underlying (latent) per species speciation rates that we want to make inferences about and is independent of the extinction process. The underlying speciation rates cannot be inferred from the distribution of parental ages of speciation events alone because this age distribution reflects both speciation and extinction processes. Specifically, older parents will be underrepresented due to extinctions. The probability that the parent species is extant at age x can be described by a survival function $S(x)$. We can then write the expected number of realized speciation events (i.e., all descendants) from a single parent species between age a and age b as

$$\textbf{(6.2)} \qquad \Lambda_{a,b} = \int_a^b S(x)h(x)dx.$$

This expectation may be interpreted as the expectation at "birth" of the parent species as it is not conditional on parental survival. The difference between these expectations representing the underlying (eq. 6.1) and realized (eq. 6.2) speciation processes is illustrated in fig. 6.1. It is important to note that, even if extinction rates are constant, speciation events that

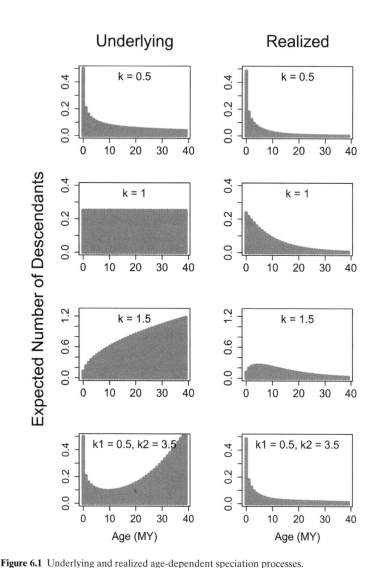

Figure 6.1 Underlying and realized age-dependent speciation processes. Left panels represent the underlying speciation process and show the expected number of descendants per 1 MY age-bins conditional on survival of the parent species (eq. 1). Right panels show the expected number of realized speciation events per 1 MY age-bins when the mortality rate of the parent species is constant (eq. 6.2). In the first three rows, the shape parameter of the Weibull hazard function, k, are set at 0.5, 1.0 or 1.5 to show scenarios where speciation is early, uniform and late with respect to the age of the parent species. The scale parameter in these plots is 4. The last row shows a "bathtub" speciation process using a "double Weibull" hazard function constructed as the sum of two Weibull components with parameters $\{k_1=0.5, \theta_1=4\}$ and $\{k_2=3.5, \theta_2=24\}$ respectively. We used a constant mortality rate of 10^{-1} MY (i.e., species duration or "life-expectancy" is exponentially distributed with mean 10 MY).

occur late in the life of the parent species will always be underrepresented among the realized speciation events because fewer parent species will reach old age (i.e., $S(x)$ is necessarily a declining function). It is therefore not possible to make inferences about underlying age-dependent specia-tion rates without simultaneously accounting for extinctions of the parent species. For example, the underlying speciation processes are very differ-ent in the top and bottom rows of fig. 6.1, but the realized distributions of parental age at speciation events are virtually identical. Nevertheless, if one has information about the extinction events or extinction rates of the parent species, it is possible to estimate the underlying speciation rates.

In the next section (Methods), we describe the data we analyzed and devise a way to estimate underlying age-dependent speciation rates using available empirical datasets.

Methods

Data

We searched the literature from 1986 to 2012 quite exhaustively for original larger paleontological datasets (at least 15 species) of clades in which di-rect ancestor-descendent relationships were proposed at the species level, where stratigraphic ranges of the same species were available, and where estimated times of branching of descendent species were reported. To date we found only 5 such datasets (table 6.1, appendix B) and fit parameters of speciation hazard functions as described in the next section. The data to be analyzed are thus the times of speciation and extinction of each species and the time of branching of all species in the dataset where known, except the earliest member of the group. In all but the macroperforate Forami-nifera dataset for which numerical stratigraphic data were provided, we had to extract species duration and age-at-speciation-event data from the time-trees as given in the publications. In some cases, authors have inferred range extensions. We used the extended speciation and extinction time points as data to be fit in such cases. Note that we otherwise used reported ages at speciation events and extinctions as given (i.e., we do not attempt to account for any uncertainty in these values). Analyses that present only cladograms are excluded as these only present sister-taxa relationships, not inferred parent-descendant relationships. We excluded pure sister-taxa re-lationships (and did not include them as data) as it is unknown which of the sister taxa is ancestral, even if they did represent direct parent-descendent

TABLE 6.1 Overview of the five datasets from the paleontological literature analyzed in this chapter. The number of known parent-descendent relationships is compared with the total number of taxa described in the publications. The next best model against which the asterisked models are compared is more than 2 AIC unit (p. 206; see also figs. 6.3–6.7).

Publication	No. taxa (known parent-descendant relationships)	Best model	Taxa
Aze et al. 2011	339(338)	Early Weibull*	Macroperforate Foraminifera
Cheetham 1986	18(17)	Late Gompertz-Makeham*	Metrarabdotos (Bryozoa)
Hulbert 1993	60(24)	Constant	Equinae
Wang 1994	27(12)	Constant	Hesperocyoninae
Wang et al. 1999	66(19)	Late Weibull	Borophaginae

relationships. This could cause bias in our inference if early or late speciation events in these excluded data are overly represented within a dataset, but we have no reason to believe such biases exist.

Models considered

We consider several different age-dependent speciation hazard functions, described below: constant, two forms of monotonically increasing/decreasing hazard functions (Weibull and Gompertz-Makeham models), and several forms of bathtub hazard functions. Note that speciation rate in our treatment depends only on parental age and not absolute time.

CONSTANT SPECIATION RATES. The most commonly used speciation model in evolutionary literature is Yule's pure birth model (Nee 2006; Yule 1924), which assumes that speciation occurs with a common constant hazard rate for all species, i.e.,

(6.3) $$h(x;\theta) = \frac{1}{\theta}$$

This model implies that parental age, x, at first speciation event, as well as time between subsequent speciation events (of the same parent), are exponentially distributed with expectation θ.

WEIBULL SPECIATION RATES. A model that is more general than Yule's pure birth model is the Weibull hazard function,

(6.4) $$h(x;k,\theta) = \frac{k}{\theta}\left(\frac{x}{\theta}\right)^{k-1}$$

where x is the species age, $k > 0$ is the shape parameter, and $\theta > 0$ is the scale parameter (see also Ezard et al. 2011a, b; Venditti et al. 2009). If $k > 1$, the speciation hazard rate will increase as a power law with species age (the second derivative is positive if $k > 2$ and negative if $1 < k < 2$, and the hazard rate increases proportionally with age if $k = 2$). If $0 < k < 1$, the speciation hazard rate will decrease and approach 0 as the species in question ages. If $k = 1$, the speciation hazard rate will be constant and speciation will be a homogeneous point-process with intensity θ^{-1}, as in the Yule model. This model implies that the age of the ancestor at its *first* speciation event is Weibull distributed but age at later speciation events are not Weibull distributed (see Tuerlinckx 2004). It also implies that the expected number of speciation events from a single parent species between age a and b given survival (eq. 6.1), is

(6.5) $$\lambda_{a,b} = \int_a^b h(x;k,\theta)dx = \frac{(b^k - a^k)}{\theta^k}.$$

Thus, a single species is expected to produce $(b/\theta)^k$ new species before age b given that it does not go extinct before this age (see fig. 6.1, right panel). That is, if we double θ the species has to survive twice as long to produce the same expected number of new species.

GOMPERTZ-MAKEHAM SPECIATION RATES.   The Gompertz-Makeham hazard function is the sum of a constant age-independent (λ) component and an age-dependent component ($\alpha e^{\beta x}$),

(6.6) $$h(x;\lambda,\alpha,\beta) = \lambda + \alpha e^{\beta x}.$$

The age-dependent component is here an exponential function, meaning that a certain increment in age will lead to a certain relative change in speciation rate, irrespective of initial age (i.e., an increase in age, x, by one unit will lead to an increase in the age-dependent component by a factor e^β). In contrast, in the Weibull model, a *relative* change in age will lead to a relative change in speciation rate (i.e., an increase in age, x, by a factor c will lead to an increase in speciation rate by a factor $c \left(\frac{c}{\theta}\right)^{k-1}$; see eq. 6.4).

BATHTUB SPECIATION RATES.   Speciation rates described by the Weibull and Gompertz-Makeham models are monotonically increasing or decreasing. But speciation rates may also first decline at early age and then increase again at higher age, or vice versa. We investigated how several potentially bathtub-shaped hazard functions fit the available data (see section on Data); a sum of two Weibull-components (see eq. 6.4), the increase-decrease-bathtub (IDB) hazard function of Hjorth (1980) and the two-parameter bathtub function of Chen (2000) as well as a modified version of Chen's function with a scaling parameter. However, none of these models was selected by the Akaike Information Criterion (AIC) (Burnham and Anderson 2002) for the best model for any of the datasets, and the fits of these more flexible models were never distinctively different from the fit of the best models (see Results and Discussion). We also experienced numerical problems (convergence failure, strong sensitivity to starting values, or failure to estimate the variance-covariance matrix of the parameter estimates) for several of the model-and-data-set combinations. The fits of these bathtub-shaped hazard functions are therefore not reported in the results.

Fitting speciation hazard functions to empirical data

It may be tempting to fit the age of the parent species at speciation directly to distributions that describe these ages, without considering extinctions of such parents. Such an exercise will cause bias in our inference because species are not immortal and do not have the same lifespans, as previously explained and illustrated in fig. 6.1. In other words, it is important to account for the extinction times of each parent species in order to properly fit speciation hazard functions to data. This is also discussed as accounting for termination (or censoring) in the statistics of recurrent events (Cook and Lawless 2007).

Our statistical model is a recurrent event model (Cook and Lawless 2007) in which a single parent species may give rise to none, one, or more new species, and where speciation hazard rates change with the age of the parent species according to the alternative hazard functions described above. This model can be formulated with an explicit likelihood function for continuous-time data (Cook and Lawless 2007), and fitted with the powerful 'survreg' function in the 'survival' package (Therneau 2015) of R (R Core Team 2014). We were not aware of this package at the time of writing this chapter and devised an approach based on discretization. We retain this discretization approach here, as we think it is instructive and easier to understand than the continuous-time formulation, which should yield very similar results.

Our approach is to first discretize parental age in short intervals, assuming that this species can go extinct only at interval borders. We then fit a Poisson maximum likelihood model to the number of observed spe-

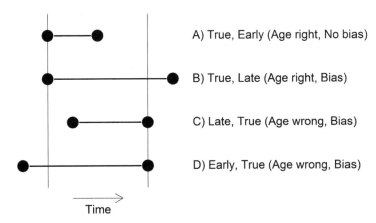

A) True, Early (Age right, No bias)

B) True, Late (Age right, Bias)

C) Late, True (Age wrong, Bias)

D) Early, True (Age wrong, Bias)

Time

Figure 6.2 Biases due to deviations from the time of true speciation and extinction. The vertical lines show the true times and the dots assumed times of speciation and extinction of the parent species in question. We show only scenarios for which at least one endpoint (speciation or extinction) is correctly known. In (A), the age of the parent species is correct and even though its early censoring or termination causes lower precision, the hazard rate estimated is not biased using our approach. In (B), the parent species is assumed to be able to give rise to offspring when it was in fact already extinct, causing bias in hazard rate estimation. In both (C) and (D) hazard rate estimation will be biased because the age of the parent species is erroneous.

ciation events in these short age intervals, where the Poisson expectations are the products of number of species extant at the beginning of the intervals and the per-species age-dependent expectations given by eq. 6.1. Thus, we condition the likelihood function on species that are extant in each of the short age intervals, thereby obtaining unbiased estimates of speciation rates in the presence of termination through extinctions (Cook and Lawless 2007). This model assumes that there are no other sources of variation in the speciation rates than the age-dependent relationship given by the speciation hazard function.

We assumed here that the times of extinction and speciation are known without error. We note, however, that if the assumed extinction of the parent species occurs before the true extinction time (fig. 6.2), we will have no bias in our model fit. However, if the inferred extinction time is after true extinction time, there will be a bias because we assume that descendent species can branch off the parent species when the latter is already truly extinct. If the speciation time of the parent species is erroneous, we will also get a bias in the model fit because the ages at speciation events are erroneous.

We compare models based on the different speciation hazard functions described above, using the AIC. We provide R code for fitting speciation hazard functions to data of species first and last appearances and speciation times in appendix A.

Results and Discussion

An early-speciation Weibull model best fits the macroperforate Foraminifera data (fig. 6.3), corroborating results from Ezard et al. (2011a), even though a different set of models are fit here. The Gompertz-Makeham model gives very similar results to the Weibull model for foraminiferans (fig. 6.3) and has a slightly higher AIC value because it is penalized for one extra parameter. A late-speciation Gompertz-Makeham model, in which speciation rate increases sharply when the parent species reaches 5–6 MY, best fits the *Metrarabdotos* bryozoans (fig. 6.4). We note, however, that from the ages of about 1 to 5 MY, the speciation hazard rate for the bryozoans is quite flat, not unlike a constant rates model. Hence, if only a part of these data is available, such that speciation events of older ancestors are not observed in the fossil record, late speciation may not be inferred. Note also that the confidence intervals are relatively wider at both early ages and late ages for the bryozoans (fig. 6.4), and wide compared with the rather small confidence intervals for foraminiferans (fig. 6.3), simply due to the different number of species in the datasets.

The remaining three datasets do not show very conclusive results with regards to age-dependent speciation. We cannot reject the null hypothesis of constant speciation in Equinae (fig. 6.5) but both the Weibull and Gompertz-Makeham models, showing higher speciation rates at greater parental ages (albeit with different forms), are less than 2 AIC units higher. The situation is similar for Hesperocyoninae (fig. 6.6), where the constant rates model and the late speciation Weibull model are only 0.2 AIC apart, and where the Gompertz-Makeham is very similar in form to the Weibull. Likewise for Borophaginae (fig. 6.7), even though the late speciation Weibull model is ranked top, the constant rates model is less than 2 AIC units higher.

Based on these few datasets, it is difficult to draw general conclusions on age-dependent speciation. The study by Ezard and colleagues (Ezard et al. 2011a) is the only one we are aware of that has fit an age-dependent speciation hazard model to paleontological data where ancestral-descendent

Macroperforate Foraminifera

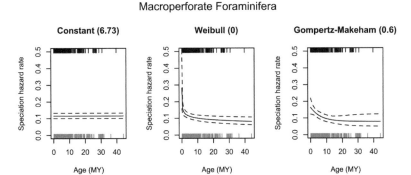

Figure 6.3 Speciation models for macroperforate foraminifera (Aze et al. 2011). Plots show fitted predictions from three alternative speciation models. Numbers in parentheses are ΔAIC values for each of the named models. Zero indicates the best fitting model. Grey ticks (bottom) are data for ages at extinction (i.e., species durations) and black ticks (top) are for ages of the parents at the time of speciation events, to which models are fit. Dotted lines show approximate 95% confidence intervals assuming a log-normal distribution of the fitted predictions (± 2 standard errors on the log-scale calculated by the delta-method (Morgan 2008)).

Metrarabdotos

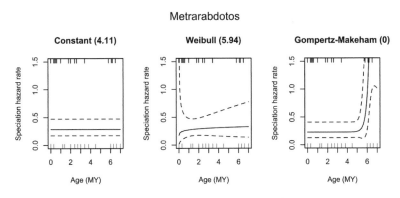

Figure 6.4 Speciation models for *Metrarabdotos* (Cheetham 1986). Conventions as in fig. 6.3. Note that the y-axes are on a different scale from figs. 6.3, 6.5–6.7.

relationships are known. That another high-quality dataset (*Metrarabdotos*) shows a completely different speciation hazard rate hints that interesting variation exists in speciation processes among clades, since patterns reflect process. Each of the age-dependent scenarios (constant, early, and late) that have been discussed in the literature best fit at least one of the five datasets, although in three of the cases (table 6.1, rows 3–5), these

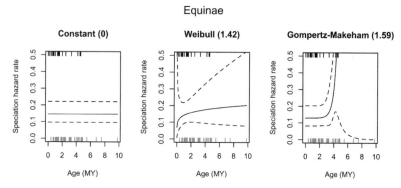

Figure 6.5 Speciation models for North American Equinae (Hulbert 1993). Conventions as in fig. 6.3.

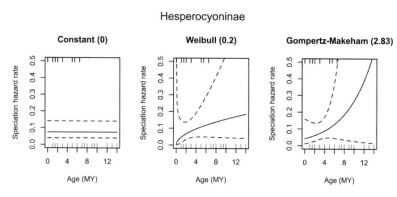

Figure 6.6 Speciation models for Hesperocyoninae (Wang 1994). Conventions as in fig. 6.3.

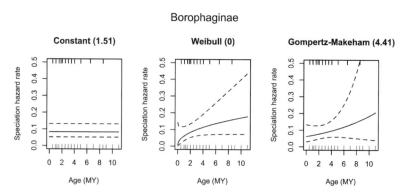

Figure 6.7 Speciation models for Borophaginae (Wang et al. 1999). Conventions as in fig. 6.3.

age-dependent models are not significantly different from a constant rates model. In these cases, several to many parent species that are part of the dataset may not have known descendants, even though these species may be present in the data (table 6.1). It is likely that if some of these relationships are resolved, we will be able to better distinguish among the models. We also note that we used speciation and extinction times as reported in the datasets, although there is surely nonnegligible uncertainty around the times of speciation and extinction in empirical datasets. Such uncertainty can be incorporated in the modeling by treating the uncertain times of speciation and extinction as unobserved variables with certain probability distributions (most easily fitted with Bayesian Markov-chain-Monte-Carlo procedures).

The evolution of macroperforate Foraminifera could have been driven by the classic forces of diversity dependence, causing their speciation hazard to drop with both age and absolute time (Ezard et al. 2011a). Although *Metrarabdotos* bryozoa show a late speciation hazard, local extirpations in older and more widespread species (see Losos and Adler 1995) does not seem like a good working hypothesis as the main cause for a higher speciation hazard for older *Metrarabdotos* species. This is because *Metrarabdotos* species in Cheetham's dataset are geographically restricted to the Caribbean. Part of this variation might alternatively reflect different taxonomic practices among systematists specializing on different groups of organisms. We note also that even though the form of the speciation hazard can be quite different among different groups, the average speciation rates hover around the same order of magnitude (about 0.05 to 0.25, except when it shoots up at older ages for the bryozoans and possibly younger ages for foraminiferans).

Origination, speciation, and extinction rates are commonly modeled as varying with respect to absolute time (e.g., Alroy 2008; Foote 2000). With sufficient data, effects of absolute time could easily be incorporated in our modeling approach. It would also be particularly interesting to incorporate species-and-time-varying covariates, such as measures of abundance or geographic spread, in the analysis. Species tend to achieve their greatest global abundance or geographic spread around the midpoint of their temporal duration (Foote et al. 2007; Liow et al. 2010; Liow and Stenseth 2007) and hence may encounter geographic and ecological situations where speciation may occur most frequently. However, to fit a model that incorporates such an effect, we would need data on species abundance or geographic spread, which are not readily available.

Previous work, using constant birth-death models and an assumption of temporally homogeneous preservation, has shown that the probability of sampling ancestor-descendant pairs each at least once in the fossil record is far from negligible (Foote 1996). Although it is not improbable to sample both ancestor and descendants in the fossil record, the question remains, what is the probability of sampling both direct ancestor and descendant during the process of speciation, given different models of lifetime-varying preservation rates? To answer this question, we need to know how long the speciation process lasts and what patterns lifetime-varying preservation rates might show. First, the duration of morphological divergence leading to stable morphospecies is purported to range from the order of 0.05 to several million years (Allmon and Smith 2011; Norris and Hull 2012), while speciation (in the sense of reproductive isolation) lasts from 103 years (Hendry et al. 2007), beyond the normal temporal resolution of the fossil record, to millions of years (Rundell and Price 2009). The duration of species depends likely on the clade in question and the manner in which we delimit the process of speciation. Second, preservation rates may depend on absolute time (Kidwell and Holland 2002) but may also depend on the age of the species. Specifically, the "hat" model of fossil occurrence trajectory (Liow et al. 2010) is a canonical temporal pattern of an average fossil taxon where there are fewer observed occurrences at the start of the taxon's existence as well as closer to the end of its existence. The bulk of observed occurrences are in the middle of the lifetime of the taxon such that a histogram of temporal fossil finds of the taxon would look Gaussian if there are enough samples. The "hat" model implies that sampling probability decreases towards the beginning of a species' lifetime. This means that it may be difficult to obtain substantial samples of individuals of a given descendant species during the time when it is diverging from its ancestor. If most speciation events also occur early in the life of the ancestors, as in the Macroperforate Foraminifera data, it may be similarly difficult to sample ancestors around the time of branching of their descendants. If species conform to a hat model of fossil occurrence and if they fit an early speciation or bathtub-shaped speciation model, it would be quite difficult to find direct ancestor-descendant pairs that would have a rich enough fossil record during the morphological divergence period.

Although age per se may not determine speciation rate (Purvis et al. 2011), age is likely to reflect geographical and other macroecological properties that change with species age. Age-dependent speciation models appear plausible and we have shown here that several such models best fit some available datasets. While assembling fossil datasets with ancestor-descendant relationships and temporal data is nontrivial, such datasets will be crucial in instructing us on what types of speciation processes prevail in nature. Only by assembling more high-quality datasets can we begin to resolve the question of the generality of different speciation models.

Appendixes

Appendix A (R code for fitting speciation hazard functions to data of species first and last appearances and speciation times) and Appendix B (five datasets used in this study; see table 6.1) are available at: [url for appendices here].

Acknowledgments

L. H. Liow thanks W. D. Allmon and M. M. Yacobucci for the invitation to speak at the session "Species and Speciation in the Fossil Record" at the Geological Society of America Meeting 2011, which inspired this work. We also thank J. Finarelli for pointing to the dog phylogenies by X. M. Wang, and M. Foote, W. Miller, W. D. Allmon, and two anonymous reviewers for constructive reviews.

References

Agapow, P. M., and A. Purvis. 2002. Power of eight tree shape statistics to detect nonrandom diversification: A comparison by simulation of two models of cladogenesis. Systematic Biology 51:866–872.

Allmon, W. D., and U. E. Smith. 2011. What, if anything, can we learn from the fossil record about speciation in marine gastropods? Biological and geological considerations. American Malacological Bulletin 29:247–276.

Alroy, J. 1998. Cope's rule and the dynamics of body mass evolution in North American mammals. Science 280:731–734.

Alroy, J. 2008. Dynamics of origination and extinction in the marine fossil record. Proceedings of the National Academy of Sciences of the United States of America 105:11536–11542.

Aze, T., T. H. G. Ezard, A. Purvis, H. K. Coxall, D. R. M. Stewart, B. S. Wade, and P. N. Pearson. 2011. A phylogeny of Cenozoic macroperforate planktonic foraminifera from fossil data. Biological Reviews 86:900–927.

Benton, M. J. 2009. The Red Queen and the Court Jester: species diversity and the role of biotic and abiotic factors through time. Science 323:728–732.

Burnham, K. P., and D. R. Anderson. 2002. Model Selection and Multimodel Inference: A Practical Information-Theoretic Approach. New York, Springer.

Cheetham, A. H. 1986. Tempo of evolution in a Neogene Bryozoan: Rates of morphologic change within and across species boundaries. Paleobiology 12:190–202.

Chen, Z. 2000. A new two-parameter lifetime distribution with bathtub shape or increasing failure rate function. Statistics & Probability Letters 49:155–161.

Cook, R. J., and J. F. Lawless. 2007. The Statistical Analysis of Recurrent Events: Statistics for Biology and Health. New York, Springer.

Coyne, J. A., and H. A. Orr. 2004. Speciation. Sunderland, Massachusetts, Sinauer Associates.

Crump, K. S., and C. J. Mode. 1968. A general age-dependent branching process. 1. Journal of Mathematical Analysis and Applications 24:494–508.

Crump, K. S., and C. J. Mode. 1969. A general age-dependent branching process. 2. Journal of Mathematical Analysis and Applications 25:8–17.

Doran, N. A., A. J. Arnold, W. C. Parker, and F. W. Huffer. 2006. Is extinction age dependent? Palaios 21:571–579.

Ezard, T. H. G., T. Aze, P. N. Pearson, and A. Purvis. 2011a. Interplay between changing climate and species' ecology drives macroevolutionary dynamics. Science 332:349–351.

Ezard, T. H. G., P. N. Pearson, T. Aze, and A. Purvis. 2011b. The meaning of birth and death (in macroevolutionary birth-death models). Biology Letters 8: 139–142.

Finnegan, S., J. L. Payne, and S. C. Wang. 2008. The Red Queen revisited: Reevaluating the age selectivity of Phanerozoic marine genus extinctions. Paleobiology 34:318–341.

Foote, M. 1996. On the probability of ancestors in the fossil record. Paleobiology 22:141–151.

Foote, M. 2000. Origination and extinction components of taxonomic diversity: Paleozoic and post-Paleozoic dynamics. Paleobiology 26:578–605.

Foote, M. 2003. Origination and extinction through the Phanerozoic: A new approach. Journal of Geology 111:125–148.

Foote, M. 2005. Pulsed origination and extinction in the marine realm. Paleobiology 31:6–20.

Foote, M., J. S. Crampton, A. G. Beu, B. A. Marshall, R. A. Cooper, P. A. Maxwell, and I. Matcham. 2007. Rise and fall of species occupancy in Cenozoic fossil molluscs. Science 318:1131–1134.

Harnik, P. G. 2011. Direct and indirect effects of biological factors on extinction risk in fossil bivalves. Proceedings of the National Academy of Sciences of the United States of America 108:13594–13599.

Hendry, A. P., P. Nosil, and L. H. Rieseberg. 2007. The speed of ecological speciation. Functional Ecology 21:455–464.

Hjorth, U. 1980. A reliability distribution with increasing, decreasing, constant, and bathtub-shaped failure rates. Technometrics 22:99–107.

Hunt, G. 2010. Evolution in fossil lineages: Paleontology and the origin of species. American Naturalist 176:S61-S76.

Hulbert, R. C., Jr. 1993, Taxonomic evolution in North American Neogene horses (subfamily Equinae): The rise and fall of an adaptive radiation. Paleobiology 19:216–234.

Jones, G. R. 2011a. Calculations for multi-type age-dependent binary branching processes. Journal of Mathematical Biology 63:33–56.

Jones, G. R. 2011b. Tree models for macroevolution and phylogenetic analysis. Systematic Biology 60:735–746.

Kalbfleisch, J. D., and R. L. Prentice. 1980. The Statistical Analysis of Failure Time Data. New York, Wiley.

Kidwell, S. M., and S. M. Holland. 2002. The quality of the fossil record: Implications for evolutionary analyses. Annual Review of Ecology and Systematics 33:561–588.

Le Rouzic, A., and Ö. Carlborg. 2007. Evolutionary potential of hidden genetic variation. Trends in Ecology & Evolution.

Liow, L. H., H. J. Skaug, T. Ergon, and T. Schweder. 2010. Global occurrence trajectories of microfossils: Environmental volatility and the rises and falls of individual species. Paleobiology 36:224–252.

Liow, L. H., and N. C. Stenseth. 2007. The rise and fall of species: Implications for macroevolutionary and macroecological studies. Proceedings of the Royal Society B-Biological Sciences 274:2745–2752.

Liow, L. H., L. Van Valen, and N. C. Stenseth. 2011. Red Queen: From populations to taxa and communities. Trends in Ecology & Evolution 26:349–358.

Losos, J. B., and F. R. Adler. 1995. Stumped by trees—a generalized null model for patterns of organismal diversity. American Naturalist 145:329–342.

Mayhew, P. J., G. B. Jenkins, and T. G. Benton. 2008. A long-term association between global temperature and biodiversity, origination, and extinction in the fossil record. Proceedings of the Royal Society B-Biological Sciences 275:47–53.

Morgan, B. J. T. 2008. Applied Stochastic Modelling. Boca Raton, Chapman & Hall.

Nee, S. 2006. Birth-death models in macroevolution. Annual Review of Ecology, Evolution, and Systematics 37:1–17.

Norris, R. D., and P. M. Hull. 2012. The temporal dimension of marine speciation. Evolutionary Ecology 26:393–415.

Phillimore, A. B., and T. D. Price. 2008. Density-dependent cladogenesis in birds. Plos Biology 6:e71.

Pigliucci, M., C. J. Murren, and C. D. Schlichting. 2006. Phenotypic plasticity and evolution by genetic assimilation. Journal of Experimental Biology 209:2362–2367.

Purvis, A., S. A. Fritz, J. Rodriguez, P. H. Harvey, and R. Grenyer. 2011. The shape of mammalian phylogeny: Patterns, processes, and scales. Philosophical Transactions of the Royal Society B-Biological Sciences 366:2462–2477.

Quental, T., and C. R. Marshall. 2010. Diversity dynamics: Molecular phylogenies need the fossil record. Trends in Ecology & Evolution 25:434–441.

R Core Team. 2014. R: A Language and Environment for Statistical Computing.

Rabosky, D. L., and R. E. Glor. 2010. Equilibrium speciation dynamics in a model adaptive radiation of island lizards. Proceedings of the National Academy of Sciences 107:22178–22183.

Ricklefs, R. E. 2006. Global variation in the diversification rate of passerine birds. Ecology 87:2468–2478.

Ricklefs, R. E., J. B. Losos, and T. M. Townsend. 2007. Evolutionary diversification of clades of squamate reptiles. Journal of Evolutionary Biology 20:1751–1762.

Rundell, R., J., and T. D. Price. 2009. Adaptive radiation, nonadaptive radiation, ecological speciation, and nonecological speciation. Trends in Ecology & Evolution 24:394–399.

Schluter, D. 2000, The Ecology of Adaptive Radiation. Oxford, Oxford University Press.

Sepkoski, J. J. 1998. Rates of speciation in the fossil record. Philosophical Transactions of the Royal Society of London Series B-Biological Sciences 353:315–326.

Steel, M., and A. McKenzie. 2001. Properties of phylogenetic trees generated by Yule-type speciation models. Mathematical Biosciences 170:91–112.

Therneau, T. 2015. A Package for Survival Analysis in S. version 2.38.

Tuerlinckx, F. 2004. A multivariate counting process with Weibull-distributed first-arrival times. Journal of Mathematical Psychology 48:65–79.

Van Valen, L. M. 1973. A new evolutionary law. Evolutionary Theory 1:1–30.

Venditti, C., A. Meade, and M. Pagel. 2009. Phylogenies reveal new interpretation of speciation and the Red Queen. Nature 463:349–352.

Wang, X. 1994. Phylogenetic systematics of the Hesperocyoninae (Carnivora, Canidae). Bulletin of the American Museum of Natural History 221:1–207.

Wang, X., R. H. Tedford, and B. E. Taylor. 1999. Phylogenetic systematics of the Borophaginae (Carnivora, Canidae). Bulletin of the American Museum of Natural History 243:1–391.

Yule, G. U. 1924. A mathematical theory of evolution based on the conclusions of Dr. J. C. Willis, FRS. Philosophical Transactions of the Royal Society B-Biological Sciences 213:21–87.

CHAPTER SEVEN

Contrasting Patterns of Speciation in Reef Corals and Their Relationship to Population Connectivity

Ann F. Budd and John M. Pandolfi

Introduction

The "species problem" has long compromised the study of long-term evolutionary patterns in scleractinian reef corals. Because of the difficulties in recognizing species, very little research has been done tracing species distributions through geologic time, reconstructing their phylogenies using the fossil record, and interpreting speciation and extinction patterns in response to long-term climate change. A number of factors have contributed to these difficulties, including high ecophenotypic plasticity (reviewed by Todd 2008), simple morphologies that often overlap among species, and patchy distributions in space and time. Recently, molecular techniques, coupled with lab and field experiments on coral reproduction, have provided a wealth of new independent data, which have improved understanding of the nature of species boundaries in corals. These new research avenues have revealed the presence of cryptic species (Knowlton 1993, 2000; Fukami et al. 2004a; Souter 2010; Ladner & Palumbi 2012) as well as differences in patterns of larval connectivity and population differentiation among species (Baums et al. 2005, 2006; Foster et al. 2012; Palumbi et al. 2012). They have also shown the potential for hybridization as a mechanism for both decreasing (homogenizing) and increasing morphological diversity; however, the degree to which the latter actually occurs in nature is still debated (Vollmer & Palumbi 2002,

2007; Willis et al. 2006; Ladner & Palumbi 2012). Increased complexity is the result of introgressive hybridization, in which a hybrid repeatedly backcrosses with one of its parent species.

Comparisons between morphological and molecular data from the same samples have yielded mixed results, despite refinement in the definition of morphological characters. In addition to our work (described below), some studies have found agreement between the two data types, whereas others have found disagreement. Examples of congruence are found in many unrelated genera including: *Pocillopora* (Flot et al. 2008a), *Porites* (Forsman et al. 2009), *Montipora* (Forsman et al. 2010) in Hawaii, and *Psammocora* across the Indo-Pacific (Stefani et al. 2008). Incongruence has been found in many other genera including: *Platygyra* (Miller & Benzie 1997; Miller & Babcock 1997), *Seriatopora* (Flot et al. 2008b), *Stylophora* (Flot et al. 2011), and Western Indian Ocean and Eastern Pacific *Pocillopora* (Souter 2010; Pinzon & LaJeunesse 2011). Mixed results have been found in "*Favia*" from Thailand (Kongjandtre et al. 2012). The discrepancies have been attributed to ecophenotypic plasticity, morphological stasis, and morphological convergence as well as to hybridization and complex metapopulation structure.

Equally germane for tracing distributions are recent studies of population connectivity based on population genetics (measures of gene flow or genetic differentiation) combined with oceanographic models, which show that connectivity patterns are responsible for a large proportion of the large-scale genetic structure and divergence among populations (Baums et al. 2005; Palumbi et al. 2012; Foster et al. 2012). Nevertheless, metapopulation structures are often found to be complex, with levels of genetic variation among populations within species sometimes being equal to those among species (e.g., *Pocillopora*; see Pinzon & LaJeunesse 2011).

In this chapter we examine how recent advances involving cryptic species and population connectivity in modern reef corals contribute to understanding species in the fossil record. To address this question, we review three examples from our own work, which represent different scleractinian families. Our results reveal significant differences in patterns of morphological variation among species within these taxa, which may be related to patterns of population differentiation. We conclude not only that the diagnostic morphological characters of species differ among families, but also that patterns of morphological variation within species and overlap among species differ among families. The causes of differences are complex, but appear to be related to reproductive biology and larval dispersal. Better knowledge and understanding of the diversity of morphological patterns in modern corals are crucial to deciphering species in the fossil record.

Cryptic Species in the *Orbicella annularis* Complex

The *Orbicella annularis* complex (formerly the *Montastraea annularis* complex; see Knowlton et al. 1992; Weil & Knowlton 1994; Lopez et al. 1999; Fukami et al. 2004a) of reef corals represents a case in which cryptic species are clearly supported using molecular, morphological, and reproductive data, and the evolutionary history of these cryptic species can be traced through geologic time. The genus name of the species complex was formally changed from "*Montastraea*" to "*Orbicella*" in Budd et al. (2012a). This complex forms large massive colonies, often >1m in diameter, and has been ecologically dominant on Caribbean reefs for >2 million years, with a depth range extending from intertidal to >80 m (Budd & Klaus 2001, 2008; Klaus & Budd 2003; Budd 2010). The geographic distribution of the complex extends across the Caribbean, Gulf of Mexico, and western Atlantic (Florida, the Bahamas, and Bermuda) and has not changed throughout its history. The complex was long thought to be one highly variable species, an "archetypal generalist" (Goreau & Wells 1967; Graus & Macintyre 1976, 1982) with seemingly limitless ability for ecophenotypic plasticity, because all of its living members have three septal cycles (24 septa) and corallite diameters of 2–3.5 mm, two features that are believed to be less plastic and therefore have been traditionally used to distinguish species in massive reef corals. However, over the past two decades, molecular work (AFLP nuclear markers, microsatellites) has shown that this one species is actually a species complex (Knowlton et al. 1992; Weil & Knowlton 1994; Lopez et al. 1999; Fukami et al. 2004a), which consists of at least three species that differ in overall colony shape: (1) *O. annularis* s.s., which forms regular-shaped smooth columns, (2) *O. faveolata*, which forms mounds and heads with keels and skirt-like edges, and (3) *O. franksi*, which forms bumpy, irregular mounds and plates. *O. faveolata* has almost no shared genotypes with *O. franksi* and *O. annularis* s.s., and *O. franksi* and *O annularis* s.s. differ in frequencies of the genotypes AAA and AA* (Fukami et al. 2004a). Moreover, lab experiments indicate gamete incompatibility between *O. faveolata* and the other two species, which themselves differ from one another in spawning time (Levitan et al. 2004).

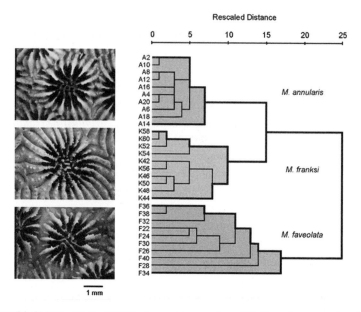

1 mm

Figure 7.1 Cluster analysis of 3-D landmark data on calices of the three Recent species of the *Orbicella annularis* complex from Panama.
Each branch of the dendrogram represents one genetically characterized colony. Colony numbers are indicated for each colony; A's indicate colonies of *O. annularis* s.s., K's indicate colonies of *O. franksi*; and F's indicate colonies of *O. faveolata*. The dendrogram clearly shows the three species in the complex to be distinct, with *O. annularis* s.s. and *O. franksi* being most similar. The most important variables in discriminating species consist of non-traditional morphological characters related to the elevation and development of the costae, the shape of the septal margin, and the length of the tertiary septa. After Budd & Klaus (2001).

Prompted by the molecular results, we have discovered several nontraditional morphological features (e.g., the overall shape of the calice and structure of the costae and corallite wall) that agree with the molecular data (fig. 7.1; Knowlton & Budd 2001; Budd & Klaus 2001): (1) *O. faveolata* has high calical relief, and thin, exsert septa and costae; (2) *O. franksi* has low calical relief, and thicker, better developed costae; (3) *O. annularis* s.s. is intermediate but closer to *O. franksi*, with moderately thick septa and costae. Moreover, our examination of transverse thin sections shows that *O. faveolata* has a parathecal corallite wall (a wall formed by dissepiments), *O. annularis* has a septothecal wall (a wall formed by coalescing septa), and, although predominantly septothecal, *O. franksi* has a combination of the two wall types. The adaptive significance of these features is unclear, but appears to be related to the interplay between upward linear extension and skeletal thickening associated with accretionary growth. Lower rates of upward extension appear to be correlated in general with increased skeletal thickening (Knowlton & Budd 2001).

Further study of the ecological distributions of the three species in the complex shows that they are sympatric, but with niche differentiation (Pandolfi & Budd 2008). Although the depth ranges of species overlap, their abundances vary along depth gradients, with each of the three species dominating a different depth zone. In addition, we have found differences in the morphological features that vary within and among species. The three species differ in wall thickness, whereas ecophenotypes of *O. faveolata* differ in development of dissepiments, and ecophenotypes of *O. faveolata* and *O. annularis* s.s. differ in corallite diameter. On the other hand, Klaus et al. (2007) found that certain morphological features that differ among species, e.g., calice relief and corallite wall thickness, decrease with water depth in *O. annularis* s.s. in Curaçao. Nevertheless, the magnitude of within-species variation is less than that of among-species variation. This correspondence between environmental variation within species and interspecies variation suggests that morphological differences among populations may be caused by natural selection and therefore adaptive.

Molecular analyses show that the nature of species boundaries within the complex varies geographically, with *O. annularis* s.s. and *O. franksi* being distinct in Panama but not in the Bahamas (Fukami et al. 2004a). Their lack of distinction in the Bahamas has been confirmed using the new morphological characters (fig. 7.2) and interpreted to be the result of an ancestral polymorphism caused by hybridization in the past (Fukami et al. 2004a). Laboratory experiments show that the three modern species can be crossed, confirming the potential for hybridization (Fukami et al. 2004a). Furthermore, our analyses of fossil samples from the late Pleistocene (~125 Ka) of the Bahamas (fig. 7.3A) reveal overlap in corallite morphology among colonies with different colony shapes, suggesting introgressive hybridization in the fossil record (Budd & Pandolfi 2004, 2010). Additional analyses of late Pleistocene and modern samples from across the Caribbean (fig. 7.3; Budd & Pandolfi 2010) show that two and often three of the modern species in the complex can be recognized in the late Pleistocene, in addition to the fossil organ-pipe species *O. nancyi* (Pandolfi 2007), and that these species are clearly distinct in central Caribbean locations (Dominican Republic, Caymans, Florida).

In contrast, analyses of samples from the late Pleistocene terraces of Barbados (>500 Ka, ~300 Ka, ~125 Ka) reveal as many as six new short-lived

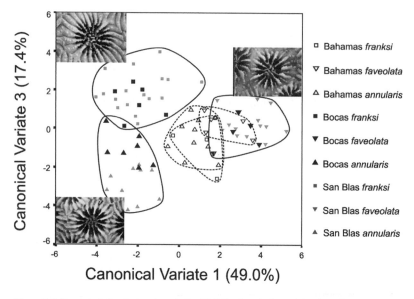

Figure 7.2 Canonical discriminant analysis of 3-D landmark data of the three Recent species of the *Orbicella "annularis"* complex from Panama (grey & black) and from the Bahamas (uncolored).
Each point on the plot represents one genetically characterized colony. Solid borders enclose species from Panama; dotted borders enclose species from the Bahamas. Relative development of major versus minor costae and septum height are strongly correlated with function one; costa length, minor septum length, and wall thickness are inversely correlated with function three. Scale bars are 1 mm. After Fukami et al. (2004a).

fossil species, in addition to *O. nancyi*, suggesting that speciation was higher along the southeastern margin of the geographic distribution of the complex (fig. 7.3; Budd & Pandolfi 2010). Although columns, mounds, and plates all occur in the terraces, analyses of corallite morphology show that only one of the three modern species in the complex (*O. faveolata*) is present. Interestingly, only after the extinction of the organ-pipe dominant *O. nancyi* (Pandolfi et al. 2001) do we see the proliferation of thin-columned *O. annularis* s.s. in Barbados in shallow-water habitats, suggesting character release and a morphological shift to columnar forms following the relaxation of competition caused by the extinction of *O. nancyi* (Pandolfi et al. 2002).

Taken together, the results for the Bahamas and Barbados indicate that lineage splitting (Barbados) and fusion (Bahamas) were concentrated at edge zones at the extreme eastern margins of the Caribbean basin having limited population connectivity and lower gene flow (fig 7.3; Budd & Pan-

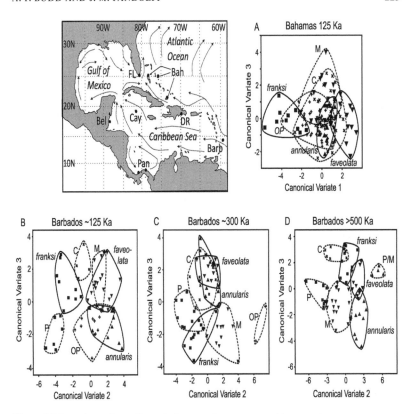

Figure 7.3 Locations of sampling relative to major oceanographic currents.
(Upper left) Map showing geographic locations of sampling relative to major oceanographic
currents (dotted): Bah, Bahamas; 'Barb', Barbados; 'Bel', Belize; 'Cay', Cayman Islands;
'DR', Dominican Republic; 'Fl', Florida; 'Pan', Panama. The Bahamas and Barbados are located at the easternmost margins of the distribution of the *Orbicella "annularis"* complex in
contrast with the other five locations. (A–D) Plots of scores on canonical variates comparing
the three Recent species the *Orbicella "annularis"* complex from Panama (solid) with Pleistocene morphospecies (dashed) in the Bahamas and Barbados. (A) ~125 Ka in the Bahamas,
(B) ~125 Ka in Barbados, (C) ~300 Ka in Barbados, and (D) >500 Ka in Barbados. Canonical
variates were selected, which had the maximum Mahalanobis distances among Recent Panama species (x axis) and among Pleistocene morphospecies (y axis). Each point represents
one colony; borders enclose the maximum variation within species or morphospecies. P, plate;
M, massive; C, column; OP, organ-pipe; P/M, an additional massive species found only in the
>500 Ka Barbados assemblages. After Budd & Pandolfi (2010).

dolfi 2010). These evolutionary patterns contrast with well-connected central locations (Dominican Republic, Cayman Islands, Florida), which contain exactly the same species in the Pleistocene as those today in Panama
and Belize. The observed difference in evolutionary patterns between the
margins and center of the distribution of the complex corresponds with the

genetic discontinuities reported today not only in the *O. annularis* complex, specifically *O. annularis* s.s. (Foster et al. 2012), but also in acroporid corals (Baums et al. 2005, 2006; Vollmer & Palumbi 2007) and in reef fish (Taylor & Hellberg 2006). In these discontinuities, eastern and western Caribbean populations are clearly separated by barriers to gene flow and the Bahamas are weakly isolated. Baums et al. (2006) found differences in population structure of *Acropora palmata* between the eastern and western Caribbean, which they attribute to differences in reproductive modes between the two regions, such that sexual recruitment is more prevalent in the east and asexual recruitment in the west. Furthermore, the wider geographic distribution that we observe during the Pleistocene in *O. faveolata* compared to *O. annularis* s.s. conforms with population genetic divergence estimates made using 10 nuclear DNA loci (7 microsatellites + 3 RFLP), which show higher gene flow and greater population connectivity in *O. faveolata* (Severance & Karl 2006). Clearly, in the case of the *O. annularis* complex, patterns in morphological and molecular data agree at the metapopulation level.

Given the agreement between morphological and genetic data, the evolutionary history of the *O. annularis* complex can also be reconstructed over millions of years of geologic time (fig. 7.4). The results indicate not only deep divergence but also that the three modern species are not sister lineages. Morphometric analyses of fossils collected from the Plio-Pleistocene of Costa Rica and Panama (~3.5–1.5 million years ago; Budd & Klaus 2001) reveal a total of ten morphospecies in the complex, which can be subdivided into three clades based on phylogenetic analysis (fig. 7.4). Modern *O. franksi* clearly belongs to a lineage in one clade (clade 1), and modern *O. faveolata* to a lineage in a second clade (clade 3). The third clade (clade 2) is extinct. *O. annularis* s.s. is ambiguous, and does not appear to have diverged until the late Pleistocene. Further analyses of fossils collected from the Mio-Pliocene of the Dominican Republic (~6.5–3.4 million years ago; Klaus & Budd 2003; Budd & Klaus 2008) result in eight morphospecies, one of which belongs to the clade containing *O. franksi* (clade 1) and three of which belong to the clade containing *O. faveolata* (clade 3; Budd 2010). These results indicate that the complex has a long evolutionary history, dating back at least 6.5 million years ago. Since its origination, the complex has consisted of 3–5 species living together at the same time, but the overall range of variation in the morphology of the complex has not changed through time. Maximum diversity occurred within the complex during the Plio-Pleistocene in association with faunal

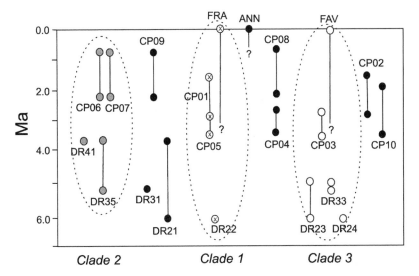

Clade 2   Clade 1   Clade 3

Figure 7.4 Morphospecies in the *Orbicella annularis* complex over the past 7 million years. Morphometric analyses reveal 10 morphospecies in the Plio-Pleistocene of Costa Rica and Panama ('CP'; Budd & Klaus 2001), and 8 morphospecies in the Mio-Pliocene of the Dominican Republic ('DR'; Klaus & Budd 2003; Budd & Klaus 2008). The three modern species are indicated as 'FRA', *O. franksi*; 'ANN', *O. annularis* s.s.; 'FAV', *O. faveolata*. Relative positions of morphospecies along the x-axis are based on a phylogenetic analysis (Budd 2010), which shows three clades with bootstrap values >40% (clade 1 with 'x', clade 2 in grey, clade 3 with no fill). Species that do not group with a clade are indicated in black. *O. franski* belongs to clade 1, *O. faveolata* belongs to clade 3, and clade 2 is extinct. *O. annularis* s.s. arose during the middle to late Pleistocene and is more closely related to *O. franksi* than to *O. faveolata*.

turnover on Caribbean reefs (Budd and Johnson 1999; Johnson et al. 2008). *O. franksi* and *O. faveolata* belong to long-lived lineages that survived turnover.

Polymorphism in *Montastraea cavernosa*

Quite different patterns have been found in the reef coral *Montastraea cavernosa*, which also forms large massive colonies, and in which morphological and molecular data disagree. *M. cavernosa* was long thought to be closely related to the *Orbicella annularis* complex, because of its plocoid colony form, but these two taxa have recently been found to belong to different families (Fukami et al. 2004b, 2008). Traditionally *M. cavernosa* is distinguished by having 4–5 septal cycles (48–60 septa) and corallite

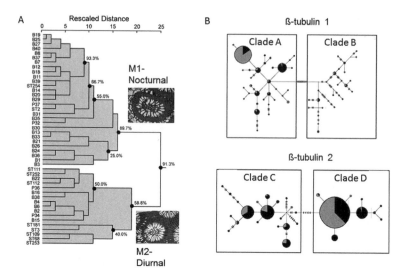

Figure 7.5 Analyses of *Montastraea cavernosa.*
(A) Cluster analysis using morphological data based on Mahalanobis distances among colonies. Each branch represents one colony. Cross-validated classification results show that the split between clusters formed using the highest rescaled distance (M1 vs M2) is 91.3% correctly classified; whereas the split between clusters formed using the second highest rescaled distance is 58.8% correctly classified. (B) Parsimony haplotype network for ß-tubulin 1 and ß-tubulin 2. The size of each circle reflects the frequency that a haplotype is observed. Notches symbolize the number of mutations between haplotypes. Haplotypes observed in morph M1 are black, and in M2 are grey. After Budd et al. (2012b).

diameters of 4–8 mm. It occurs across a depth range similar to the *O. annularis* complex, but has relatively lower abundances and a broader spatial distribution extending beyond that of the *O. annularis* complex to West Africa and Brazil. Recently it too has been hypothesized to consist of a species complex, because two morphologically distinct feeding morphs co-occur (Lasker 1976, 1979, 1980, 1981): (1) a diurnal morph with smaller polyps [=M2 in our analysis below] and (2) a nocturnal morph with large polyps [=M1 in our analysis below]. In Puerto Rico, the diurnal morph has been found to be more abundant in shallow reef environments, whereas the nocturnal morph is more abundant in deep reef environments (Ruiz 2004). Our morphometric analyses of samples collected from across the Caribbean and as far east as West Africa (Belize, Panama, Puerto Rico, São Tomé) confirm the existence of two distinct morphs; M1 = large corallites, M2 = small corallites (fig. 7.5; Budd et al.

2012b). The most important features distinguishing the two morphs are corallite size, wall thickness, and the development of the tertiary septa. In our samples, no differences in morphology were detected overall among geographic locations, or among water depths in Belize, so although polyp size may be related to feeding behavior, the observed morphological polymorphism does not appear to be caused by geographic or environmental variation (Budd et al. 2012b).

Haplotype networks derived from DNA sequence data collected on the same samples reveal two distinct clades within each of two loci (i.e., two copies of the β-tubulin gene referred to as β-tubulin 1 and β-tubulin 2); however, each of the molecular clades is composed of a mixture of the two morphs (fig. 7.5; Budd et al. 2012b). Genotype frequencies and two-locus genotype assignments indicate genetic exchange across clades, and $\emptyset_{st}$ values (which use allele frequencies to measure differentiation among populations) show no genetic differentiation between morphs at different locations. Unlike the *O. annularis* complex, both morphs share nearly all of the same genotypes. Differences in frequency of one of these genotypes (AADD) have been detected between morphs in association with São Tomé, but they may have been caused by sampling bias. Taken together, our morphological and genetic results do not provide evidence for cryptic species in *M. cavernosa*, but indicate instead that this species has an unusually high degree of polymorphism over a wide geographic area, and that gene flow is high among populations (Budd at al. 2012). Furthermore, again unlike the *O. annularis* complex (described above), Nunes et al. (2009) found lower levels of genetic diversity in peripheral populations of *M. cavernosa* in Brazil and West Africa than in central populations in the Caribbean.

Morphometric analyses of fossil colonies of *M. cavernosa*–like corals from the Plio-Pleistocene of Costa Rica and Panama (~3.5–1.5 million years ago) reveal four distinct fossil morphotypes, with morphometric differences (i.e., Mahalanobis distances) among the three well-sampled fossil morphotypes roughly equivalent to those between the two modern morphs (fig. 7.6; Budd et al. 2012b). These results suggest that the three fossil morphotypes are also polymorphs of *M. cavernosa*. Similarly, analyses of *M. cavernosa*–like corals from the Mio-Pliocene of the Dominican Republic (~6.5–3.4 million years ago) also reveal four fossil morphotypes (Budd 1991), and Johnson (2007) reported three fossil morphotypes from the late Oligocene of Antigua (~25–26 million years ago). Study of morphological differences between the two modern morphs and these fossil

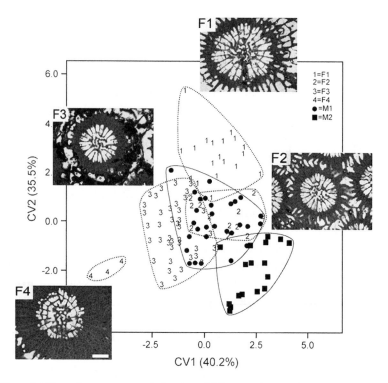

Figure 7.6 Plots of scores on canonical variates (CV) comparing the six *Montastraea caver-nosa* morphotypes (2 Recent: M1, M2; 4 Plio-Pleistocene: F1–F4).
Each point on the plot represents one colony; areas enclosed in dots include each fossil morphotype (photos), and those enclosed in solid curves include each modern morphotype. All photos at the same scale; scale bar in F4 is 2 mm. After Budd et al. (2012b).

morphotypes indicates that *M. cavernosa* has a long duration dating back >25 million years, and that it has been polymorphic throughout its duration (Budd et al. 2012b).

Speciation in *Favia fragum*

The third example involves speciation in the reef coral *Favia fragum* (Carlon and Budd 2002; Carlon et al. 2011), a hermaphroditic brooder, which is unrelated to either the *Orbicella annularis* complex or *Montastraea cavernosa* (Fukami et al. 2004b). Hermaphrodites have male and female reproductive cells, sometimes resulting in self-fertilization. Brooders release sperm into the external water column, but they retain their eggs and have internal fertilization, which may limit dispersal. Unlike the other two examples, *Favia fragum* forms small colonies (usually <6.5 cm in diameter), and occurs only in intertidal and shallow subtidal (generally <5 m depth) environments. Its geographic distribution, however, is comparable to that

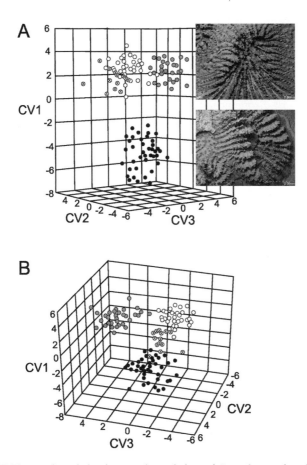

Figure 7.7 Phenotypic variation in natural populations of *Favia fragum* plotted as scores along three major canonical variates (CVs). (A) Perspective emphasizing variation in CV 1. (B) Perspective emphasizing variation in CVs 1 and 3. Grey circles = short ecomorph, reef habitat, lagoon; circles with x = short ecomorph, seagrass habitat, lagoon; unfilled circles = short ecomorph, exposed reef; black circles = tall ecomorph, seagrass, lagoon. CV1 explained 69.4% of the variation and is strongly correlated with costa height. CV2 explained 26.3% of the variation and is strongly correlated with septal length. CV 3 explained 4.3% of the variation and is strongly correlated with overall size. Scale bars are 2 mm. After Carlon et al. (2011).

of the *O. annularis* complex. Two genetically distinct ecomorphs occur in the Bocas del Toro region of Panama with only a narrow zone of overlap (in other words, a narrow contact zone): a "tall" morph that is restricted to seagrass habitats, and a "short" morph that is more abundant on coral reefs. Study of allele frequencies in five allozyme loci shows that the two morphs are in the process of speciating, but are only partially isolated, with a limited amount of gene flow still occurring between the two morphs (Carlon and Budd 2002). This pattern of ecological overlap has been interpreted as an example of parapatric speciation (Carlon and Budd 2002; Carlon et al. 2011), in which the ranges of two speciating populations do not significantly overlap but are immediately adjacent to one another. The results suggest divergence with gene flow (recently reviewed by Pinho & Hey 2010). In the present case, population isolation appears to be a consequence of a high degree of self-fertilization and inbreeding within ecomorphs, combined with limited outcrossing (mixed mating; Carlon and Lippé 2011).

Morphometric analyses show that the two ecomorphs have distinct and nonoverlapping morphologies, the most important features being related to calical elevation and development of costae (fig. 7.7; Carlon et al. 2011). Calices are more elevated (tall and thin costae) in more disturbed environments, which would "lift" polyps off of the main colony surface and away from any accumulating sediment. Estimates of additive genetic variation obtained using a marker-based approach and common garden experiments reveal significant narrow-sense heritability in these features and relatively small genotype-environment (G × E) interactions, indicating that the morphological differences are under genetic control (Carlon et al. 2011). Better understanding of the growth and functional significance of these features will help further to determine whether the divergence between the two morphs can be explained by divergent natural selection between morphs (with the tall morph being better adapted to seagrass habitats), or alternatively by the neutral forces of genetic drift.

Conclusion

Clearly, coral species differ in patterns of morphological variation and distinctiveness, and are structured differently in these three examples. In the *Orbicella annularis* complex (family Merulinidae sensu Budd et al. 2012a), modern species are deeply diverged and are not sister lineages.

They overlap in both morphology and spatial/temporal distribution. Morphological overlap may be attributed in part to evolutionary convergence (independent evolution of similar features that have the same function), as well as to introgressive hybridization. Evolutionary convergence is particularly problematic at all taxonomic levels in reef corals due to constraints imposed by their simple morphology (Budd et al. 2010). In *Montastraea cavernosa* (family Montastraeidae sensu Budd et al. 2012a), one highly variable species forms numerous morphological polymorphs, which are sympatric and genetically indistinct. This one polymorphic species has an unusually wide spatial/temporal distribution. In *Favia fragum* (family Mussidae sensu Budd et al. 2012a), speciation is currently in progress with no morphological overlap between species, and only a narrow zone of ecological overlap between species.

Taken together, these results indicate that making general predictions that could be applied to the fossil record is a complex undertaking because of all of the factors that need to be considered. Nevertheless, despite extensive ecophenotypic plasticity and morphological overlap among closely related species, we have found that morphological differences among species do tend to be statistically significant, and that in many cases, genetic and morphological data agree. These results argue in favor of a multivariate statistical approach comparing populations at both the intra- and interspecific level sampled in the same environment as well as along an environmental gradient. Several nontraditional corallite-level morphological traits need to be analyzed simultaneously, and adequate sampling of populations is essential (with numbers of corallites per colony and numbers of colonies per population being determined quantitatively, for example using cumulative sampling curves). Moreover, environmental data associated with each population are needed for evaluating trends along environmental gradients. By using such an approach (multivariate, population based, with environmental correlates), morphospecies can be recognized that agree with molecular data. Comparisons between such morphospecies can then be used to determine sets of qualitative characters that can be used in subsequent paleoecological investigations.

One factor that may be partially responsible for the differences among the three systems treated herein is reproductive biology. The *O. annularis* complex consists of hermaphroditic broadcast spawners, *M. cavernosa* consists of gonochoric broadcast spawners, and as noted above, *F. fragum* is a hermaphroditic brooder (Kerr et al. 2011). In contrast to brooders (defined above), broadcaster spawners release both sperm and eggs into the external water column, and tend to disperse more widely than brooders. In contrast to hermaphrodites (defined above), gonochores have only one kind of reproductive cell, either male or female. Although these reproductive traits are not preserved in fossils, molecular phylogenetic analyses of modern corals (i.e., mapping reproductive traits onto molecular trees) suggest that gonochorism is an ancestral trait in the Anthozoa, with hermaphrodites arising in three large, distantly related lineages of scleractinian reef corals, such that two-thirds of all reef corals today are hermaphroditic broadcast spawners (Kerr et al. 2011). The patterns we observe in the *O. annularis* complex, which include concordance between molecular and morphological data at the metapopulation level, appear to be related to the dispersal and population connectivity characteristic of hermaphroditic broadcast spawners, which dominate reefs today and through much of the Cenozoic. In contrast, due to inbreeding and limited dispersal, the brooder *F. fragum* is composed of locally adapted demes and highly structured populations, which have low amounts of genetic and morphological variation (Carlon and Lippé 2011). Given the association with brooding, this pattern is less common today and may have also been in the geologic past. Finally, the broad distribution and highly polymorphic nature of both molecular and morphological variation in the gonochore *Montastraea cavernosa* remains an enigma, but appears to be associated with the extreme longevity of this lineage (>25 million years) and the fact that molecular phylogenetic analyses indicate it to be ancestral to many modern scleractinians (Fukami et al. 2004b). Such polymorphism may have been more common in the geologic past than today (especially the Mesozoic), and requires further study in order to better interpret patterns in the fossil record.

Rates of evolutionary change and population connectivity are important species attributes, not just for the appreciation of the evolution of past life, but also for the future of present-day coral reef ecosystems. Recent work is showing the important role that species life history traits play in their susceptibility to extinction in both ancient and modern seas (van Woesik et al. 2012), and our understanding of susceptibility to extinction is strengthened when evolutionary history and phylogenetic information is added (Huang 2012). The important role of past and present evolution in the continued future maintenance of biodiversity is increasingly being appreciated (e.g., Fukami et al. 2004b; Budd & Pandolfi 2010; Huang 2012), making taxonomic and evolutionary studies important components of modern approaches to marine management.

Acknowledgments

We thank Warren Allmon, Bob Elias, and Jon Hendricks for reviewing the manuscript. This research was supported by NSF grants EAR97–25273 and DEB-0343208 to A.F.B. and by Smithsonian Institution Marine Science Network and Biodiversity grants, a Smithsonian Tropical Research Institute Tupper Fellowship, and an Australian Research Council Centre of Excellence grant to J.M.P. and others.

References

Baums IB, Miller MW, Hellberg ME (2005). Regionally isolated populations of an imperiled Caribbean coral, *Acropora palmata*. Molecular Ecology 14: 1377–1390.

Baums, IB, Miller MW, Hellberg ME. (2006). Geographic variation in clonal structure in a reef building Caribbean coral, *Acropora palmata*. Ecological Monographs 76: 503–519.

Budd AF (1991). Neogene Paleontology in the Northern Dominican Republic. 11. The Family Faviidae (Anthozoa: Scleractinia). Part I. Bulletins of American Paleontology 101 (338): 5–83, pls. 1–29.

Budd AF (2010). Tracing the long-term evolution of a species complex and its relationship to environmental change: Examples from the *Montastraea "annularis"* complex. Palaeoworld 19: 348–356.

Budd AF, Johnson KG (1999). Origination preceding extinction during Late Cenozoic turnover of Caribbean reefs. Paleobiology 25: 188–200.

Budd AF, Klaus JS (2001). The origin and early evolution of the *Montastraea "annularis"* species complex (Anthozoa: Scleractinia). Journal of Paleontology 75: 527–545.

Budd AF, Klaus JS (2008). Early evolution of the *Montastraea "annularis"* species complex (Anthozoa: Scleractinia): Evidence from the Mio-Pliocene of the Dominican Republic. In: RH Nehm, and AF Budd (editors), 2008. Evolutionary Stasis and Change in the Dominican Republic Neogene, Springer, New York, 85–124.

Budd AF, Pandolfi JM (2004). Overlapping species boundaries and hybridization within the *Montastraea "annularis"* reef coral complex in the Pleistocene of the Bahama Islands. Paleobiology 30: 396–425.

Budd AF, Pandolfi JM (2010). Evolutionary novelty is concentrated at the edge of species distributions. Science 328: 1558–1561.

Budd AF, Romano SL, Smith ND, Barbeitos MS (2010). Rethinking the phylogeny of scleractinian corals: A review of morphologic and molecular data. Integrative and Comparative Biology 50: 411–427

Budd AF, Fukami H, Smith ND, Knowlton N (2012a). Taxonomic classification of the reef coral family Mussidae (Cnidaria: Anthozoa: Scleractinia). Zoological Journal of the Linnean Society 166: 465–529.

Budd AF, Nunes FLD, Weil E, Pandolfi JM (2012b). Polymorphism in a common Atlantic reef coral (*Montastraea cavernosa*) and its long-term evolutionary implications. Evolutionary Ecology 26: 265–290.

Carlon DB, Budd AF (2002). Incipient speciation across a depth gradient in a scleractinian coral? Evolution 56: 2227–2242.

Carlon DB, Budd AF, Lippé C, Andrew RL (2011). The quantitative genetics of incipient speciation: Heritability and genetic correlations of skeletal traits in populations of diverging *Favia fragum* ecomorphs. Evolution 65: 3428–3447.

Carlon DB, Lippé C (2011). Estimation of mating systems in Short and Tall ecomorphs of the coral *Favia fragum*. Molecular Ecology 20: 812–828.

Flot J-F, Magalon H, Cruaud C, Couloux A, Tillier S (2008a). Patterns of genetic structure among Hawaiian corals of the genus *Pocillopora* yield clusters of individuals that are compatible with morphology. Comptes Rendus Biologies 331: 239–247.

Flot J-F, Licuanan W, Nakano Y, Payri C, Cruaud C, Tillier S (2008b). Mitochondrial sequences of *Seriatopora* corals show little agreement with morphology and reveal the duplication of a tRNA gene near the control region. Coral Reefs 27: 789–794.

Flot J-F, Blanchot J, Charpy L, Cruaud C, Licuanan W, Nakano Y, Payri C, Tillier S (2011). Incongruence between morphotypes and genetically delimited species in the coral genus *Stylophora*: Phenotypic plasticity, morphological convergence, morphological stasis, or interspecific hybridization? BMC Ecology 11: 22.

Forsman ZH, Concepcion GT, Haverkort RD, Shaw RD, Maragos JE, Toonen RJ (2010). Ecomorph or endangered coral? DNA and microstructure reveal Hawaiian species complexes: *Montipora dilatata/flabellata/turgescens* and *M. patula/verrilli*. PLoS One 5(12): e15021(1–10).

Forsman ZH, Barshis DJ, Hunter CL, Toonen RJ (2009) Shape-shifting corals: Molecular markers show morphology is evolutionarily plastic in *Porites*. BMC Evolutionary Biology 9: 45.

Foster NL, Paris CB, Kool JT, Baums IB, Stevens JR, Sanchez JA, Bastildas C, Agudelo C, Bush P, Day O, Ferrari R, Gonzalez P, Gore S, Guppy R, McCartney MA, McCoy C, Mendes J, Srinivasan A, Steiner S, Vermeij MJA, Weil E, Mumby PJ (2012). Connectivity of Caribbean coral populations: Complementary insights from empirical and modelled gene flow. Molecular Ecology 21: 1143–1157.

Fukami H, Budd AF, Levitan DR, Jara J, Kersanach R, Knowlton N (2004a). Geographic differences in species boundaries among members of the *Montastraea annularis* complex based on molecular and morphological markers. Evolution 58: 324–337.

Fukami H, Budd AF, Paulay G, Solé-Cava A, Chen CA, Iwao K, Knowlton N

(2004b). Conventional taxonomy obscures deep divergence between Pacific and Atlantic Corals. Nature 427: 832–835.

Fukami H, Chen CA, Budd AF, Collins A, Wallace C, Chuang Y-Y, Chen C, Dai C-F, Iwao K, Sheppard C, Knowlton N (2008). Mitochondrial and nuclear genes suggest that stony corals are monophyletic but most families of stony corals are not (Order Scleractinia, Class Anthozoa, Phylum Cnidaria). PLoS One 3(9): e3222(1–9).

Goreau TF, Wells JW (1967). The shallow-water Scleractinia of Jamaica: Revised list of species and their vertical distribution range. Bulletin of Marine Science 17:442–453.

Graus RR, Macintyre IG (1976). Control of growth form in colonial corals: Computer simulation. Science 193:895–897.

Graus RR, Macintyre IG (1982). Variations in the growth forms of the reef coral *Montastraea annularis* (Ellis and Solander): A quantitative evaluation of growth response to light distribution using computer simulation. In: Rützler K, Macintyre IG (eds.). The Atlantic barrier reef ecosystem at Carrie Bow Cay, Belize. 1. Structure and communities. Smithsonian Contributions to Marine Sciences 12:441–464.

Huang D (2012). Threatened reef corals of the world. PLoS One 7(3): e34459.

Johnson KG (2007). Reef-coral diversity in the Late Oligocene Antigua Formation and temporal variation of local diversity on Caribbean Cenozoic reefs. In: Hubmann B, Piller WE (eds.), Fossil Corals and Sponges. Proceedings of the 9th International Symposium on Fossil Cnidaria and Porifera. Österr. Akad. Wiss., Schriftenr. Erdwiss. Komm. 17: 471–491. Wien.

Johnson KG, Jackson JBC, Budd AF (2008). Caribbean reef development was independent of coral diversity over 28 million years. Science 319(5869): 1521–1523.

Kerr AM, Baird AH, Hughes TP (2011). Correlated evolution of sex and reproductive mode in corals (Anthozoa: Scleractinia). Proceedings of the Royal Society B 278: 75–81.

Klaus JS, Budd AF (2003). Comparison of Caribbean coral reef communities before and after Plio-Pleistocene faunal turnover: Analyses of two Dominican Republic reef sequences. Palaios 18:3–21.

Klaus JS, Budd AF, Heikoop JM, Fouke BW (2007). Environmental controls on corallite morphology in the reef coral *Montastraea annularis*. Bulletin of Marine Science 80: 233–260.

Knowlton N (1993). Sibling species in the sea. Annual Review of Ecology and Systematics 24: 189–216.

Knowlton N (2000). Molecular genetic analyses of species boundaries. Hydrobiologica 420: 73–90.

Knowlton N, Budd AF (2001). Recognizing coral species past and present. In: Jackson JBC, Lidgard S, McKinney FK (eds.), Evolutionary Patterns: Growth, Form, and Tempo in the Fossil Record. University of Chicago Press, Chicago, 97–119.

Knowlton N, Weil E, Weigt LA, Guzmán HM (1992). Sibling species in *Montastraea annularis*, coral bleaching, and the coral climate record. Science 255: 330–333.

Kongjandtre N, Ridgeway T, Cook LG, Huelsken T, Budd AF, Hoegh-Guldberg O (2012). Taxonomy and species boundaries in the coral genus *Favia* Milne Edwards and Haime, 1857 (Cnidaria: Scleractinia) from Thailand revealed by morphological and genetic data. Coral Reefs 31: 581–601.

Ladner JT, Palumbi SR (2012). Extensive sympatry, cryptic diversity, and introgression throughout the geographic distribution of two species complexes. Molecular Ecology 21: 2224–2238.

Lasker HR (1976). Intraspecific variability of zooplankton feeding in the hermatypic coral *Montastraea cavernosa*. In: Mackie GW (ed.), Coelenterate Ecology and Behavior, Plenum Press, New York, 101–109.

Lasker HR (1979). Light dependent activity patterns among reef corals: *Montastraea cavernosa*. Biology Bulletin 156: 196–211.

Lasker HR (1980). Sediment rejection by reef corals: The roles of behavior and morphology in *Montastraea cavernosa* (Linnaeus). Journal of Experimental Marine Biology and Ecology 47: 77–87.

Lasker HR (1981). Phenotypic variation in the coral *Montastraea cavernosa* and its effects on colony energetics. Biology Bulletin 160: 292–302.

Levitan DR, Fukami H, Jara J, Kline D, McGovern TM, McGhee KE, Swanson CA, Knowlton N (2004). Mechanisms of reproductive isolation among sympatric broadcast-spawning corals of the *Montastraea annularis* species complex. Evolution 58: 308–323.

Lopez JV, Kersanach R, Rehner SA Knowlton N (1999). Molecular determination of species boundaries in corals: Genetic analysis of the *Montastraea annularis* complex using amplified fragment length polymorphisms and a microsatellite marker. Biological Bulletin 196: 80–93.

Miller KJ, Benzie JAH (1997). No clear genetic distinction between morphological species within the coral genus *Platygyra*. Bulletin of Marine Science 61: 907–917

Miller K, Babcock R (1997). Conflicting morphological and reproductive species boundaries in the coral genus *Platygyra*. Biological Bulletin 192: 98–110.

Nunes F, Norris RD, Knowlton N (2009). Implications of isolation and low genetic diversity in peripheral populations of an amphi-Atlantic coral. Molecular Ecology 18: 4283–4297.

Palumbi SR, Vollmer S, Romano S, Oliver T, Ladner J (2012). The role of genes in understanding the evolutionary ecology of reef building corals. Evolutionary Ecology 26: 317–335.

Pandolfi JM (2007). A new, extinct Pleistocene reef coral from the *Montastraea* 'annularis' species complex. Journal of Paleontology 81: 472–482.

Pandolfi JM, Budd AF (2008). Morphology and ecological zonation of Caribbean reef corals: The *Montastraea* 'annularis' species complex. Marine Ecology Progress Series 369: 89–102.

Pandolfi JM, Jackson JBC, Geister J (2001). Geologically sudden natural extinction of two widespread Late Pleistocene Caribbean reef corals. In: Jackson JBC, Lidgard S, Mc-Kinney FK (eds.), Evolutionary Patterns: Growth, Form, and Tempo in the Tossil Record. University of Chicago Press, Chicago, 120–158.

Pandolfi JM, Lovelock CE, Budd AF (2002). Character release following extinction in a Caribbean reef coral species complex. Evolution 56: 479–501.

Pinho C, Hey J (2010). Divergence with gene flow: Models and data. Annual Review of Ecology, Evolution, and Systematics 41: 215–230.

Pinzon JH, LaJeunesse TC (2011). Species delimitation of common reef corals in the genus *Pocillopora* using nucleotide sequence phylogenies, population genetics, and symbiosis ecology. Molecular Ecology 20: 311–325.

Ruiz H (2004). Morphometric examination of corallite and colony variability in the Caribbean coral *Montastraea cavernosa* (Linnaeus 1766). Ms.C. thesis (advisor Dr. Ernesto Weil). Department of Marine Sciences, University of Puerto Rico Mayaguez.

Severance EG, Karl SA (2006). Contrasting population genetic structures of sympatric, mass-spawning Caribbean corals. Marine Biology 150: 57–68.

Souter P (2010). Hidden genetic diversity in a key model species of coral. Marine Biology 157: 875–885.

Stefani F, Benzoni F, Pichon M, Cancelliere C, Galli P (2008). A multidisciplinary approach to the definition of species boundaries in branching species of the coral genus *Psammocora* (Cnidaria, Scleractinia). Zoologica Scripta 37: 71–91.

Taylor MS, Hellberg ME (2006). Comparative phylogeography in a genus of coral reef fishes: Biogeographic and genetic concordance in the Caribbean. Molecular Ecology 15: 695–707.

Todd PA (2008). Morphological plasticity in scleractinian corals. Biological Reviews 83: 315–337.

van Woesik R, Franklin EC, O'Leary J, Mcclanahan TR, Klaus JS, Budd AF (2012). Hosts of the Plio-Pleistocene past reflect modern-day coral vulnerability. Proceedings of the Royal Society B: Biological Sciences 279: 2448–2456.

Vollmer SV, Palumbi S (2002). Hybridization and the evolution of reef coral diversity. Science 296: 2023–2025.

Vollmer SV, Palumbi S (2007). Restricted gene flow in the Caribbean staghorn coral *Acropora cervicornis*: Implications for the recovery of endangered reefs. Journal of Heredity 98: 40–50.

Weil E, Knowlton N (1994). A multi-character analysis of the Caribbean coral *Montastraea annularis* (Ellis and Solander 1786) and its two sibling species, *M. faveolata* (Ellis and Solander 1786) and *M. franksi* (Gregory 1895). Bulletin of Marine Science 55: 151–175.

Willis BL, van Oppen MJ, Miller DJ, Vollmer SV, Ayre DJ (2006). The role of hybridization in the evolution of reef corals. Annual Review of Ecology Evolution and Systematics 37: 489–517.

CHAPTER EIGHT

Towards a Model for Speciation in Ammonoids

Margaret M. Yacobucci

Introduction

E very fossil buff is familiar with ammonoid cephalopods or "ammonites,"[1] whose iconic coiled shells can be found in rock shops around
the world. Ammonoids have been used as biostratigraphic markers for
over two centuries, and therefore the temporal and spatial context of ammonoid clades is very well known. These data have revealed extraordinarily rapid rates of speciation and extinction among ammonoid groups.
The drivers behind this evolutionary volatility are not clear. In this chapter, I present a theoretical model for speciation in ammonoid cephalopods that synthesizes what we know about ammonoid paleobiology with
modern concepts of ecological speciation and the role of developmental
variability in fueling evolutionary change. First, I review key aspects of
ammonoid paleobiology that are relevant to understanding their rapid
speciation rates. Next, I outline contemporary ideas about speciation, including species concepts relevant to paleontologists, the geographic context of the speciation process, ecological speciation, and the roles of natural
and sexual selection (especially in sympatric speciation). These ideas are
integrated to produce a general model for how speciation may occur in
many ammonoid clades.

---

1. The term "ammonoid" here refers to any member of Subclass Ammonoidea. "Ammonite" technically refers to members of Suborder Ammonitina but also used informally
(especially among geologists) to refer to any ammonoid fossil.

Ammonoid Evolution

Among all the mollusk clades, the extinct ammonoid cephalopods show perhaps the most dramatic patterns of macroevolution (Neige et al. 2009). Living cephalopods include two extant orders showing quite disparate anatomies, the internally shelled or shell-less Coleoidea (squid, octopus, and cuttlefish) and the externally shelled Nautiloidea (modern *Nautilus* and *Allonautilus* as well as numerous extinct forms). Four additional externally shelled orders are entirely extinct: Actinoceratoidea, Endoceratoidea, Bactritoidea, and Ammonoidea. Ammonoidea is derived from Bactritoidea and considered to be the sister group to Coleoidea (Engeser 1996). Ammonoids arose from a bactritoid ancestor in the Late Silurian or early Devonian (Erben 1960, 1964, 1965; Dzik 1984; Korn 2001; Klug and Korn 2004), and became extinct at or just above the Cretaceous-Paleogene boundary (Landman et al. 2012).

Ammonoids varied widely in their body size and shell form, both between and within species (De Baets et al. 2013). The smallest ammonoids reached final adult size at less than 1 cm shell diameter while the largest species could be more than 2 m across. Sexual dimorphism was also common among ammonoids, expressing itself as differences in shell size and/ or shell shape (Davis et al. 1996). General shell shapes ranged from highly compressed, discus-like forms to wide and globular outlines (fig. 8.1). The degree of shell coiling also varied, with some shells looking like a coiled rope while others overlapped most of the preceding whorl to create a narrow umbilicus. While most ammonoid species coiled in a planispiral fashion, Suborder Ancyloceratina includes a variety of heteromorph ammonoids, with shell forms varying from straight cones to paperclip and snail-like shapes to highly irregular and/or open coiled shells. The shells of many ammonoids were ornamented with complex combinations of ribs, tubercles, and spines. Extensive intraspecific variation in shell form has also been documented in numerous ammonoid species. All these variations in shell size and form have typically been related to habitat and mode of life (Westermann 1996; Ritterbush and Bottjer 2012).

Like all living cephalopods, ammonoids underwent direct development, with no larval stage (Landman et al. 1996). Rather, the innermost portion of the shell, the ammonitella, formed within the egg. The hatchling ammonoid would have been relatively small, as measured ammonitella diameters range from 0.5 to 2.6 mm, with most less than 1.5 mm (Landman et al.

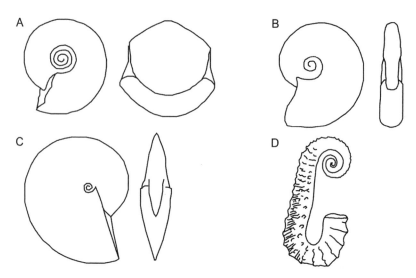

Figure 8.1 Some ammonoid shell shape variations.
Images redrawn and modified from Arkell et al. 1957, p. L83, fig. 125, and p. L84, fig. 126.
A. Cadicone (depressed, evolute / loosely coiled shell)
B. Platycone (compressed shell with planar flanks)
C. Oxycone (compressed, involute / tightly coiled, disc-shaped shell)
D. Heteromorph (nonplanispiral coiling; also note ribs and tubercles ornamenting shell)

1996; Wani 2011). These small hatchlings, at least in some ammonoid groups, may have been planktotrophic, indicating a good dispersal capability. Others may have quickly settled into a nektobenthic mode of life, limiting dispersal. Some evidence exists for masses of ammonoid eggs, deposited on the sea floor or floating in the water column, which may suggest a semelparous mode of reproduction (reproducing just once before dying), as is seen in many extant cephalopods (Landman et al. 1996 and references therein). Age at maturity has been estimated for several ammonoid groups, using a range of methods (e.g., septal spacing, oxygen isotopes, epizoans growing on living ammonoids); most estimates range from 1 to 8 years (Bucher et al. 1996). By comparison, modern *Nautilus* species reach maturity in about 5 to 11 years (Landman and Cochran 1987) while modern coleoid cephalopods mature more quickly, with ages at maturity ranging from 90 days in the cuttlefish *Sepiella inermis* to 3.8 years in the deep water octopus *Bathypolypus arcticus* (Wood and O'Dor 2000).

Ammonoids display a distinctively volatile pattern of evolution, with repeated episodes of rapid diversification followed by frequent large extinction events. The Jurassic-Cretaceous ammonoid suborder Ammonitina shows a particularly volatile evolutionary pattern, with family-level volatility levels an order of magnitude higher than those of bivalve mollusks (Gilinsky 1994, 1998; Yacobucci 2005). A frequently cited explanation for this evolutionary pattern involves the shallow, epeiric sea habitat of many ammonites. Such settings were subject to frequent environmental perturbations such as dysoxic events and sea level fluctuations; these factors have long been thought to be important controls on the evolutionary dynamics of ammonites (Wiedmann 1988; Hallam 1989; House 1989, 1993; Wiedmann and Kullmann 1996; Yacobucci 1999; Sandoval et al. 2001). Many key environmental variables within epeiric seas are typically arranged as spatial gradients (e.g., water depth, temperature, salinity, dissolved oxygen) while others are distributed as more discrete patches (e.g., nutrients, substrate types).

With their often ornate shells and wide geographic distributions, ammonoids have an excellent fossil record and demonstrably high preservation probability (Foote and Sepkoski 1999). Combined with their rapid rates of origination and extinction, the rich fossil record has resulted in ammonoid species being used extensively as biostratigraphic markers. Their biostratigraphic utility, however, has perhaps hindered a more contemporary approach to their systematics. Most publications that name and describe new ammonoid species use outdated methods of phylogenetic analysis and taxon recognition based on stratophenetics, with little sense of a biological or phylogenetic species concept. Rather, species may be initially separated by their stratigraphic occurrence, with anatomical features then identified as diagnostic for those groupings (Donovan 1994). Little attempt has been made to link nominal ammonoid species with a particular species concept (such as the biological or phylogenetic species concepts; see below). Few ammonoid studies have employed contemporary phylogenetic systematic approaches (Rouget et al. 2004; Neige et al. 2007; Yacobucci 2012), and even these often include an emphasis on stratigraphic positions (e.g., Pardo et al. 2008). On the other hand, the vast number of described ammonoid taxa (with over 2000 named genera) provides a rich database documenting potentially phylogenetically significant morphological traits, and many studies have explored temporal patterns of evolution within individual clades.

Role of Developmental Timing in Ammonoid Evolution

Researchers have demonstrated growing interest in the role of developmental changes in promoting speciation (Naisbit et al. 2003; West-Eberhard

2003, 2005; Minelli and Fusco 2012). Variations in developmental timing and expression of developmental regulatory genes can readily produce new innovations and intrapopulational polyphenism, and these can lead to premating isolation (Grant et al. 2006; Kobayashi et al. 2006; Maan et al. 2006; Minelli and Fusco 2012; Boehne et al. 2013; Gunter et al. 2013). Investigation of such factors, however, has been limited, and little attempt has been made to apply these concepts to fossil groups. This lack of study is all the more surprising given the long history of documenting changes in developmental timing among many extinct clades, especially ammonoids.

The study of developmental variations and heterochrony in ammonoids extends back decades and is still an active area of inquiry (Dommergues et al. 1986; Landman 1988; Kennedy 1989; Landman and Geyssant 1993; Gerber et al. 2007; Gerber 2011; Korn 2012; De Baets et al. 2015b; Neige and Rouget 2015). The term heterochrony has been used by ammonoid paleontologists broadly to refer to evolutionary changes in the timing and rate of growth. As with any fossil group, determining ontogenetic age is challenging. While overall shell size or number of chambers can serve as imperfect proxies for age, ammonoid shells also show changes at hatching (e.g., primary constriction, changes in ornamentation, initiation of growth lines) and at maturation (e.g., close spacing of septa, changes in body chamber shape and ornamentation) that allow identification of juvenile vs. sexually mature individuals and shell portions (Davis et al. 1996; Landman et al. 1996; De Baets et al. 2015a; Klug et al. 2015). Further complicating studies of heterochrony in ammonoids is the tendency of workers to blur the observation of a heterochronic pattern with the inference of the evolutionary process of heterochrony. Documenting the former does not necessarily prove operation of the latter; other processes could conceivably produce a heterochronic pattern (e.g., maturity at smaller size in later species).

Both paedomorphosis (the retention of ancestral juvenile traits in the adult descendant) and peramorphosis ("overmaturation" of descendants past the ancestral adult form) have been described in ammonoids. Within the paedomorphic pattern, progenesis (early sexual maturation) is most common (fig. 8.2), though examples of neoteny (slowed growth) have also been cited. Progenetic dwarfs have been described from the Devonian (Korn 1992, 1995a, 1995b), Carboniferous (Swan 1988; Stephen et al. 2002), Permian (Frest et al. 1981; Glenister and Furnish 1988); Jurassic (Cariou and Sequeiros 1987; Marchand and Dommergues 1988; Landman et al. 1991; Meister 1993; Mignot et al. 1993; Dommergues 1994; Linares and Sandoval

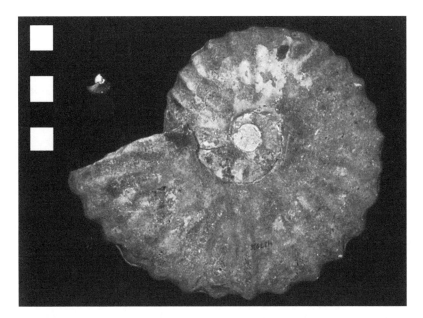

Figure 8.2 Example of progenesis, a type of heterochrony in which maturation rate is accelerated, producing an adult descendant that resembles a juvenile ancestor.
Metoicoceras praecox (USNM 427936) and its progenetic dwarf offshoot *Cryptometoicoceras
mite* (USNM 423766) from the Late Cenomanian of North America. The 1.2 cm diameter
C. mite specimen is a mature adult, based on crowding of last septa and change in ornamen-
tation on body chamber (Kennedy and Cobban 1990b). It strongly resembles juveniles of
M. praecox, with which it co-occurs. Scale bar in centimeters. USNM: National Museum of
Natural History, Smithsonian Institution, Washington.

1996; Neige et al. 1997; Parent 1997, 1998), and Cretaceous (Kennedy 1977;
Wright and Kennedy 1980; Kennedy 1988; Landman 1989; Kennedy and
Cobban 1990a,b; Landman et al. 1991; Wright et al. 1996; Kennedy et al.
2001; Courville and Cronier 2003; Harada and Tanabe 2005) Periods, and
across several major ammonoid clades (Goniatitina, Clymeniina, Ammonitina, Ancyloceratina). Peramorphosis is less common than paedomorphosis (Landman and Geyssant 1993), and is often seen within taxa
that also show paedomorphic changes, producing a mosaic form of heterochrony (Dommergues 1987; Linares and Sandoval 1996; Neige et al.
1997; Parent 1998; Swan 1998; Stephen et al. 2002; Courville and Cronier
2003).

Researchers have made various connections between the phenomenon
of heterochrony and other aspects of ammonoid paleobiology. Differences
between sexual dimorphs (i.e., macroconchs and microconchs) have been

related to heterochronic shifts. Tintant (1963), Guex (1981), and Parent (1997) all suggested microconchs were produced by progenesis or neoteny. Neige (1992) also identified progenesis as the source of some microconchs among Jurassic ammonites, but additionally recognized hypomorphosis as a contributing process in some taxa. These studies do not clearly explain how heterochronic evolutionary change would affect only microconchs (presumed males) and not macroconchs (presumed females) of the same species. Perhaps the genetic growth program was sex-linked such that only males were subject to heterochronic evolutionary change. This notion is supported by the observation that modern cephalopods are gonochoristic (i.e., sex is genetically determined, although cases of intersexuality are known [Hoving et al. 2006]). Others have related continuous intraspecific shell variation to heterochronic processes, suggesting that rounder forms retained their juvenile geometry longer than more compressed forms and could therefore be considered "paedomorphic" (Hammer and Bucher 2006). Again, one might argue whether such a term can apply to variations seen within a single species.

Numerous authors have argued that certain heterochronic changes would be adaptively favored in particular environments. For instance, progenesis could be advantageous in unstable environments, as rapid maturation would allow individuals to exploit abundant juvenile resources. Neoteny (slowed growth), in contrast, would be favored in more stable environments (Gould 1977; McKinney and McNamara 1991). Paleontologists have applied these general concepts to specific case studies of ammonoids, the best of which clearly delineate independent evidence for environmental parameters. Mancini (1978) argued that the progenetic dwarfs of the Cretaceous Grayson Formation of Texas were better adapted than their larger ancestors to live on the unusually soft substrates present at that time. Alternatively, Enay and Gygi (2001) suggested that the Jurassic progenetic dwarfs they investigated would have been more tolerant of dysoxic bottom waters, as smaller body sizes require less oxygen and more rapid maturation allows populations to take advantage of brief intervals of higher oxygen content. In a similar vein, Mignot et al. (1993) argued that paedomorphosis within the Early Jurassic ammonoid *Hildoceras* was an adaptive response to suboptimal environmental conditions, and Zaton (2008) suggested that the range of mature body sizes seen in Jurassic tulitids were due to varying environmental conditions. Monnet et al. (2003) argued that paedomorphic changes resulting in smaller adult body sizes during the Late Cenomanian were driven by environmental perturbations such as sea level rise, temperature increase, and productivity changes. Stevens (1988) suggested that large adult body sizes (as are produced in certain heterochronic shifts such as neotenic or hypermorphic gigantism) might be expected in cold, deep-water environments. Landman and Geyssant (1993) reviewed 167 reported cases of heterochrony in ammonoids, relating the different heteromorphic processes to different modes of life (e.g., nektobenthic, oceanic, megaplanktonic). While paedomorphosis still predominated, neritic nektobenthic forms were more likely to show peramorphosis than other ecologies. Vertical migrators showed the highest rate of progenesis.

The prevalence of heterochrony among ammonoid clades has also been related to diversification rates and the production of higher taxa. For example, Korn (1995b) argued that the diversification of several Late Devonian goniatite and clymeniid clades was driven by sea level fluctuations that favored accelerated maturation and reproductive rates during times of relative sea level fall. Landman (1989) noted that repeated instances of progenesis produced different ammonoid species that had nearly identical juvenile forms, but diverged at maturity. Landman et al. (1991) emphasized that the various Jurassic and Cretaceous progenetic species were not merely sexually mature juveniles, but also had unique mature traits that make them diagnosable taxa. They argued that this "novel combination of juvenile, adult, and unique features may endow progenetic species with the evolutionary potential to play a role in the origin of higher taxa" (Landman et al. 1991, 409). Yacobucci (1999) linked the rapid endemic radiation of acanthoceratid ammonoids in the Late Cretaceous Western Interior Seaway of North America to their developmental flexibility, noting the prevalence of progenetic offshoots within this group.

In addition to heterochronic processes, various other forms of developmental flexibility have been documented in ammonoids and used for systematic purposes. In ammonoids, a large number of characters change during the course of ontogeny and the timing of appearance and disappearance of traits through ontogeny is quite variable. In contrast, other metazoans show less change through ontogeny and the ontogenetic timing of traits is more fixed. Because of the prevalence of these developmentally controlled characters, ammonoid paleontologists have been pioneers in using them in systematics. The systematic description and differentiation of ammonoid species from similar and co-occurring forms often includes reference to developmentally based characters. One species of a genus might reach maturity at a smaller size than another. Features of ornamentation like ribs and tubercles may occur only on one portion of the shell, indicating a developmental shift in the shell's growth program. Traits like the density of ribs or the shape or pattern of spacing of tubercles may change during growth. The adult suture may remain relatively simple in one species while developing more complexity through ontogeny in a close relative. Characters like these are routinely used to diagnose and differentiate closely related ammonoid species. By contrast, the use of juvenile traits and aspects of developmental timing is less common in the systematic study of extant animals, leading to calls for greater awareness of such traits by advocates for the role of development in speciation (Minelli and Fusco 2012).

Significance of Homeomorphy in Ammonoid Evolution

Homeomorphy, the repetition of shell forms in more or less distantly related groups, is pervasive in ammonoids (Schindewolf 1940; Haas 1942; Arkell et al. 1957; Kennedy and Cobban 1976; Saunders and Swan 1984; Dommergues et al. 1989; Dommergues 1994; Donovan 1994; Guex 2001; Monnet et al. 2011). It is understood among ammonoid workers that homeomorphy is to be expected when describing new species, and many taxonomic descriptions of ammonoid taxa therefore include sections on how to distinguish the new group from homeomorphs. Examples of homeomorphy have been particularly well-documented from the Jurassic (e.g., Dommergues et al. 1984; Dommergues and Mouterde 1987; Cariou et al. 1990; Meister 1993; El Hariri et al. 1996; Dommergues 2002; Cecca and Rouget 2006; Schweigert et al. 2012) and the Cretaceous (e.g., Reyment 1955; Obata 1975; Jeletzky and Stelck 1981; Delanoy and Poupon 1992; Maeda 1993; Kennedy and Wright 1994; Delanoy and Busnardo 2007; Bujtor 2010) Periods (fig. 8.3).

The repeated parallel evolution of similar shell shapes, ornamentation types, and suture patterns within ammonoid clades has been related to heterochrony by several workers. Dommergues and colleagues (1989) argued that homeomorphy of shell forms in various Jurassic ammonoids was due to heterochronic processes that recurrently produced similar shell morphologies (disk-shaped oxycones, globular sphaerocones). Both Landman (1989) and Dommergues (1994) specifically cited iterative progenesis as the mechanism producing smaller-bodied species that resembled the juveniles of older or co-occurring ammonoid species. These progenetic trends repeated several times, producing similar-looking species—home-

Tetrahoplites   *Calycoceras*

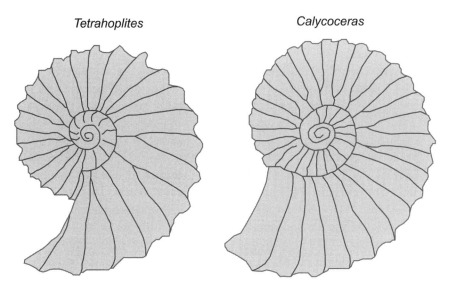

Figure 8.3 Example of homeomorphy in Cretaceous ammonoids.
Tetrahoplites (Hoplitaceae, Hoplitidae, Lower Albian) and *Calycoceras* (Acanthocerataceae,
Acanthoceratidae, Cenomanian) show similar shell sizes, coiling, and ornamentation, despite
being only distantly related. Line drawings created by tracing over photographs of specimens.

omorphs—in each iteration. Similarly, Meister (1993) suggested that paedomorphosis by neoteny was responsible for producing homeomorphic
suboxyconic shell forms in multiple groups of Early Jurassic phyllocera-
tine ammonoids.

Selection favoring certain morphs in certain habitats has long been seen
as the most likely process driving the recurrent evolution of ammonoid homeomorphs. Seilacher and Gunji (1993) argued that certain shell shapes
would be adapted to particular water depths, and therefore that parallel
evolution of similar shell forms could be expected within shallow epeiric seas. Similar arguments relating homeomorphic shell forms to water
depth have been made by Bayer and McGhee (1984), Cecca and Pochettino (2000), Courville (2007), and Bujtor (2010). These evolutionary
trends have been related to sea level cycles. Courville (2007), for instance,
proposed that Cenomanian-Turonian ammonites can be divided into a
cosmopolitan fauna adapted to life in open platform and shelf habitats
and groups of endemic ammonites that diversified in epeiric seaways
(such as the Trans-Saharan Seaway of West Africa) during sea level highs.
Each time sea level rose, a new group of seaway endemics evolved from

open shelf ancestors. These endemics display homeomorphic adaptations to seaway habitats, with the same shell forms and ornaments recurring in each sea level cycle.

As an alternative to adaptation to particular water depths, Monnet et al. (2012) noted that repeated trends to larger shell size and increased shell coiling in Middle Triassic ammonoids might best be explained as a manifestation of Cope's rule—the often-cited trend of increasing adult body size within a clade. De Baets et al. (2012) suggested that the opposite trend, towards smaller embryonic/hatchling size in at least three separate lineages of Devonian ammonoids, might represent adaptations for increased fecundity and higher mobility of hatchlings within the water column. These changes would have been favored during the Devonian "nekton revolution," when free-swimming predators diversified (Klug et al. 2010).

Guex (2000, 2001) has argued that environmental stress may be the root cause of homeomorphy in ammonoids. "Major evolutionary jumps in ammonoids occur during severe extinction events, and are characterized by the sudden appearance of simple, primitive-looking forms which are atavistic with respect to their more complex immediate ancestors" (Guex 2000, 115). For Guex, environmental stress preferentially caused more complex ammonoid forms to die out, while simpler forms that resemble distant ancestors evolved to take their place. In this view, homeomorphs were more likely to occur during or immediately after times of environmental perturbation and heightened turnover, and to show atavistic or ancestral traits.

Current Speciation Models

With this background on ammonoid evolutionary paleobiology, we can now turn our attention to contemporary ideas on how new species form. Butlin et al. (2008) suggest three key questions to ask at each stage of the speciation process: what is the spatial context, what is the driving force for divergence, and what is the (genetic) basis of reproductive isolation? In this section, I review and discuss each of these issues. First, though, we must address what we mean by "species" in the fossil record.

Species Concepts

Allmon and Smith (2011) noted that paleontology as a field has not given very much thought to how the term "species" is used; does it correspond with biological species, a morphological cluster, a temporally defined grouping? Working with the fossil remains of long-dead organisms certainly places limits on what paleontologists can recognize and document, limits that neontologists do not usually face. However, in practice, the operational definitions of species used by modern and fossil workers may not be that different.

Literally dozens of different definitions of "species" exist in the scientific literature. Here I focus on the few species concepts most commonly referenced by paleontologists. The biological species concept (BSC) is certainly the most widely cited (e.g., it is the species definition found in most introductory textbooks). The BSC states that species are groups of interbreeding natural populations that are reproductively isolated from other groups (Mayr 1942, 1995). Coyne and Orr (2004) provide an overview of some of the pros and cons of the BSC. Of most obvious concern to paleontologists is the impossibility of applying the BSC (at least directly) to fossil species. However, most other species definitions try to capture the idea highlighted by the BSC—that species are distinct, isolated gene pools, each therefore with its own unique evolutionary history (Allmon, chapter 3, this volume; Miller, chapter 2, this volume).

Many paleontologists tend naturally to prefer a species concept that provides them a voice in evolutionary biology. Perhaps for that reason, George Gaylord Simpson's evolutionary species concept (ESC; Simpson 1951, 1961) is cited by many paleontologists (e.g., Miller 2006); it helps that Simpson was himself a paleontologist with a deep awareness of the nature of the fossil record. The ESC defines a species as a lineage of organisms that maintains its identify from other such lineages and has its own evolutionary tendencies and historical fate (Simpson 1961; Wiley 1978). The ESC has been criticized by some because it may, for many groups, end up as synonymous with the BSC, and because it suffers from imprecision about which "evolutionary tendencies and historical fate" are significant enough to deem a lineage a species (Coyne and Orr 2004).

Phylogeneticists often cite one of the several phylogenetic species concepts (PSC; Cracraft 1989; de Queiroz and Donoghue 1988). The PSCs focus on recognizing diagnosable, monophyletic groups that are basal (i.e., cannot be further subdivided into monophyletic groups). The PSCs provide a more operational species definition that the BSC or ESC—if one can assemble phylogenetic data and perform a rigorous phylogenetic analysis, one should be able to identify species directly from the results. They are still rooted, however, in evolutionary theory, and in particular in the conception of species as unique evolutionary entities.

In daily practice, of course, the majority of paleontologists still use a traditional morphospecies concept: species are groups of organisms that are morphologically similar to each other and can be diagnosed by specific morphological traits. The morphospecies concept has rightfully been criticized for using arbitrary cutoffs for how similar is similar "enough" to define a new species, and for the idiosyncratic selection of "key" characters by different workers. However, it is still the most practicable method for recognizing and delineating both fossil and modern species (Allmon, chapter 3, this volume); a modern ecologist assessing biodiversity in a habitat patch, for instance, will identify species morphologically with a key. The paleontological morphospecies concept is also rooted in the BSC, ESC, and PSCs: it is implied that morphological similarity should reflect evolutionary proximity (Raup and Stanley 1978, 130). Hence, fossil morphospecies are intended to approximate unique evolutionary lineages (Allmon, chapter 3, this volume; Miller, chapter 2, this volume).

Speciation Models

A spectrum of speciation models has been proposed, distinguished by their geographic context and/or the degree of gene flow permitted between diverging populations. Under the allopatric speciation model, the diverging population is geographically separated from its parent population, with no gene flow occurring between them (fig. 8.4). The extent of the geographic separation needed for speciation to occur varies with the mobility and dispersal ability of the organisms involved. Several workers have shown, for instance, that allopatry may occur within small geographic scales in shallow marine settings (Meyer et al. 2005; Krug 2011). "Microallopatric" speciation involves separation of populations into microhabitats, which can be located quite close to each other in an absolute sense (Smith 1955, 1965; Mayr 1947; Fitzpatrick et al. 2008). However, the fundamental isolating barrier is still related to the physical distance between microhabitats, as opposed to situations in which the isolating barrier is not related to physical distance but to some biological process, such as attraction to different hosts.

Allopatry is widely accepted as the most common mode of speciation; Coyne and Orr (2004) go so far as to formally define it as the null hypothesis for speciation studies, i.e., to demonstrate that another mode operated in a specific case, one must first show that allopatry was impossible. Others have argued that this requirement may be too strict (Johannesson

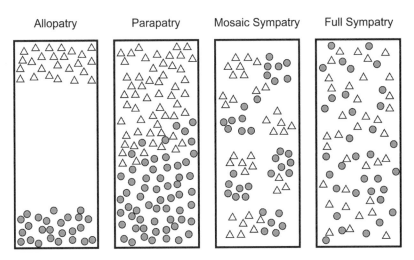

Figure 8.4 Schematic spatial distribution of populations in allopatry, parapatry, mosaic sympatry, and full sympatry.
Triangles and circles represent two definable populations. Redrawn based on Mallet et al. 2009, fig. 2, p. 2334.

2001; Bolnick and Fitzpatrick 2007; Marie Curie Speciation Network 2012).

Parapatry refers to a set of populations that are arrayed along an ecological gradient or cline (fig. 8.4). The varying ecological parameter could be biotic (e.g., food resources, predator concentrations) or abiotic (e.g., temperature, salinity, water depth). Adjacent populations may show some geographic overlap and gene flow, while more distantly separated populations will show much less gene flow. Parapatric speciation is typically envisioned to result from ecological divergence, as populations adapt to their local environments. While direct evidence for parapatric speciation has been limited, theoretical modeling shows it is possible to produce a new species by divergent selection along a gradient even in the face of some gene flow (Coyne and Orr 2004; Gavrilets 2004; Nosil 2008; Pinho and Hey 2010; Keller and Seehausen 2012).

Populations in sympatry show a large overlap in geographic range and full gene flow (fig. 8.4). Sympatric speciation was originally defined spatially, as divergence between two coexisting populations in the absence of any geographic barrier between them (Poulton 1904; Mayr 1942). The meaning of the term has since been shifted to refer to speciation in the presence of a high degree of gene flow between populations, or to some combination of the spatial and genetic concepts (Rice and Hostert 1993; Gavrilets 2003; Coyne and Orr 2004; Butlin et al. 2008; Fitzpatrick et al. 2008; Mallet et al. 2009; Bird et al. 2012). Some classic examples of sympatric speciation, such as insect host races, have been redefined as "microallopatric" (Smith 1955, 1965) or "heteropatric" (Getz and Kaitala 1989), indicating a broadly overlapping geographic range but different ecological niches that physically prevent populations from encountering each other. This variation in how sympatry is defined has contributed to the ongoing controversy surrounding its relative importance in producing biodiversity. Indeed, some have recommended that the term sympatry no longer be used (Fitzpatrick et al. 2008, 2009; Coyne 2011).

Mayr (1942, 1963) argued strongly against the possibility of speciation in sympatric populations—he saw gene flow as too large an obstacle to the development of reproductive isolation, even in the face of strong divergent selection. Another problem with sympatric speciation is that incipient species cannot occupy the same ecological niche, or one population will simply out-compete the other. The main challenge of sympatric speciation, then, is that the process must simultaneously prevent recombination from erasing genetic differences between incipient species and also prevent ecological competition from eliminating one or more coexisting populations (Johannesson 2001; Coyne and Orr 2004). Hence, reproductive isolation via assortative mating must evolve within the group while the population simultaneously evolves ecological differences to prevent competition. Typical models of sympatric speciation therefore involve adaptation of subpopulations to specific ecological niches along with sexual selection to prevent interbreeding between subpopulations.

The term "magic traits" was coined by Gavrilets (2004) to denote single characteristics that address both ecological differentiation and reproductive isolation simultaneously, thereby solving the sympatric speciation dilemma. Magic traits are determined by "magic genes" that are subject to divergent selection and also pleiotropically produce nonrandom mating (Servedio et al. 2011). Servedio et al. (2011) argue that the mechanism can be natural or sexual selection, as long as the product is divergence, and therefore that magic traits may be more prevalent in speciation events than has previously been thought.

It has been argued that allopatry and sympatry are two extreme end members of a continuum of degree of gene flow (Butlin et al. 2008; Mallet et al. 2009). It is unlikely for speciation to take place in pure sympatry, within a fully panmictic population (Coyne and Orr 2004; Fitzpatrick et al. 2008; 2009; Marie Curie Speciation Network 2012), though it is not impossible (Bolnick and Fitzpatrick 2007; Coyne 2011; Bird et al. 2012). Rather, most speciation may involve both an early allopatric stage and a later sympatric stage before reproductive isolation is complete (Rundle and Schluter 2004; Rundle and Nosil 2005; Butlin et al. 2008; Aguilée et al. 2011; Marie Curie Speciation Network 2012). Alternatively, most speciation may fall within a parapatric regime, with reduced but present gene flow, or in mosaic sympatry, with randomly distributed habitat patches in a region occupied by different populations (fig. 8.4; Mallet 2008; Mallet et al. 2009). Because speciation is rarely instantaneous, but may take tens of thousands of years, various environmental and geographic shifts are likely to take place before full reproductive isolation is established (Norris and Hull 2012).

Several attempts have been made to develop tests for recognizing allopatric vs. sympatric speciation. Hindering these efforts is a lack of consensus on what exactly must be assessed in such a test. Mallet et al. (2009) argued that the allopatry vs. sympatry debate is really about the relative importance of nonbiotic factors vs. natural selection in driving speciation. Krug (2011) suggested that the focus on geographic context is misleading, as we are actually interested in how reproductive isolation arises: via natural selection, sexual selection, or random drift.

Despite these concerns, many recent assessments of allopatric vs. sympatric speciation still focus on geography. Most neontological models, experiments, and observations on speciation have studied terrestrial organisms, especially plants, insects, and vertebrates. The potential role of geographic barriers in promoting speciation is especially evident for terrestrial groups. Marine organisms, on the other hand, are frequently broadcast spawners and therefore have high dispersal potentials, making allopatric speciation seem, at least at first glance, to be more challenging. However, not-so-obvious geographic barriers have been well-documented for many marine groups, ranging from the size and temperature of ocean basins (Claremont et al. 2011) to water currents and substrates affecting larval settling (Cowen and Sponaugle 2099) to physical barriers on a small scale due to microhabitat preference (e.g., degree of wave energy on a rocky coastline; Claremont et al. 2012). The consensus view is that allopatric speciation is the norm in the ocean (Claremont et al. 2012), though some examples of sympatric divergence also exist (Krug 2011).

Coyne and Orr (2004) expressed perhaps the strictest test for sympatric speciation. To demonstrate that sympatric speciation has occurred, one must show that the species in question are currently in sympatry, are reproductively isolated from each other, are each other's closest sister taxa, *and* that any previous history of allopatry is very unlikely. The last criterion could be seen as an impossible obstacle for proponents of sympatric speciation to overcome. Coyne and Orr (2004) also suggested that, when studying larger, clade-level patterns, if the percentage of endemic species in a clade is correlated with the degree of mobility of the organisms, allopatric speciation is the likely mechanism (although they do not fully address why such a correlation must be causal).

Other geographic tests include that of Losos and Schluter (2000), who noted that allopatric speciation is supported if new species seem to arise only when the habitat area in which they are found is sufficiently large to allow geographic isolation. Barraclough and Vogler (2000) proposed a quantitative test to identify allopatric vs. sympatric speciation. For multiple pairs of sister species, construct a plot of geographic range overlap vs. age of divergence. A positive slope, where geographic overlap increases with the age of a clade, indicates allopatry while a negative slope, where geographic overlap decreases with the age of a clade, indicates sympatry. Coyne and Orr (2004) note that this method works only if postspeciation range shifts are rare and/or small, and suggest focusing on just the youngest species pairs.

As a final note, one paleontological observation that cannot tell us much about the particular mechanism(s) of speciation is punctuated equilibrium, in which speciation events appear instantaneous (or nearly so) in the fossil record, with species then maintaining relative stasis through most of their stratigraphic range. This punctuated equilibrium pattern, which is frequently seen in the fossil record of animals and some other groups, is "agnostic" with respect to speciation mechanism (Gould 2002, 780). Punctuated equilibrium was originally explained as what allopatric speciation would look like as preserved in the fossil record, but it is also compatible with sympatric speciation. As Gould (2002) put it, punctuated equilibrium "simply requires that any asserted mechanism of speciation, whatever its mode or style, be sufficiently rapid and localized to appear as a punctuation *when scaled into geological time*" (Gould 2002, 780, emphasis in the original).

Isolating Barriers

Modern biologists have emphasized a distinction between physical barriers, such as geographic isolation, and the biological barriers that prevent gene flow. Coyne and Orr (2004) defined isolating barriers as the biological features of organisms that prevent gene flow among sympatric populations. It is these features that cause divergence of separate populations into two species. Geographic barriers may lead to the development of isolating barriers, but under Coyne and Orr's definition, are not themselves sufficient to produce divergence.

Isolating barriers are generally divided into premating and postmating barriers. Postmating barriers, such as hybrid inviability and sterility, are understandably difficult to assess in most fossil taxa, so here I focus on premating barriers. Perhaps most often cited is habitat isolation, in which populations live in different habitats and therefore develop adaptations separately for each habitat. Niche differentiation by itself is not a barrier, but if individuals in different habitats do not encounter and mate with each other (or hybrids have reduced fitness because they are unsuitable to either parent's habitat), reproductive isolation can follow. Habitat isolation can involve spatially separated habitats or habitats that are found as patches within the same general area.

Other premating barriers include allochronic isolation (populations breed at different times, preventing gene flow among them), mechanical isolation (reproductive structures are physically incompatible), and behavioral isolation (a lack of attraction prevents successful courtship or mating). Behavioral isolation would play a significant role in organisms that show relatively complex social interactions and sexual selection. Riesch et al. (2012) have proposed cultural differences as another isolating barrier that might be expected in such groups. Culture in social animals refers to behaviors that are transmitted by social learning rather than by genes; such traits include vocalizations, feeding preferences, foraging strategies, grooming behaviors, etc. In some social animal groups, cultural differences between populations could represent barriers that lead to assortative mating and, in tandem with ecological divergence, reproductive isolation (e.g., Riesch et al. [2012] on divergence in *Orcinus orca* [killer whale] populations).

Inferring the isolating barriers that led to speciation is not straightforward. Coyne and Orr (2004) rightly note that currently observed isolating barriers between two species may not be the barriers that led to speciation in the first place, but rather may have developed after speciation was completed. Ideally, therefore, one would wish to reconstruct the order in which isolating barriers evolved. Whether premating or postmating barriers are more likely to evolve first is also an open question (Marie Curie Speciation Network 2012).

Ecological and Parallel Speciation

Ecological speciation refers to the evolution of reproductive isolation via divergent natural selection (Schluter 2001, 2009; Rundle and Nosil 2005; Hendry et al. 2007; Nosil 2012). This model of speciation, championed by Dolph Schluter of the University of British Columbia and colleagues, among others, has gained in favor over the past decade or so (Schluter 1996a,b, 1998, 2000, 2001, 2009; Funk 1998; Coyne and Orr 2004; Rundle and Nosil 2005; Grant and Grant 2008; Nosil 2012). It typically is thought to involve populations that undergo divergent selection to occupy different ecological niches. This divergence produces anatomical, behavioral, and/or physiological differences that then lead to reproductive isolation. Sexual dimorphism within populations can increase the range of ecologically relevant variation that fuels speciation (Butler et al. 2007). Ecological speciation can happen in allopatry, parapatry, or sympatry. Specific examples in the literature, though, often involve ecological divergence along gradients; such a model would represent parapatric speciation.

How, precisely, reproductive isolation is acquired under ecological speciation can vary. For instance, isolation may merely be a side consequence of selection on other traits (Schluter's (2001) "by-product mechanism"), a result of assortative mating by populations occupying different niches, or due to reduced hybrid fitness (involving direct selection for isolation; Schluter 2001). This model, then, involves both habitat isolation and often subsequent behavioral isolation. One line of evidence for ecological speciation is when multiple populations of an ancestral species independently evolve reproductive isolation in correlation with the environment (Rundle and Nosil 2005).

Workers studying several model systems of ecological speciation have noted that the ecological divergence of populations seemed to repeat itself in multiple events (Schluter and Nagel 1995; Johannesson 2001; Jones et al. 2012). For example, a single fish species colonizing a lake might diverge into a benthic and a pelagic form. When the same fish colonized another lake, the same benthic/pelagic divergence took place, with the resulting morphs looking very much like those found in the first lake. This phenomenon, in which different ecotypes repeatedly and independently evolve reproductive isolation in sympatry, has been dubbed parallel speciation (Schluter and Nagel 1995; Johannesson 2001). Butlin et al. (2008) argued that the process of parallel speciation may not require the parallel independent evolution of a new allele each time, but rather parallel independent selection for the same allele in different populations. The relevant allele would have been present in the ancestral population, reflecting a degree of genetic variability (Butlin et al. 2008). Confirming this idea, Jones et al. (2012) found that, in the classic stickleback fish model, the same set of loci were important in multiple transitions from marine to freshwater populations.

Roles of Selection and Developmental Regulation

Regardless of speciation model, speciation is widely seen as driven by natural selection and/or sexual selection. Sobel et al. (2009), for instance, argued that the ecological speciation concept was not helpful, as *any* speciation event will involve divergent natural selection at some stage. The relative roles of natural versus sexual selection in speciation have been debated (Coyne and Orr 2004). Hendry (2009) contended that reporting biases favor the publication of examples of speciation by natural selection over those driven by sexual selection. Others have argued that the two types of selection are most effective when they both operate within a diverging lineage (Marie Curie Speciation Network 2012). While nonselective genetic drift within small populations isolated in founder events was emphasized by Mayr (1942, 1963) as a key mechanism producing reproductive isolation, drift has fallen out of favor as examples of selection-fueled speciation have accumulated (Coyne and Orr 2004). As the Marie Curie Speciation Network (2012) argued, drift is not likely to be strong enough to produce a new species by itself, but it may play a role in the early stages of speciation, especially in situations involving small populations invading new habitats. Others still advocate for further study of the role of genetic drift in speciation (Templeton 2008; Rundell and Price 2009). While acknowledging that founder speciation, in which a very small number of individuals becomes isolated from a much larger parent population and is therefore subject to strong genetic drift and rapid speciation, is relatively rare, Templeton (2008) noted that " 'rare' does not mean 'unimportant' in evolution" (470). It should also be noted that a more important role for drift may lie in evolution above the species level, specifically in species sorting, as the number of evolutionary "individuals" (that is, species within clades) is frequently quite small (Gould 2002). Drift becomes a more powerful influence when the number of individuals is small.

A modification of the more typical selectionist view would allow for additional mechanisms besides the conventional accumulation and selective sorting of mutations in coding loci. Of particular interest here is the importance of developmental variations in fueling speciation. For example, McCune (1982), writing about fish species flocks, argues that speciation may rely more on developmental recombinations of existing traits than the appearance of new traits:

> If genetic-developmental potentials are shared (primitively) by large groups, the generation of diversity, especially in speciose groups, may come less from the accumulation of mutations in reproductively isolated populations than from the selective unmasking and mixing of already present and cryptically accumulating genetic potentials. (McCune 1982, 325)

While Hoekstra and Coyne (2007) argued emphatically that natural selection on expressed structural genes is the key mechanism driving adaptive radiations, others have countered that changes to cis-regulatory regions that control the expression of genes during development are critical to speciation (Carroll 2008; Frankel et al. 2011; Jones et al. 2012; Wittkopp and Kalay 2012). Jones et al. (2012), for instance, found evidence that changes to regulatory and coding regions are both important in stickleback fish speciation, but regulatory changes predominate.

More generally, Minelli and Fusco (2012) have argued that differences in timing of maturation and variations in when the breeding season occurs can lead to reproductive isolation, and hence that to understand speciation, one must consider changes occurring across the organism's entire life cycle, not just changes seen in reproductive adults. One important question is whether intraspecific variation in developmental timing leads to interspecific heterochrony (Spicer et al. 2011; Tills et al. 2011). The term "heterokairy" was coined by Spicer and Burggren (2003) to refer to variations in the developmental timing within a species (as opposed to heterochrony, which involves timing changes between species). Heterokairy often refers specifically to nongenetic variation, although Tills et al. (2011) found that intraspecific variations in developmental timing within the freshwater snail *Radix* had a genetic basis. In this snail, the traits showing developmental variability within a species were the same traits used to distinguish different species. Developmental flexibility, then, could be an important driver of speciation.

Rates of Speciation

How long does it take to make a new species? This question can be taken in two ways: (1) how long does it take to achieve reproductive isolation

between two diverging populations, and (2) how long is the interval between branching events in a clade? Evidence from the fossil record as well as
modern anatomy and genetics shows that punctuated change at speciation
events is common (Gould 2002; Pagel et al. 2006; Pachut and Anstey 2009;
Hunt 2010). The genetic and anatomical changes that separate two diverging species may accumulate over as little as 1,000 generations (Hunt et al.
2008; Hunt 2010). The total duration of a speciation event may typically
range from 10,000 to 100,000 years; however, certain groups show much
more rapid speciation (Coyne and Orr 2004).

We can also ask how frequently a given clade is likely to experience speciation events. Some clades (such as ammonoids) show much higher speciation rates than others. Speciation rate may be linked to specific biological traits, like sexual dimorphism and low dispersal rates (which promote
reproductive isolation), or highly variable anatomical features associated
with niche divergence and ecological isolation (Coyne and Orr 2004).
Theoretically, we would expect sympatric speciation to be faster than allopatric speciation (because the intermediate stage of speciation in sympatry is unstable and the transition time will therefore be short), and speciation driven by sexual selection to be faster than speciation driven by
natural selection (Bush and Smith 1998; McCune and Lovejoy 1998; Johannesson 2001; Coyne and Orr 2004). Rosenblum et al. (2012) argued
with their "ephemeral speciation model" that species may, in fact, form
rapidly and readily, but only rarely persist for any length of time. Variables that affect the probability of persistence, then, would be the factors
controlling observed speciation rates.

Adaptive radiations, in which species within a clade diversify readily
and rapidly, have been tied to the combination of new ecological opportunities and strong sexual selection (Wagner et al. 2012). Gavrilets and Vose
(2009) found that theoretical modeling supported the conclusion that:

> strong ecologically based spatially heterogeneous selection coupled with lim-
> ited migration, genetically based habitat choice and genetically based mate
> choice can indeed result in rapid phenotypic and ecological diversification and
> the emergence of multiple species reproductively isolated by a variety of mech-
> anisms. (Gavrilets and Vose 2009, 120)

Note that, under this view, adaptive radiations can occur in a variety
of geographic contexts and with a range of isolating barriers. Of greatest
importance are a strong selection regime and genetically rooted ecological and mating preferences.

A Model for Ammonoid Speciation

Can we take what is known about speciation from neontological studies and apply it to an extinct group, the ammonoids? In order to develop a model for ammonoid speciation, it is necessary to synthesize key biological processes known to be important in ammonoid evolution—such as developmental flexibility, heterochrony, and homeomorphy—with the apparent linkage between ammonoid diversity and sea level cycles. These phenomena can then be set into the context of contemporary speciation models. To summarize the key findings from the literature reviewed above:

- Ammonoids have an excellent fossil record and demonstrably high preservation probability.
- Ammonoids display a wide range of body sizes, shell shapes, and shell ornamentation.
- Ammonoids show extremely high rates of origination compared to most marine animals.
- This volatile pattern of ammonoid evolution has frequently been linked to sea level changes and the extent of epeiric seaways.
- Many ammonoid clades experienced heterochronic evolution, often connected to environmental variations and possibly driving diversification.
- Ammonoid species are frequently diagnosed by traits that represent variations in developmental timing.
- Homeomorphy is pervasive among ammonoids, often linked to heterochronic processes, and may be a result of repeated invasions into similar epeiric habitats.

Findings from contemporary research on speciation include:

- Microallopatric, parapatric, and sympatric speciation are all viable alternatives to the traditional allopatric model.
- Traits that can be simultaneously linked with ecological differentiation and reproductive isolation are especially important in sympatric and parapatric models.
- Isolating barriers can include ecological, behavioral, and cultural differences.
- Speciation may be driven by divergent natural selection to occupy different ecological niches. These processes may occur repeatedly in similar habitats, producing a pattern of parallel evolution.

- Changes in developmental regulatory genes and other types of developmental flexibility may play important roles in driving speciation.
- Rapid diversification events are often associated with both ecological niche divergence and sexual selection.

Integrating these concepts and observations produces the following proposed speciation model for ammonoids:

- The ancestral ammonoid species moves into a new habitat, such as a newly formed epeiric seaway created by a sea level rise (fig. 8.5.A).
- Small random changes in the flexible developmental program of individuals produce variable adult sizes and shell forms (fig. 8.5.B).
- These variable morphs sort into different ecological niches and/or occupy distinct microhabitats within the epeiric seaway (fig. 8.5.C).
- Assortative mating and disruptive selection result in reproductive isolation and divergence. If these subpopulations persist, one or more new, endemic species may be produced (fig. 8.5.D).
- Finally, if a related ammonoid species later moves into a similar new epeiric habitat, it will undergo the same sort of process. Developmental constraints on shell form will result in the production of anatomical variants similar to earlier endemic radiations (i.e., homeomorphs), which will then sort themselves into similar microhabitats (fig. 8.5.E).

Note that under this model, speciation is implied to be sympatric or microallopatric. By tying both reproductive isolation and ecological differentiation to a single causal mechanism (developmental flexibility), the model addresses the major challenge of speciation in sympatry. Different anatomical variants are produced in situ, and then separate into microhabitats. These microhabitats may be patchily distributed within the same general region (e.g., different benthic substrates), or may result in a fossil record that combines multiple microhabitats in a single location (e.g., ammonoids occupying different portions of the water column). The model assumes that ammonoid morphology should match specific ecological niches consistently, that is, a particular mode of life is reflected in shell anatomy and size. The model also requires that the ecological niches to which ammonoid morphs adapt are consistently available through space and time.

This model of speciation emphasizes the importance of both biological processes (developmental flexibility) and environmental factors (sea level change and a mosaic of microhabitats) in explaining high diversification

Figure 8.5 Cartoons illustrating the proposed ammonoid speciation model.
A. The ancestral ammonoid species moves into a newly formed seaway.
B. Changes in the developmental program of individuals produce a variety of shell forms and adult sizes.
C. These variable morphs sort into different ecological niches as they occupy distinct micro-habitats within the seaway.
D. Assortative mating and disruptive selection result in reproductive isolation and divergence.
E. Invasion of a similar seaway will result in a repeat of the process; developmental constraints on shell form produce homeomorphs of the first seaway.

rates among ammonoids. Neither by itself is sufficient to explain ammonoid evolution. The inherent developmental flexibility of ammonoids can produce a great diversity of forms, but these will persist and diverge only when environmental conditions allow it. Sea level rises that produce new shallow marine habitat area may represent a particularly important environmental change driving ammonoid diversification. However, Holland (2012) documented that not all sea level rises are equal. While sea level rises necessarily increase the total area of flooded continent, they need not increase shallow marine habitat area, depending on what depth range of habitat one considers (e.g., 0–25 m, 75–100 m). Hence, a clade's response to sea level change will be contingent on the specifics of the case: what the starting sea level was, the particular bathymetric profile of that region, and the larger paleogeographic context. These subtleties may help to explain conflicting specific ammonoid case studies that show diversification peaks during transgressions vs. regressions.

Conclusions

Ammonoid cephalopods show a distinctive suite of evolutionary characteristics, including frequent heterochrony, homeomorphy, and a high origination rate that is often linked to sea level cycles. The model for ammonoid speciation presented in this chapter integrates these observations on ammonoids with contemporary understandings of the speciation process, in particular the importance of natural and sexual selection as well as developmental variability in fueling speciation. When shallow, epeiric seaways formed due to sea level rise, ammonoid clades could invade these new habitats and diverge into separate niches as developmental variations arise and differentiate due to assortative mating and disruptive selection. Speciation would occur in sympatry or microallopatry. Given developmental constraints, ammonoid clades diverging within different seaways would produce a similar range of anatomical variants, producing homeomorphs. This model of speciation links the inherent developmental flexibility of ammonoids with environmental variability to explain ammonoids' propensity to speciate readily.

To test this speciation model requires a variety of approaches. Of primary importance is developing better phylogenetic and biogeographic contexts for radiating ammonoid clades (Yacobucci 2015). Such information is essential for studying ammonoid evolution, but is sorely lacking for most clades. Detailed morphological studies that document developmental variation and constraint are also necessary. We need to document how anatomical variants are linked to different environmental settings and ecological niches. We also need to investigate cases of homeomorphic parallelism in order to determine whether parallel shell forms could be linked specifically to developmental constraints. Finally, in this age of evo-devo it is remarkable that we still know very little about how developmental regulatory genes control molluscan shell growth through ontogeny. What exactly turns on and off tubercles and ribs during development? What is the genetic basis for heterochronic changes like progenetic dwarfism? Can changes in developmental regulatory genes produce "magic traits"? With a broad vision and interdisciplinary approach, we can gain a much better understanding of ammonoid evolution.

Acknowledgments

I wish to thank N. Landman, W. Allmon, and T. Avruch for helpful conversations, suggestions, and editorial guidance, and three reviewers for their detailed suggestions to improve the manuscript. I dedicate this chapter to William A. Cobban and to Stephen J. Gould, two paleontologists whose advice and writings have profoundly influenced my view of ammonoid evolution.

References

Aguilée, R., Lambert, A., and Claessen, D. 2011. Ecological speciation in dynamic landscapes. *Journal of Evolutionary Biology* 24: 2663–2677.

Allmon, W. D., and Smith, U. E. 2011. What, if anything, can we learn from the fossil record about speciation in marine gastropods? Biological and geological considerations. *American Malacological Bulletin* 29: 247–276.

Anderson, F. E. 2000. Phylogeny and historical biogeography of the loliginid squids (Mollusca: Cephalopoda) based on mitochondrial DNA sequence data. *Molecular Phylogenetics and Evolution* 15(2): 191–214.

Arkell, W. J., Furnish, W. M., Kummel, B., Miller, A. K., Moore, R. C., Schindewolf, O. H., Sylvester-Bradley, P. C., and Wright, C. W. 1957. *Treatise on Invertebrate Paleontology, Part L, Mollusca 4, Cephalopoda, Ammonoidea.* Boulder and Lawrence, Geological Society of America and University of Kansas Press.

Barraclough, T. G., and Vogler, A. P. 2000. Detecting the geographical pattern of speciation from species-level phylogenies. *American Naturalist* 155: 419–434.

Bayer, U., and McGhee, G. R., Jr. 1984. Iterative evolution of Middle Jurassic ammonite faunas. *Lethaia* 17: 1–16.

Bird, C. E., Fernandez-Silva, I., Skillings, D. J., and Toonen, R. J. 2012. Sympatric speciation in the post "Modern Synthesis" era of evolutionary biology. *Evolutionary Biology* 39: 158–180.

Blakey, R. C. 2011. Paleogeography and geologic evolution of North America. http://cpgeosystems.com/nam.html.

Boehne, A., Heule, C., Boileau, N., and Salzburger, W. 2013. Expression and sequence evolution of aromatase cyp19a1 and other sexual development genes in East African cichlid fishes. *Molecular Biology and Evolution* 10: 2268–2285.

Bolnick, D. I., and Fitzpatrick, B. M. 2007. Sympatric speciation: Models and empirical evidence. *Annual Reviews in Ecology, Evolution, and Systematics* 38: 459–487.

Bonacum, J., Landman, N. H., Mapes, R. H., White, M. M., White, A. J., and Irlam, J. 2011. Evolutionary radiation of present-day *Nautilus* and *Allonautilus*. *American Malacological Bulletin* 29: 77–93.

Bucher, H., Landman, N. H., Klofak, S. M., and Guex, J. 1996. Mode and rate of growth in ammonoids. Pp. 407–461 *in* Landman, N. H., Tanabe, K., and Davis, R. A. (eds.). *Ammonoid Paleobiology*. New York, Plenum Press.

Bush, G. L., and Smith, J. J. 1998. The genetics and ecology of sympatric speciation: A case study. *Researches on Population Ecology* 40: 175–187.

Bujtor, L. 2010. Systematics, phylogeny and homeomorphy of the Engonoceratidae HYATT, 1900 (Ammonoidea, Cretaceous) and revision of *Engonoceras duboisi* LATIL, 1989. *Carnets de Geologie* Article No. CG2010_A08.

Butler, M. A., Sawyer, S. A., and Losos, J. B. 2007. Sexual dimorphism and adaptive radiation in *Anolis* lizards. *Nature* 447: 202–205.

Butlin, R. K., Galindo, J., and Grahame, J. W. 2008. Sympatric, parapatric, or allopatric? The most important way to classify speciation? *Philosophical Transactions of the Royal Society of London Series B-Biological Sciences* 363: 2997–3007.

Cariou, E., and Sequeiros, L. 1987. Callovian *Taramelliceras* (Ammonitina, Taramelliceratinae): Discovery of the ancestral forms and probable progenetic origin of the genus. *Geobios* 20: 495–516.

Cariou, E., Elmi, S., and Mangold, C. 1990. *Securisites*, new genus (Ammonitina, Jurassic) and its phylogenetic position in the family Oppeliidae: An example of iterative evolution. *Comptes Rendus de l'Academie des Sciences Serie II Mecanique Physique Chimie Sciences de l'Univers Sciences de la Terre* 315: 1267–1273.

Carroll, S. B. 2008. Evo-devo and an expanding evolutionary synthesis: A genetic theory of morphological evolution. *Cell* 134: 25–36.

Cecca, F. and Pochettino, M. 2000. The Early Kimmeridgian genus *Metastreblites* Olóriz, 1978 (Ammonoidea, Oppeliidae) from Rocca Drago (western Sicily, Italy): Homeomorphy and iterative evolution within the Subfamily Streblitinae. *Geobios* 33: 97–107.

Cecca, F., and Rouget, I. 2006. Anagenetic evolution of the early Tithonian ammonite genus *Semiformiceras* tested with cladistic analysis. *Palaeontology* 49: 1069–1080.

Claremont, M., Williams, S. T., Barraclough, T. G., and Reid, D. G. 2011. The geographic scale of speciation in a marine snail with high dispersal potential. *Journal of Biogeography* 38: 1016–1032.

Claremont, M., Reid, D. G., and Williams, S. T. 2012. Speciation and dietary specialization in *Drupa*, a genus of predatory marine snails (Gastropoda: Muricidae). *Zoologica Scripta* 41: 137–149.

Courville, P. 2007. Échanges et colonisations fauniques (Ammonitina) entre Téthys et Atlantique sud au Crétacé supérieur: voies atlantiques ou sahariennes? *Carnets de Géologie, Mémoire* 2007/02: 16–19.

Courville, P., and Cronier, C. 2003. Ontogenetic heterochronies: A tool to study both variability and phyletic relationships? Example: *Nigericeras*, Ammonitina of the African Upper Cretaceous. *Comptes Rendus Palevol* 2: 535–546.

Cowen, R. K., and Sponaugle, S. 2009. Larval dispersal and marine population connectivity. *Annual Review of Marine Science* 1: 443–466.

Coyne, J. A. 2011. Speciation in a small space. *Proceedings of the National Academy of Sciences* 108: 12975–12976.

Coyne, J. A., and Orr, H. A. 2004. *Speciation.* Sunderland, Massachusetts, Sinauer Associates, Inc.

Cracraft, J. 1989. Speciation and its ontology: The empirical consequences of alternative species concepts for understanding patterns and processes of differentiation. Pp. 28–59 *in* Otte, D., and Endler, J (eds.). *Speciation and Its Consequences* Sunderland, Massachusetts, Sinauer Associates, Inc.

Davis, R. A., Landman, N. H., Dommergues, J-L. Marchand, D., and Bucher, H. 1996. Mature modifications and dimorphism in ammonoid cephalopods. Pp. 463–539 *in* Landman, N. H., Tanabe, K., and Davis, R. A. (eds.). *Ammonoid Paleobiology.* New York, Plenum Press.

De Baets, K., Klug, C., Korn, D., and Landman, N. H. 2012. Early evolutionary trends in ammonoid embryonic development. *Evolution* 66: 1788–1806.

De Baets, K., Klug, C., and Monnet, C. 2013. Intraspecific variability through ontogeny in early ammonoids. *Paleobiology* 39: 75–94.

de Queiroz, K., and Donoghue, M. J. 1988. Phylogenetic systematics and the species problem. *Cladistics* 4: 317–338.

De Baets, K., Landman, N. H., and Tanabe, K. 2015a. Ammonoid embryonic development. *In* Klug, C., Korn, D., De Baets, K., Kruta, I., and Mapes, R. H. (eds.). *Ammonoid Paleobiology: From Anatomy to Ecology.* Topics in Geobiology, vol. 43. Dordrecht, Springer.

De Baets, K., Bert, D., Hoffmann, R., Monney, C., Yacobucci, M. M., and Klug, C. 2015b. Ammonoid intraspecific variability. *In* Klug, C., Korn, D., De Baets, K., Kruta, I., and Mapes, R. H. (eds.). *Ammonoid Paleobiology: From Anatomy to Ecology.* Topics in Geobiology, vol. 43. Dordrecht, Springer.

Delanoy, G., and Busnardo, R. 2007. *Anglesites* gen. nov. (Ammonoidea, Ancyloceratina), a new genus of heteromorphic ammonites from the upper Barremian from South-East of France. *Geobios* 40: 801–807.

Delanoy, G., and Poupon, A. 1992. About the genus *Lytocrioceras* Spath, 1924: (Ammonoidea, Ancyloceratina). *Geobios* 25: 367–382.

Dommergues, J.-L. 1987. L'evolution chez les Ammonitina du Lias moyen (Carixian, Domerien basal) en Europe occidentale. *Documents des Laboratoires de Géologie de la Faculté des Sciences de Lyon* 98: 1–297.

Dommergues, J.-L. 1994. The Jurassic ammonite *Coeloceras*: An atypical example of dimorphic progenesis elucidated by cladistics. *Lethaia* 27: 143–152.

Dommergues, J-L. 2002. Les premiers Lytoceratoidea du Nord-Ouest de l'Europe (Ammonoidea, Sinemurien inferieur, France): Exemple de convergence evolutive vers les morphologies "capricornes." *Revue de Paleobiologie* 21: 257–277.

Dommergues, J.-L., and Mouterde, R. 1987. The endemic trends of Liassic ammonite faunas of Portugal as the result of the opening up of a narrow epicontinental basin. *Palaeogeography Palaeoclimatology Palaeoecology* 58: 129–138.

Dommergues, J.-L., Cariou, E., Contini, D., Hantzpergue, P., Marchand, D., Meister, C., and Thierry, J. 1989. Homéomorphies et canalisations évolutives: Le role de l'ontogenèse. Quelques exemples pris chez les ammonites du Jurassique. *Geobios* 22: 5–48.

Dommergues, J.-L., David, B., and Marchand, D. 1986. Ontogeny and phylogeny: Paleontological applications. *Geobios* 19: 335–356.

Dommergues, J.-L., Mouterde, R., and Rivas, P. 1984. A false polymorphism: *Dubariceras*, new genus of the Ammonita from the Mesogean Carixian. *Geobios* 17: 831–839.

Donovan, D. T. 1994. History of classification of Mesozoic ammonites. *Journal of the Geological Society, London* 151: 1035–1040.

Dzik, J. 1984. Phylogeny of the Nautiloidea. *Palaeontologia Polonica* 45: 1–267.

El Hariri, K., Neige, P., and Dommergues, J.-L. 1996. Rib morphometrics of Pliensbachian Harpoceratinae (Ammonitina) from the High Atlas (Morocco). Comparison with specimens from the Central Apennines (Italy). *Comptes Rendus de l'Academie des Sciences, Serie II A, Sciences de la Terre et des Planetes* 322: 693–700.

Enay, R. and Gygi, R. A. 2001. Les ammonites de la zone à Bifurcatus (Jurassique supérieur, Oxfordien) de Hinterstein, près de Oberehrendingen (canton d'Argovie, Suisse). *Eclogae geologicae Helvetiae* 94: 447–487.

Engeser, T. 1996. The position of the Ammonoidea within the Cephalopoda. Pp. 3–19 *in* Landman, N. H., Tanabe, K. and Davis, R. A. (eds.) *Ammonoid Paleobiology*. New York, Plenum Press.

Erben, H. K. 1960. Primitive Ammonoidea aus dem Unterdevon Frankreichs und Deutschlands. *Neues Jahrbuch für Geologie und Paläontologie, Abhandlungen* 110: 1–128.

Erben, H. K. 1964. Die Evolution der ältesten Ammonoidea (Lieferung I). *Neues Jahrbuch für Geologie und Paläontologie, Abhandlungen* 120: 107–212.

Erben, H. K. 1965. Die Evolution der ältesten Ammonoidea (Lieferung II). *Neues Jahrbuch für Geologie und Paläontologie, Abhandlungen* 122: 275–312.

Fitzpatrick, B. M., Fordyce, J. A., and Gavrilets, S. 2008. What, if anything, is sympatric speciation? *Journal of Evolutionary Biology* 21: 1452–1459.

Fitzpatrick, B. M., Fordyce, J. A., and Gavrilets, S. 2009. Pattern, process, and geographic modes of speciation. *Journal of Evolutionary Biology* 22: 2342–2347.

Foote, M., and Sepkoski, J. J. Jr. 1999. Absolute measures of the completeness of the fossil record. *Nature* 398: 415–417.

Frankel, N., Erezyilmaz, D. F., McGregor, A. P., Wang, S., Payre, F., and Stern, D. L. 2011. Morphological evolution caused by many subtle-effect substitutions in regulatory DNA. *Nature* 474: 598–603.

Frest, T. J., Glenister, B. F., and Furnish, W. M. 1981. Pennsylvanian-Permian cheiloceratacean ammonoid families Maximitidae and Pseudohaloritidae. *Journal of Paleontology* 55, suppl. no. 3: 1–46.

Funk, D. J. 1998. Isolating a role for natural selection in speciation: Host adaptation and sexual isolation in *Neochlamisus bebbianae* leaf beetles. *Evolution* 52: 1744–1759.

Gavrilets, S. 2003. Models of speciation: What have we learned in 40 years? *Evolution* 57: 2197–2215.

Gavrilets, S. 2004. *Fitness Landscapes and the Origin of Species.* Princeton, Princeton University Press.

Gavrilets, S., and Vose, A. 2009. Dynamic patterns of adaptive radiation: evolution of mating preferences. Pp. 102–129 *in* Butlin, R., Bridle, J., and Schluter, D. (eds.). *Speciation and Patterns of Diversity.* Ecological Reviews Series. Cambridge, Cambridge University Press.

Gerber, S. 2011. Comparing the differential filling of morphospace and allometric space through time: The morphological and developmental dynamics of Early Jurassic ammonoids. *Paleobiology* 37: 369–382.

Gerber, S., Neige, P., and Eble, G. J. 2007. Combining ontogenetic and evolutionary scales of morphological disparity: A study of early Jurassic ammonites. *Evolution and Development* 9: 472–482.

Getz, W. M., and Kaitala, V. 1989. Ecogenetic models, competition, and heteropatry. *Theoretical Population Biology* 36: 34–58.

Gilinsky, N. L. 1994. Volatility and the Phanerozoic decline of background extinction. *Paleobiology* 20: 445–458.

Gilinsky, N. L. 1998. Evolutionary turnover and volatility in higher taxa. Pp. 162–184 *in* McKinney, M. L., and Drake, J. A. (eds.). *Biodiversity Dynamics: Turnover of Populations, Taxa, and Communities.* New York, Columbia University Press.

Glenister, B. F., and Furnish, W. M. 1988. Terminal progenesis in Late Paleozoic ammonoid families. Pp. 51–66 *in* J. Wiedmann and J. Kullmann (eds.). *Cephalopods: Present and Past.* Stuttgart, Schweizerbart'sche Verlagsbuchhandlung.

Gould, S. J. 1977. *Ontogeny and Phylogeny.* Cambridge, Massachusetts, Harvard University Press.

Gould, S. J. 2002. *The Structure of Evolutionary Theory.* Cambridge, Massachusetts, Belknap Press of Harvard University Press.

Grant, P. R., and Grant, B. R. 2008. *How and Why Species Multiply: The Radiation of Darwin's Finches.* Princeton, Princeton University Press.

Grant, P. R., Grant, B. R., and Abzhanov, A. 2006. A developing paradigm for the development of bird beaks. *Biological Journal of the Linnean Society* 88: 17–22.

Guex, J. 1981. Quelques cas de dimorphisme chez les ammonoids du Lias Inferieur. *Bulletin des Laboratoires de Géologie, Minéralogie, Géophysique et du Musée géologique de l'Université de Lausanne* 258: 239–248.

Guex, J. 2000. *Paronychoceras* gen. n., un nouveau genre d'ammonites (Cephalopoda) du Lias superieur. *Bulletin de la Société Vaudoise des Sciences Naturelles* 87: 115–124.

Guex, J. 2001. Environmental stress and atavism in ammonoid evolution. *Eclogae geologicae Helvetiae* 94: 321–328.

Gunter, H. M., Fan, S., Xiong, F., Franchini, P., Fruciano, C., and Meyer, A. 2013. Shaping development through mechanical strain: The transcriptional basis for diet-induced phenotypic plasticity in a cichlid fish. *Molecular Ecology* 22: 4516–4531.

Haas, O. 1942. Recurrence of morphologic types and evolutionary cycles in Mesozoic ammonites. *Journal of Paleontology* 16: 643–650.

Hallam, A. 1989. The case for sea-level change as a dominant causal factor in mass extinction of marine invertebrates. *Royal Society of London Philosophical Transactions, ser. B* 325: 437–455.

Hammer, Ø., and Bucher, H. 2006. Generalized ammonoid hydrostatics modelling, with application to *Intornites* and intraspecific variation in *Amaltheus*. *Paleontological Research* 10:91–96

Harada, K., and Tanabe, K. 2005. Paedomorphosis in the Turonian (Late Cretaceous) collignoniceratine ammonite lineage from the north Pacific region. *Lethaia* 38: 47–57.

Hendry, A. P. 2009. Ecological speciation? Or the lack thereof? *Canadian Journal of Fisheries and Aquatic Sciences* 66: 1383–1398.

Hendry, A. P., Nosil, P., and Rieseberg, L. H. 2007. The speed of ecological speciation. *Functional Ecology* 21: 455–464.

Hoekstra, H. E., and Coyne, J. A. 2007. The locus of evolution: Evo devo and the genetics of adaptation. *Evolution* 61: 995–1016.

Holland, S. M. 2012. Sea level change and the area of shallow-marine habitat: Implications for marine biodiversity. *Paleobiology* 38: 205–217.

Hoving, H. J. T., Roeleveld, M. A. C., Lipinski, M. R., and Videler, J. J. 2006. Nidamental glands in males of the oceanic squid *Ancistrocheirus lesueurii* (Cephalopoda: Ancistrocheiridae)—sex change or intersexuality? *Journal of Zoology* 269(3): 341–348.

House, M. R. 1989. Ammonoid extinction events. *Royal Society of London Philosophical Transactions, ser. B* 325: 307–326.

House, M. R. 1993. Fluctuations in ammonoid evolution and possible environmental controls. Pp. 13–34 *in* House, M. R. (ed.). *The Ammonoidea: Environment, Ecology, and Evolutionary Change.* London, Clarendon Press.

Hunt, G. 2010. Evolution in fossil lineages: Paleontology and *The Origin of Species. American Naturalist* 176: S61-S76.

Hunt, G., Bell, M. A., and Travis, M. P. 2008. Evolution toward a new adaptive optimum: Phenotypic evolution in a fossil stickleback lineage. *Evolution* 62: 700–710.

Jeletzky, J. A., and Stelck, C. R. 1981. *Pachygrycia*, a new *Sonneratia*-like ammonite from the Lower Cretaceous (Earliest Albian?) of Northern Canada. *Geological Survey of Canada Paper* 80–25.

Johannesson, K. 2001. Parallel speciation: A key to sympatric divergence. *Trends in Ecology and Evolution* 16: 148–153.

Jones, F. C., Grabherr, M. G., Chan, Y. F., Russell, P., Mauceli, E., Johnson, J., Swofford, R., Pirun, M., Zody, M. C., White, S., Birney, E., Searle, S., Schmutz, J., Grimwood, J., Dickson, M. C., Myers, R. M., Miller, C. T., Summers, B. R., Knecht, A. K., Brady, S. D., Zhang, H., Pollen, A. A., Howes, T., Amemiya, C., Broad Institute Genome Sequencing Platform and Whole Genome Assembly Team, Lander, E. S., Di Palma, F., Lindblad-Toh, K., and Kingsley, D. M. 2012. The genomic basis of adaptive evolution in threespine sticklebacks. *Nature* 484: 55–61.

Keller, I., and Seehausen, O. 2012. Thermal adaptation and ecological speciation. *Molecular Ecology* 21: 782–799.

Kennedy, W. J. 1977. Ammonite evolution. Pp. 251–304 *in* Hallam, A. (ed.). *Patterns of Evolution, As Illustrated by the Fossil Record.* Amsterdam, Elsevier.

Kennedy, W. J. 1988. Mid-Turonian ammonite faunas from northern Mexico. *Geological Magazine* 125: 593–612.

Kennedy, W. J. 1989. Thoughts on the evolution and extinction of Cretaceous ammonites. *Proceedings of the Geological Association* 100(3): 251–279.

Kennedy, W. J., and Cobban, W. A. 1976. Aspects of ammonite biology, biogeography, and biostratigraphy. *Special Papers in Palaeontology* 17.

Kennedy, W. J., and Cobban, W. A. 1990a. Cenomanian ammonite faunas from the Woodbine Formation and lower part of the Eagle Ford Group, Texas. *Palaeontology* 33: 75–154.

Kennedy, W. J., and Cobban, W. A. 1990b. Cenomanian micromorph ammonites from the Western Interior of the USA. *Palaeontology* 33: 379–422.

Kennedy, W. J., and Wright, C. W. 1994. The affinities of *Nigericeras* Schneegans, 1943 (Cretaceous, Ammonoidea). *Geobios* 27: 583–589.

Kennedy, W. J., Cobban, W. A., and Landman, N. H. 2001. A revision of the Turonian members of the ammonite subfamily Collignoniceratinae from the United

States Western Interior and Gulf Coast. *Bulletin of the American Museum of Natural History* 267: 1–148.

Klug, C., and Korn, D. 2004. The origin of ammonoid locomotion. *Acta Palaeontologica Polonica* 49: 235–242.

Klug, C., Kröger, B., Kiessling, W., Mullins, G. L., Servais, T., Frýda, J., Korn, D., and Turner, S. 2010. The Devonian nekton revolution. *Lethaia* 43:465–477.

Klug, C., Zatoń, M., Parent, H., Hostettler, B., and Tajika, A. 2015. Mature modifications and sexual dimorphism. *In* Klug, C., Korn, D., De Baets, K., Kruta, I., and Mapes, R. H. (eds.). *Ammonoid Paleobiology: From Anatomy to Ecology.* Topics in Geobiology, vol. 43. Dordrecht, Springer.

Kobayashi, N., Watanabe, M., Kijimoto, T., Fujimura, K., Nakazawa, M., Ikeo, K., Kohara, Y., Gojobori, T., and Okada, N. 2006. *magp4* gene may contribute to the diversification of cichlid morphs and their speciation. *Gene* 373: 126–133.

Korn, D. 1992. Heterochrony in the evolution of Late Devonian ammonoids. *Acta Palaeontologica Polonica* 37: 21–36.

Korn, D. 1995a. Impact of environmental perturbations on heterochronic development in Palaeozoic ammonoids. Pp. 245–260 *in* McNamara, K. J. (ed.). *Evolutionary Change and Heterochrony.* New York, Wiley.

Korn, D. 1995b. Paedomorphosis of ammonoids as a result of sealevel fluctuations in the Late Devonian Wocklumeria Stufe. *Lethaia* 28: 155–165.

Korn, D. 2001. Morphometric evolution and phylogeny of Palaeozoic ammonoids: Early and Middle Devonian. *Acta Geologica Polonica* 51(3): 193–215.

Korn, D. 2012. Quantification of ontogenetic allometry in ammonoids. *Evolution and Development* 14: 501–514.

Krug, P. J. 2011. Patterns of speciation in marine gastropods: A review of the phylogenetic evidence for localized radiations in the sea. *American Malacological Bulletin* 29: 169–186.

Landman, N. H. 1988. Heterochrony in ammonites. Pp. 159–182 *in* McKinney, M. L. (ed.). *Heterochrony in Evolution.* New York, Plenum Press.

Landman, N. H. 1989. Iterative progenesis in Upper Cretaceous ammonites. *Paleobiology* 15: 95–117.

Landman, N. H., and Cochran, J. K. 1987. Growth and longevity of Nautilus. Pp. 401–420 *in* Saunders, W. B., and Landman, N. H. (eds.). *Nautilus: The Biology and Paleobiology of a Living Fossil.* New York, Plenum Press.

Landman, N. H., and Geyssant, J. R. 1993. Heterochrony and ecology in Jurassic and Cretaceous ammonites. *Geobios* 15: 247–255.

Landman, N. H., Dommergues, J-L., and Marchand, D. 1991. The complex nature of progenetic species: Examples from Mesozoic ammonites. *Lethaia* 24: 409–421.

Landman, N. H., Tanabe, K., and Shigeta, Y. 1996. Ammonoid embryonic development. Pp. 343–405 *in* Landman, N. H., Tanabe, K., and Davis, R. A. (eds.). *Ammonoid Paleobiology.* New York, Plenum Press.

Landman, N. H., Garb, M. P., Rovelli, R., Ebel, D. S., and Edwards, L. E. 2012. Short-term survival of ammonites in New Jersey after the end-Cretaceous bolide impact. *Acta Palaeontologica Polonica* 57: 703–715.

Leckie, R. M., Yuretich, R. F., West, O. L. O., Finkelstein, D., and Schmidt, M. 1998. Paleoceanography of the southwestern Western Interior Sea during the time of the Cenomanian–Turonian boundary (Late Cretaceous). *In* Dean, W. E., and Arthur, M. A. (eds.). *Stratigraphy and Paleoenvironments of the Cretaceous Western Interior Seaway, USA.* SEPM Concepts in Sedimentology and Paleontology 6:101–126.

Linares, A., and Sandoval, J. 1996. The genus *Haplopleuroceras* (Erycitidae, Ammonitina) in the Betic Cordillera, southern Spain. *Geobios* 29: 287–305.

Losos, J. B., and Schluter, D. 2000. Analysis of an evolutionary species-area relationship. *Nature* 408: 847–580.

Naisbit, R. E., Jiggins, C. D., and Mallet, J. 2003. Mimicry: Developmental genes that contribute to speciation. *Evolution and Development* 5(3): 269–280.

Maan, M. E., Haesler, M. P., Seehausen, O., and Van Alphen, J. J. M. 2006. Heritability and heterochrony of polychromatism in a Lake Victoria cichlid fish: Stepping stones for speciation? *Journal of Experimental Zoology* 306B(2): 168–176.

Maeda, H. 1993. Dimorphism of late Cretaceous false-puzosiine ammonites, *Yokoyamaoceras* Wright and Matsumoto, 1954 and *Neopuzosia* Matsumoto, 1954. *Transactions and Proceedings of the Palaeontological Society of Japan, New Series* 169: 97–128.

Mallet, J. 2008. Hybridization, ecological races, and the nature of species: Empirical evidence for the ease of speciation. *Philosophical Transactions of the Royal Society of London Series B-Biological Sciences* 363: 2971–2986.

Mallet, J., Meyer, A., Nosil, P., and Feder, J. L. 2009. Space, sympatry, and speciation. *Journal of Evolutionary Biology* 22: 2332–2341.

Mancini, E. A. 1978. Origin of the Grayson micromorph fauna, Upper Cretaceous of North Central Texas, USA. *Journal of Paleontology* 52: 1294–1314.

Marchand, D. and Dommergues, J.-L. 1988. Rhythmes évolutifs et hétérochronies due développement: Examples pris parmi les ammonites Jurassiques. Pp. 67–78 *in* Wiedmann, J., and Kullmann, J. (eds.). *Cephalopods: Present and Past.* Stuttgart, Schweizerbart'sche Verlagsbuchhandlung.

Marie Curie Speciation Network. 2012. What do we need to know about speciation? *Trends in Ecology and Evolution* 27: 27–39.

Mayr, E. 1942. *Systematics and the Origin of Species.* New York, Columbia University Press.

Mayr, E. 1947. Ecological factors in speciation. *Evolution* 1:263–288.

Mayr, E. 1963. *Animal Species and Evolution.* Cambridge, Massachusetts, Belknap Press.

Mayr, E. 1995. Species, classification, and evolution. Pp. 3–12 *in* Arai, R., Kato, M., and Doi, Y. (eds.). *Biodiversity and Evolution.* Tokyo, National Science Museum Foundation.

McCune, A. R. 1982. *Early Jurassic Semionotidae (Pisces) from the Newark Supergroup: Systematics and evolution of a fossil species flock.* Ph.D. dissertation, Yale University.

McCune, A. R., and Lovejoy, N. R. 1998. The relative rate of sympatric and allopatric speciation in fishes: Test using DNA sequence divergence between sister species and among clades. Pp. 172–185 *in* Howard, D. J., and Berlocher, S. H. (eds.). *Endless Forms—Species and Speciation.* Oxford, Oxford University Press.

McKinney, M. L., and McNamara, K. J. 1991. *Heterochrony: The Evolution of Ontogeny.* New York, Plenum Press.

Meister, C. 1993. L'évolution parallèle des Juraphyllitidae euroboréaux et téthysiens au Pliensbachien: Le rôle des contraintes internes et externes. *Lethaia* 26: 123–132.

Meyer, C. P., Geller, J. B., and Paulay, G. 2005. Fine scale endemism on coral reefs: Archipelagic differentiation in turbinid gastropods. *Evolution* 59: 113–125.

Mignot, Y., Elmi, S., and Dommergue, J.-L. 1993. Croissance et miniaturization de quelques *Hildoceras* (Cephalopoda) en liaison avec des environnments contraignant de la Téthys toarcianne. *Geobios Memoire Special* 15: 305–312.

Miller, W., III. 2006. What every paleontologist should know about species: New concepts and questions. *Neues Jahrbuch für Geologie und Paläontologie Monatshefte* 2006: 557–576.

Minelli, A., and Fusco, G. 2012. On the evolutionary developmental biology of speciation. *Evolutionary Biology* 39: 242–254.

Monnet, C., Bucher, H., Escarguel, G., and Guex, J. 2003. Cenomanian (early Late Cretaceous) ammonoid faunas of Western Europe. Part II: diversity patterns and the end-Cenomanian anoxic event. *Eclogae geologicae Helvetiae* 96: 381–398.

Monnet, C., De Baets, K., and Klug, C. 2011. Parallel evolution controlled by adaptation and covariation in ammonoid cephalopods. *BMC Evolutionary Biology* 11: article no. 115.

Monnet, C., Bucher, H., Guex, J., and Wasmer, M. 2012. Large-scale evolutionary trends of Acrochordiceratidae Arthaber, 1911 (Ammonoidea, Middle Triassic) and Cope's rule. *Palaeontology* 55: 87–107.

Neige, P. 1992. *Mise en place du dimorphisme (sexuel) chez les Ammonoides: Approche ontogénétique et interpretation hétérochronique.* Diplome d'Etudes Approfondies (D.E.A.), Université de Bourgogne.

Neige, P., and Rouget, I. 2015. Evolutionary trends within Jurassic ammonoids. *In* Klug, C., Korn, D., De Baets, K., Kruta, I., and Mapes, R. H. (eds.). *Ammonoid Paleobiology: From Macroevolution to Paleogeography.* Topics in Geobiology, vol. 44. Dordrecht, Springer.

Neige, P., Marchand, D., and Laurin, B. 1997. Heterochronic differentiation of sexual dimorphs among Jurassic ammonite species. *Lethaia* 30: 145–155.

Neige, P., Rouget, I., and Moyne, S. 2007. Phylogenetic practices among scholars of fossil cephalopods, with special reference to cladistics. Pp. 3–14 *in* Landman,

N. H., Davis, R. A., and Mapes, R. H. (eds.). *Cephalopods—Present and Past: New Insights and Fresh Perspectives*. Berlin, Springer.

Neige, P., Brayard, A., Gerber, S., and Rouget, I. 2009. Les Ammonoïdes (Mollusca, Cephalopoda): avancées et contributions récentes à la paleobiology évolutive. *Comptes rendus-Palevol* 8: 167–178.

Norris, R. D., and Hull, P. M. 2012. The temporal dimension of marine speciation. *Ecology and Evolution* 26: 393–415.

Nosil, P. 2008. Speciation with gene flow could be common. *Molecular Ecology* 17: 2103–2106.

Nosil, P. 2012. *Ecological Speciation*. Oxford Series in Ecology and Evolution. Oxford, Oxford University Press.

Obata, I. 1975. Lower Cretaceous ammonites from the Miyako Group: *Diadochoceras* from the Miyako Group. *Bulletin of the National Science Museum Series C (Geology)* 1: 1–10.

Pachut, J. F., and Anstey, R. L. 2009. Inferring evolutionary modes in a fossil lineage (Bryozoa: Peronopora) from the Middle and Late Ordovician. *Paleobiology* 35: 209–230.

Pagel, M., Venditti, C., and Meade, A. 2006. Large punctuational contribution of speciation to evolutionary divergence at the molecular level. *Science* 314: 119–121.

Pardo, J. D., Huttenlocker, A. K., and Marcot, J. D. 2008. Stratocladistics and evaluation of evolutionary modes in the fossil record: An example from the ammonite genus *Semiformiceras*. *Palaeontology* 51: 767–773.

Parent, H. 1997. Ontogeny and sexual dimorphism of *Eurycephalites gottschei* (Tornquist) (Ammonoidea) of the Andean Lower Callovian (Argentine-Chile). *Geobios* 30: 407–419.

Parent, H. 1998. Upper Bathonian and lower Callovian ammonites from Chacay Melehué (Argentina). *Acta Palaeontologica Polonica* 43: 69–130.

Pinho, C., and Hey, J. 2010. Divergence with gene flow: Models and data. *Annual Reviews in Ecology, Evolution, and Systematics* 41: 215–230.

Poulton, E. B. 1904. What is a species? *Proceedings of the Entomological Society of London* 1903: lxxvii–cxvi.

Raup, D. M., and Stanley, S. M. 1978. *Principles of Paleontology*, 2nd ed. New York, W. H. Freeman and Company.

Reyment, R. A. 1955. Some examples of homeomorphy in Nigerian Cretaceous ammonites. *Geologiska Foreningens i Stockholm Forhandlingar* 77: 567–594.

Rice, W. R., and Hostert, E. E. 1993. Laboratory experiments on speciation: What have we learned in 40 years? *Evolution* 47: 1637–1653.

Riesch, R., Barrett-Lennard, L. G., Ellis, G. M., Ford, J. K. B., and Deecke, V. B. 2012. Cultural traditions and the evolution of reproductive isolation: Ecological speciation in killer whales? *Biological Journal of the Linnean Society* 106: 1–17.

Ritterbush, K. A., and Bottjer, D. J. 2012. Westermann Morphospace displays ammonoid shell shape and hypothetical paleoecology. *Paleobiology* 38: 424–446.

Rosenblum, E. B., Sarver, B. A. J., Brown, J. W., Des Roches, S., Hardwick, K. M., Hether, T. D., Eastman, J. M., Pennell, M. W., and Harmon, L. J. 2012. Goldilocks meets Santa Rosalia: An ephemeral speciation model explains patterns of diversification across time scales. *Evolutionary Biology* 39: 255–261.

Rouget, I., Neige, P., Dommergues, J.-L. 2004. L'analyse phylogénétique chez les ammonites: État des lieux et perspectives. *Bulletin de la Société Géologique de France* 175: 507–512.

Rundle, H. D., and Nosil, P. 2005. Ecological speciation. *Ecology Letters* 8: 336–352.

Rundle, H. D., and Schluter, D. 2004. Natural selection and ecological speciation in sticklebacks. Pp. 192–209 *in* Dieckmann, U., Doebeli, M., Metz, J. A. J., and Tautz, D. (eds.). *Adaptive Speciation*. Cambridge Studies in Adaptive Dynamics. Cambridge, Cambridge University Press.

Sandoval, J., O'Dogherty, L., and Guex, J. 2001. Evolutionary rates of Jurassic ammonites in relation to sea-level fluctuations. *Palaios* 16: 311–335.

Saunders, W. B., and Swan, A. R. H. 1984. Morphology and morphological diversity of mid-Carboniferous Namurian ammonoids in time and space. *Paleobiology* 10: 195–228.

Schindewolf, O. H. 1940. Konvergenz bei Korallen und Ammoniten. *Fortschritte der Geologie und Paläontologie* 12: 387–491.

Schluter, D. 1996a. Ecological causes of adaptive radiation. *American Naturalist* 148: S40–S64.

Schluter, D. 1996b. Ecological speciation in postglacial fishes. *Philosophical Transactions of the Royal Society of London Series B-Biological Sciences* 351: 807–814.

Schluter, D. 1998. Ecological causes of speciation. Pp. 114–129 *in* Howard, D. J., and Berlocher, S. H. (eds.). *Endless Forms: Species and Speciation*. Oxford, Oxford University Press.

Schluter, D. 2000. *The Ecology of Adaptive Radiation*. Oxford, Oxford University Press.

Schluter, D. 2001. Ecology and the origin of species. *Trends in Ecology and Evolution* 16: 372–380.

Schluter, D. 2009. Evidence for ecological speciation and its alternative. *Science* 323: 737–741.

Schluter, D., and Nagel, L. M. 1995. Parallel speciation by natural selection. *American Naturalist* 146: 292–301.

Schweigert, G., Zeiss, A., and Westermann, G. E. G. 2012. The *Gravesia* homeomorphs from the latest Kimmeridgian of Mombasa, Kenya. *Revue de Paleobiology*, special issue 11: 13–25.

Seilacher, A., and Gunji, P. Y. 1993. Morphogenetic countdowns in heteromorph shells. *Neues Jahrbuch für Geologie und Paläontologisches Abhandlungen* 190: 237–265.

Servedio, M. R., Van Doorn, G. S., Kopp, M., Frame, A. M., and Nosil, P. 2011. Magic traits in speciation: "Magic" but not rare? *Trends in Ecology and Evolution* 26: 389–397.

Simpson, G. G. 1951. The species concept. *Evolution* 5: 285–298.

Simpson, G. G. 1961. *Principles of Animal Taxonomy.* New York, Columbia University Press.

Sinclair, W., Newman, S. J., Vianna, G. M. S., Williams, S., and Aspden, W. J. 2011. Spatial and genetic diversity in populations on the east and west coasts of Australia: The multi-faceted case of *Nautilus pompilius* (Mollusca, Cephalopoda). *Reviews in Fisheries Science* 19(1): 52–61.

Smith, H. M. 1955. The perspective of species. *Turtox News* 33: 74–77.

Smith, H. M. 1965. More evolutionary terms. *Systematic Zoology* 14: 57–58.

Sobel, J. M., Chen, G. F., Watt, L. R., and Schemske, D. W. 2009. The biology of speciation. *Evolution* 64: 295–315.

Spicer, J. I., and Burggren, W. W. 2003. Development of physiological regulatory systems: Altering the timing of crucial events. *Zoology* 106: 91–99.

Spicer, J. I., Rundle, S. D., and Tills, O. 2011. Studying the altered timing of physiological events during development: It's about time . . . or is it? *Respiratory Physiology and Neurobiology* 178: 3–12.

Stephen, D. A., Manger, W. L., and Baker, C. 2002. Ontogeny and heterochrony in the middle Carboniferous ammonoid *Arkanites relictus* (Quinn, McCaleb, and Webb) from northern Arkansas. *Journal of Paleontology* 76: 810–821.

Stevens, G. R. 1988. Giant ammonites, a review. Pp. 141–166 *in* Wiedmann, J., and Kullmann, J. (eds.). *Cephalopods: Present and Past.* Stuttgart, Schweizerbart'sche Verlagsbuchhandlung.

Swan, A. R. H. 1988. Heterochronic trends in Namurian ammonoid evolution. *Palaeontology* 31: 1033–1051.

Tills, O., Rundle, S. D., Salinger, M., Haun, T., Pfenninger, M., and Spicer, J. I. 2011. A genetic basis for intraspecific differences in developmental timing? *Evolution and Development* 13: 542–548.

Tintant, H. 1963. Les Kosmoceratidés du Callovien inférieur et moyen d'Europe occidentale. *Publications de l'Université de Dijon* 29: 1–500.

Voss, N. A. 1985. Systematics, biology, and biogeography of the cranchiid cephalopod genus *Teuthowenia. Bulletin of Marine Science* 36(1): 1–85.

Wagner, C. E., Harmon, L. J., and Seehausen, O. 2012. Ecological opportunity and sexual selection together predict adaptive radiation. *Nature* 487: 366–370.

Wani, R. 2011. Sympatric speciation drove the macroevolution of fossil cephalopods. *Geology* 39: 1079–1082.

West-Eberhard, M. J. 2003. *Developmental Plasticity and Evolution.* Oxford: Oxford University Press.

West-Eberhard, M. J. 2005. Developmental plasticity and the origin of species differences. *Proceedings of the National Academy of Sciences (USA)* 102: 6543–6549.

Westermann, G. E. G. 1996. Ammonoid life and habit. Pp. 607–707 *in* Landman, N. H., Tanabe, K., and Davis, R. A. (eds.). *Ammonoid Paleobiology.* New York, Plenum Press.

Wiedmann, J. 1988. Plate tectonics, sea level changes, climate, and the relationship to ammonite evolution, provincialism, and mode of life. Pp. 737–765 *in* Wiedmann, J., and Kullmann, J. (eds.). *Cephalopods: Present and Past*. Stuttgart, Schweizerbart'sche Verlagsbuchhandlung.

Wiedmann, J., and Kullman, J. 1996. Crises in ammonoid evolution. Pp. 795–813 *in* Landman, N. H., Tanabe, K., and Davis, R. A. (eds.). *Ammonoid Paleobiology*. New York, Plenum Press.

Wiley, E. O. 1978. The evolutionary species concept reconsidered. *Systematic Zoology* 27: 17–26.

Wittkopp, P. J., and Kalay, G. 2012. *Cis*-regulatory elements: Molecular mechanisms and evolutionary processes underlying divergence. *Nature Reviews Genetics* 13: 59–69.

Wood, J. B., and O'Dor, R. K. 2000. Do larger cephalopods live longer? Effects of temperature and phylogeny on interspecific comparisons of age and size at maturity. *Marine Biology* 136: 91–99.

Wright, C. W., and Kennedy, W. J. 1980. Origin, evolution, and systematics of the dwarf acanthoceratid *Protacanthoceras* Spath, 1923 (Cretaceous Ammonoidea). *Bulletin of the British Museum of Natural History, Geology* 34: 65–108.

Wright, C. W., with Calloman, J. H., and Howarth, M. K. 1996. Cretaceous Ammonoidea. Pp. 1–362 *in* Kaesler, R. L. (ed.). *Treatise on Invertebrate Paleontology, Part L, Mollusca 4, Cretaceous Ammonoidea*. Boulder and Lawrence, Geological Society of America and University of Kansas Press.

Yacobucci, M. M. 1999. Plasticity of developmental timing as the underlying cause of high speciation rates in ammonoids: An example from the Cenomanian Western Interior Seaway of North America. Pp. 59–76 *in* Olóriz, F., and Rodríguez-Tovar, F. J. (eds.). *Advancing Research on Living and Fossil Cephalopods*. Proceedings, IV International Symposium Cephalopods—Present and Past. Plenum Press, New York.

Yacobucci, M. M. 2005. Multifractal and white noise evolutionary dynamics in Jurassic-Cretaceous Ammonoidea. *Geology* 33: 97–100.

Yacobucci, M. M. 2012. Meta-analysis of character utility and phylogenetic information content in cladistic studies of ammonoids. *Geobios* 45: 139–143.

Yacobucci, M. M. 2015. Macroevolution and paleobiogeography of Jurassic-Cretaceous ammonoids. *In* Klug, C., Korn, D., De Baets, K., Kruta, I., and Mapes, R. H. (eds.). *Ammonoid Paleobiology: From Macroevolution to Paleogeography*. Topics in Geobiology, vol. 44. Dordrecht, Springer.

Zatoń, M. 2008. Taxonomy and palaeobiology of the Bathonian (Middle Jurassic) tulitid ammonite *Morrisiceras*. *Geobios* 41: 699–717.

Species of Decapoda (Crustacea) in the Fossil Record: Patterns, Problems, and Progress

Carrie E. Schweitzer and Rodney M. Feldmann

Introduction

S pecies are the material basis upon which all paleontological studies of biodiversity must be grounded. Higher taxa are conceptual constructs whose legitimacy is based solely upon the quality of the physical evidence, that is, the species they embrace. Yet we sometimes have considerable difficulty validating species within fossil decapod crustaceans: the shrimp, lobsters, crabs, and their relatives. The problem is not so much with the definition of species per se as it is with describing and distinguishing species given the vagaries of the fossil record. In a masterful essay, Weller (1961) discussed the issues surrounding the definition of species in the fossil record as well as in the modern world. Examination of systematic literature confirms that his succinct definition remains as valid and inclusive now as it was 50 years ago.

> A species is a formal unit of biologic taxonomy, to be identified by a specific name, consisting of a natural continuing population of individuals presumably closely related to each other and generally similar morphologically, that is distinct and distinguishable from all other contemporaneous populations, and separated from related ancestral and descendant populations at some convenient but arbitrarily selected boundaries. (Weller 1961, 1192–1193)

The fossil record of Decapoda is obscured in a number of ways. Decapod remains consist of tens or hundreds of individual elements, are variably calcified, and are often dissociated in the fossil record so that fossil species are usually described on partial remains. Different species, even within the same family or genus, may well be defined on different anatomical parts, thus making comparison difficult. Decapod remains, whether corpses or exuviae, are frequently scavenged for their nutritive value, which has the effect of greatly reducing the material that is preserved. Many species are described on only one partial specimen. Given the scarcity of decapod fossils, defining the geological and geographical range of a taxon, the range of intraspecific morphological variation and the presence and mode of expression of sexual dimorphism is limited or impossible in many species. These and other issues are the subject of the discussion to follow.

However, there is much good news. The complete known fossil record of the Decapoda has been compiled (Schweitzer et al. 2010, as a basis). Proxy characters (Schweitzer 2003), preserved hard parts that track taxonomically important soft parts, make it possible to identify species in the fossil record that are consonant with extant congeners. The approach of using proxy characters has been validated by recent phylogenetic analyses employing both fossil and extant species (Karasawa et al. 2008, 2011, 2013). Further, there has been a surge of interest in the study of fossil decapods in the past 40 years that has resulted in greatly increasing the number of species recognized and expanding the areas of investigation from Europe and North America into the rest of the world. Coupled with these developments, improvements in preparation and illustration techniques have permitted elucidation of morphological features on decapod fossils that were simply not observed or observable in the 19th and early 20th centuries. All of this introduces optimism in the definition of fossil species and interpretation of their paleoecology, biogeography, and phylogeny.

Recognizing Fossil Decapod Species

Fundamentally, there is no difference between the manner in which fossil and living species of decapods are named and described: it is most frequently based upon external morphology (figs. 9.1, 9.2). Examination of recent literature on systematics of living decapods in which new species are

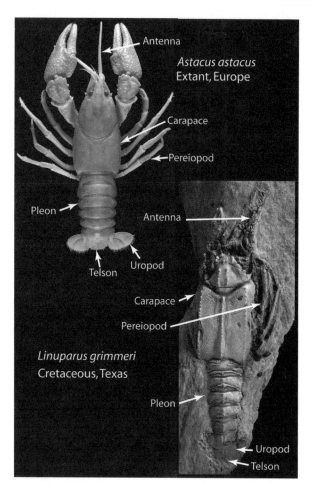

Figure 9.1 Lobster-like forms found both in fossils and extant animals.
Lobster-like forms, indicating the same structures recognized in extant forms (upper left, *Astacus astacus* Linnaeus, 1758, Senckenberg Forschungsinstitut und NaturMuseum, Frankfurt, catalog no. 13096, a crayfish) preserved in fossils (lower right, *Linuparus grimmeri* Stenzel, 1945, United States National Museum of Natural History, Smithsonian Institution, acc. # 259571, a spiny lobster). *Astacus* photo by Sven Tänkner.

named (e.g., Ahyong 2012; Davie and Ng 2013) demonstrates that the species concept most frequently used is based upon external morphology. A few employed molecular techniques combined with external morphology to define species (Gouws et al. 2001), and others have used a combination of morphology, genetics, and estimates of divergence times to define new

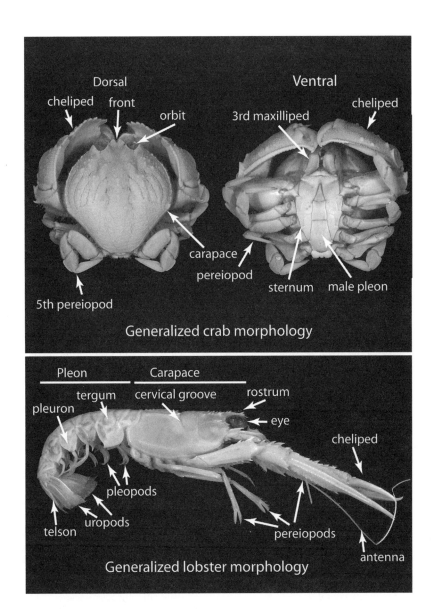

Figure 9.2 Generalized morphology of brachyurans (crabs) and lobsters. Crab is *Cycloes granulosa* De Haan, 1837, male, United States National Museum of Natural History, Smithsonian Institution (USNM), catalog no. 29926, and lobster is *Nephrops norvegicus* (Linnaeus, 1758), commonly known as scampi or shrimp scampi, USNM 152172.

species (Anker et al. 2007). In each case, the other techniques were ancillary and supportive to the morphological descriptions of the new species. Very few studies rely entirely on genetics and phylogeny (Thoma et al. 2009; Mantelatto et al. 2009, for example), and these are largely to sort out generic-level relationships among species. A major exception exists, in the area of so-called cryptic species, in which morphology appears to be almost identical. Cryptic species differ in behavioral habits or subtle color patterns and in genetic sequences (Matthews 2006). It seems unlikely that such cryptic species will be recognizable in the fossil record with current methodology.

The major difference between morphological species descriptions of ancient and extant forms is that biologists use some characters that are rarely preserved in the fossil record, such as gill structure, reproductive organs, and mouth parts. Higher-level classification, such as at the genus and family level, is often accomplished by neontologists by using gonopores, gonopods, and other elements of the sternum and pleon. In the past, these were not commonly used by paleontologists (for example, see the diagnoses in Glaessner [1969], the first edition of the *Treatise*). However, better fossil preparation techniques, using micro-jack tools not available in the 19th century for example, and a general recognition that these are important for classification have led to widespread use of these features when they are preserved (i.e., Karasawa et al. 2008, 2011, 2013; Van Bakel et al. 2012).

Unfortunately, some material simply does not have these structures preserved, such as the Jurassic brachyuran material from Europe (i.e., Schweitzer and Feldmann 2009). For this type of material, Schweitzer (2003) applied the concept of "proxy characters," those characters not often described or mentioned by neontologists but that tracked often unpreservable morphology such as antennae, antennules, mouthparts, and genitalia. Structures of the orbits, the rostrum, the lateral and posterior margins of the carapace, and the groove patterns (fig. 9.2) have been found to be sufficiently distinctive for family-level classification.

Using proxy characters as well as morphological characters used by biologists when preserved, classification of fossil taxa appears to be relatively stable and concordant with neontological classification. Indeed, placement of fossil taxa based upon these proxy characters has been supported by phylogenetic analyses including both fossil and extant decapods (Karasawa and Schweitzer 2006; Karasawa et al. 2008, 2011, 2013).Reconciling biological and paleontological methods of recognizing species and higher-level taxa has thus become more congruent in the past 30 years.

Problems with Recognizing Fossil Decapoda Species

Despite the advances in decapod species and higher-level taxon recognition outlined above, several specific problems arise when working with the fossil decapod record.

The problem of single specimens

Many fossil decapod species are named based upon a single specimen, the holotype, yielding no chance to observe the range of variation in morphology, sexual dimorphism, or differences between juveniles and adults. Often only part of the animal is preserved, the dorsal carapace. Single-specimen status makes the taxon a true singleton. This status is generally regarded as less useful for assessing diversity and other comprehensive studies of patterns in paleobiology as compared to those species which are known from multiple specimens and multiple time periods. Singletons are usually omitted from diversity studies (Aberhan and Kiessling 2012). However, single-specimen species are rampant in the decapod record and cannot be excluded from comprehensive studies because to do so would greatly diminish the utility of such studies by excluding a large proportion of the decapod record. For example, four out of eight species within the long-ranging brachyuran family Homolodromiidae (middle Jurassic [Bathonian]–Holocene) (Schweitzer and Feldmann 2010a) are single specimen species.

The problem of recognizing sexual dimorphism

Sexual dimorphism among male and female brachyurans (crabs) has long been recognized in the shape of the pleon and in some aspects of the sternum. Male crabs and sometimes lobsters can possess large claws used for mating display (Feldmann 1998). Sexual dimorphism is also common among the Gebiidae de Saint Laurent, 1979, and Axiidea de Saint Laurent, 1979, the ghost and mud shrimp. However, this phenomenon has been recognized only recently in the fossil record (Schweitzer Hopkins and Feldmann 1997; Schweitzer, Feldmann, et al. 2006). Members of these two infraorders display clear secondary sexually dimorphic characteristics in the shape and size of the major chelae. Juveniles and adults exhibit differences in chela morphology; the juvenile chelae are less differentiated than the adult and even when male, appear intermediate between male and female chelae. Recognition of these dimorphic characters has led to the synonymy of several fossil species of ghost shrimp (Schweitzer Hopkins and Feldmann 1997).

Sexual dimorphism has also been recognized in the dorsal carapace of the Raninidae, in which males sometimes display larger and more ornamented lateral spines (Feldmann and Schweitzer 2007). These features can be observed in extant and fossil members of the family. In addition, widespread dimorphism can be observed in the narrow pleons of the Raninidae, which is subtle but can also be quantified and observed in fossils (Feldmann and Schweitzer 2007). Several species of the extinct genus *Lophoranina* Fabiani, 1910, currently recognized from the same locality, may in fact be males and females, for example, based upon these observations.

Lobsters display sexual dimorphism in a variety of ways. Scyllaridae Latreille, 1825, and Palinuridae Latreille, 1802, are almost always reported as being achelate on all appendages, but examination of extant, preserved specimens in the United States National Museum (March 2011) shows that females may have pseudochelate closures on the fifth pereiopod, probably for egg handling, which is not often reported in diagnoses or neontological literature on these animals. Schweigert (2001) reported dimorphism in the extinct genus *Cycleryon* Glaessner, 1965, a member of Eryonidae de Haan, 1841, in which the fifth pereiopods are achelate in males and chelate in females. It has long been known that relative breadth of the pleon of *Homarus americanus* H. Milne Edwards, 1837, is broader in females than in males (Herrick 1909), and that the pleonal pleurae are sharply terminated in males and rounded in females of glypheoid lobsters (Étallon 1859; Forest and de Saint Laurent 1989; Feldmann and de Saint Laurent 2002). Thus, sexual dimorphism must be considered when naming new species of fossil lobsters.

The problem of differential cuticle preservation

Recent studies on the microstructure of decapod cuticle have shown that the cuticle of the same animal can appear remarkably different depending on whether the exocuticle or endocuticle is preserved. The two layers of cuticle often display quite different levels of ornamentation. The exocuticle can be very heavily ornamented, with relatively smooth endocuticle (fig. 9.3A), and the opposite pattern can also be true, wherein the exocuticle smooths out the ornamentation better seen on the endocuticle. Molds of the interior of the carapace often appear very different as compared to

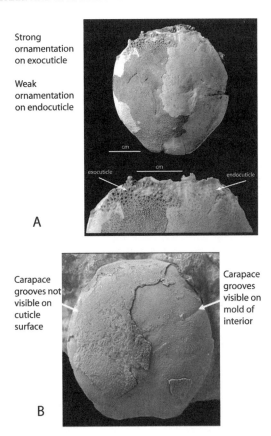

Strong ornamentation on exocuticle

Weak ornamentation on endocuticle

A

Carapace grooves not visible on cuticle surface

Carapace grooves visible on mold of interior

B

Figure 9.3 Differences in carapace ornamentation depending on cuticular preservation. **A**, *Jurellana tithonia* Schweitzer and Feldmann, 2010b, Naturhistorisches Museum Wien, catalog no. 1990/0041/2518, holotype, Tithonian, Austria. **B**, *Cyrtorhina fusseli* Blow and Manning, 1996, Charleston Museum, catalog no. 18558, Eocene, South Carolina.

specimens with any layer of cuticle preserved (fig. 9.3B). Thus, extreme caution must be taken in using ornamentation to differentiate among species of a genus, especially if there are few other characteristics upon which to base the species.

The problem of geographic isolation; or "it's a new locality, so it must be a new species!"

Of the 3679 decapod species arrayed in 1065 genera known in the fossil record, only about 200 are known from multiple, broadly geographically spaced localities (examination of data compiled from Schweitzer et al. 2010 and updated to December 2014). Most species are named from a single locality or set of localities separated by a few kilometers to tens of kilometers at most, generally in the same formation.

The apparent geographic isolation of species may be explained in part because mode of preservation within different rock units yields different preservational styles. Concretionary preservation, Lagerstätten, and crack-out-type preservation may result in specimens that are so different in appearance that they are interpreted as representing different species (fig. 9.4). The apparent geographic restriction might suggest that decapod species did not migrate through space and time and that invasion patterns cannot be examined as, for example, has been done for Ordovician rocks of the Cincinnati, Ohio, region (Stigall 2012). Diversity and biogeographic patterns at the species level cannot be examined if species are restricted to one locality. At this time, it seems judicious to conduct diversity, biogeographic, and other comprehensive studies on the genus level (Schweitzer 2001; Schweitzer et al. 2002; Fraaije 2003; Feldmann and Schweitzer 2006).

The problem of multiple species of the same genus in the same locality

Many of the most species-rich localities in the decapod record contain multiple species of the same genus (Rathbun 1935; Schweitzer and Feldmann 2009). In these cases, the species do not appear to be sexual dimorphs or juveniles and adults. Oversplitting of genera into too many species at the same locality may be occurring. However, multiple species living in the same environment does seem to have precedent in modern oceans. Multiple species of *Callinectes* Stimpson, 1862, for example, inhabit the east coast of the USA (as reported by Williams 1984), and their ranges overlap to a remarkable degree. Davie (2002: 448–449) reported that *Caphyra* Guérin-Méneville, 1832, a portunoid crab, with six species reported from Australia, included five on the Great Barrier Reef, Queensland; many were reported to inhabit the same genus of coral. Given these modern examples of co-occurring congeners, investigation of niche partitioning in extinct decapod genera is underway.

The problem of collector interest and bias

The decapod species record is likely strongly biased by collector bias, collector interest, and possibly by collector knowledge. Decapods within rocks often do not look like decapods or perhaps like anything interesting

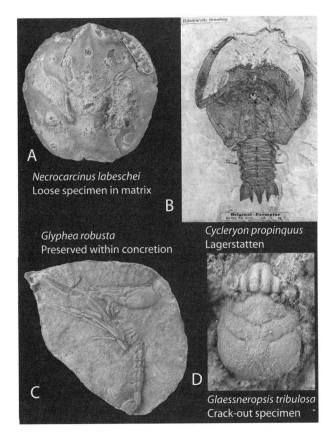

Figure 9.4 Four common preservational styles for Decapoda.
A, Specimens found loose in or eroded from matrix; example, *Necrocarcinus labeschei* (Eudes-Deslongchamps, 1835), Sedgwick Museum, Cambridge University, catalog no. 23152. Specimens, because they are usually three-dimensional, often retain portions of the ventral surface. **B**, Konservat-lagerstätten; example, *Cycleryon propinquus* (Schlotheim, 1822), from Solnfohen-type limestone deposits, Late Jurassic, Germany, Carnegie Museum of Natural History, Pittsburgh, catalog no. 34359. Specimen is flattened so that dorsal and ventral features are superimposed. **C**, Concretionary preservation, *Glyphea robusta* Feldmann and McPherson, 1980, Geological Survey of Canada, Eastern Paleontology Division, Ottawa, catalog no. 61398, holotype. Specimen is three-dimensional, but only about half of the morphology can be seen. **D**, Crack-out faunas, *Glaessneropsis tribulosa* Schweitzer and Feldmann, 2009, Naturhistorisches Museum Wien, catalog no. 1990/0041/272, holotype, Tithonian limestones, Austria. Only the dorsal surface is typically seen in these types of occurrences. Rarely is the entire specimen preserved or recoverable.

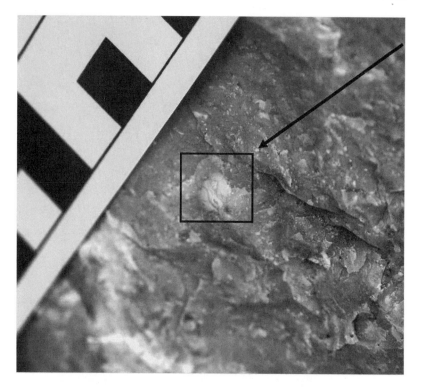

Figure 9.5 Tiny Jurassic brachyuran, probably *Goniodromites* sp., embedded in matrix.
Scale in upper left is in centimeters.

at all (fig. 9.5). This is especially true of such taxa as the Jurassic brachyurans, which we have been told on many occasions by collaborators and
other geologists and paleontologists do not look like crabs. These types of
faunas are usually so-called crack-out faunas, which are collected by sit-
ting down and smashing rocks for several days. These faunas are difficult
to see, find, and collect.

Many fossil decapods are very small, under a few centimeters. For example, one of us (RMF) visited a well-known locality for large *Chaceon*
Manning and Holthuis, 1989, fossils in Argentina, a species that is typically 10–15 cm across. Hundreds had been collected, but totally missed
were the hundreds of other concretions, 2–3 cm in size, which yielded an
entirely new fauna that had been overlooked in favor of the larger, showy
Chaceon (Schweitzer and Feldmann 2000a,b,c; 2001a).

Collector interest is also a factor in the number of decapod species
named per stage and in various geographic areas. As a specific example,

work on more derived groups in the decapod fossil record accelerated more recently, probably due to the fact that the rocks containing them, including those of Eocene, Oligocene, or Miocene age, are not always where the main "science" was happening in the 1800s (Germany, Britain). Large exposures of rocks of these ages are in Italy, Hungary, Spain, Argentina, Chile, New Zealand, and Japan, which have received intensive study much more recently. Amateur collectors in the Pacific Northwest of North America, both American and Canadian, have led to large increases in the number of Paleogene and Neogene species known from that region in recent decades (e.g., Tucker 1998; Schweitzer and Feldmann 2000a, b, c; Schweitzer et al. 2003). Collector interest in the Late Jurassic has also been intense, not only in the 19th century but also recently (e.g., Schweigert 2001; Schweitzer and Feldmann 2009). Interest in the Eocene of Italy by Italian workers has increased the number of Eocene taxa from the country by hundreds (i.e., Beschin et al. 2007). Thus, collector interest has had a profound effect on the number of recognized species.

The problem of missing or destroyed material

Many named species, mostly from the 19th century, cannot be verified because the type or other illustrated material has been lost or destroyed. For example, fifteen species have been synonymized with *Eryma bedeltum* (Quenstedt, 1857) (Schweitzer et al. 2010: 23). However, it may be difficult to determine the full extent of the number of species-level synonymies that should exist, due to the destruction of many European collections and specimens during the wars of the first half of the 20th century. Original type material, which was often the only material of the species that existed, was destroyed in many cases. In other cases, many of the specimens were held in private collections that have since been lost. Thus, in many instances species are currently recognized that may be synonymous with others, but there remains no real method by which to test this notion because the type and only specimens were destroyed.

The problem of species named for different morphological parts

Many decapod species are named based only upon claws. Especially susceptible were the ghost and mud shrimp, commonly reported as *Callianassa* Leach, 1814 sensu lato, prior to 1935. Species in almost all other callianassid genera were named after 1987, undoubtedly due to the work of Raymond Manning and Darryl Felder (1991) on revising the American

Callianassidae Dana, 1852, providing characteristics of the merus, carpus, and manus that could possibly be observed in fossils, and also restricting the genus *Callianassa* considerably. Their work called attention to the existence of ghost and mud shrimp genera other than *Callianassa* and revolutionized the identification of fossil ghost shrimp. For those *Callianassa* spp. named prior to 1935, unless more material is recovered, we may never know the true generic placement of the species. Further, many of those species may be sexual dimorphs.

A further issue is the problem that arises when claws receive a species name and the carapace of a potentially conspecific decapod receives a different name. If specimens are recovered in which the claw and carapace are preserved together, a synonymy can be made. More commonly, the relationship between claws and carapace must be inferred by their occurrence at the same locality or in the same rock unit.

A related problem is establishment of a new species based upon fragmentary or poorly preserved material, especially common in 19th- and early 20th-century literature. Many of the genera and species in Van Straelen (1925) are fragmentary and poorly preserved, and the type material is lost. Thus, the species are valid but there is no way to evaluate them.

The problem of look-alikes

Our experience has shown that there are several families of brachyurans (crabs) with rectangular, relatively feature-less dorsal carapaces. It can be very difficult to classify these fossils in the absence of sternal or abdominal elements, especially when more than one taxon of this type occurs in the same formation (Schweitzer and Feldmann 2001b; Feldmann et al. 2010; Feldmann, Schweitzer, Casadío, and Griffin et al. 2011). Other families retain genera that are quite similar, such as Carpiliidae, and Eocene localities can yield many species within different genera of a single family (Feldmann, Schweitzer, Bennett et al. 2011). It can be easy to overlook these differences and misidentify species or underestimate species diversity.

Brief History of Species Recognition

In spite of issues mentioned above, the cumulative curve of named fossil decapod species continues to rise (fig. 9.6). Although it has shown times of slowing, notably during the period before, during, and after World War II, the curve shows no signs of leveling off, and in fact, in the past 40 years it has risen considerably faster than it ever has. This recent period has seen substantial interest in the decapod crustaceans. Improved preparation techniques and the ability to collect in areas that were for various reasons inaccessible (political, undeveloped, etc.) has led to an explosion in described taxa. The integration of Eastern Europe into the European Union, providing access for Eastern workers to the West (e.g., Franțescu 2011; Hyžný and Schlögl 2011) as well as easier access to collections in China (Feldmann, Schweitzer et al. 2012), Iran (Feldmann et al. 2007; Yazdi et al. 2010), and Cuba (Schweitzer, Feldmann, et al. 2006) are good examples. The numbers of named families and genera have followed a slightly different pattern from that of species, each showing a slight flattening after an initial burst in the 19th century, with a renewed blossoming of taxon naming over the last 20 or so years (fig. 9.6).

For comparison to modern numbers within Decapoda, De Grave et al. (2009) reported 162 families of decapods from the fossil record out of a total (extinct and extant) of 233; herein, we report 175 families known from the fossil record. De Grave et al. reported 2725 genera, extinct and extant. We report 1065 genera known from the fossil record, extinct species and extant species with a fossil record included.

The numbers are far different for species, as is perhaps predictable. As of December 2014, 3679 decapod species were reported from the fossil record, including some that also were known to be extant. De Grave et al. (2009) reported 17,635 species, both extant and extinct. Noting that the De Grave et al. number is slightly undervalued because it is three years older than the December 2014 number, it is clear that the fossil record is much less diverse than the modern record, considering that the fossils, not counting the three taxa known from the Paleozoic, are spread over about 250 million years. Perhaps then, it should not be surprising that the number of fossil species being described is still climbing!

Within families, the number of fossil species is quite variable. Palinuridae, which has a record spanning 245 million years, has an approximately equal number of fossil and extant species. Raninidae de Haan, 1839, spanning approximately 112 million years, has about 4.5 times as many fossil species as extant ones (table 9.1). Thus, it is not possible to determine just from examining numbers of species within families whether we are overestimating (i.e., splitting) or underestimating (i.e., lumping) the number of fossil species within Decapoda.

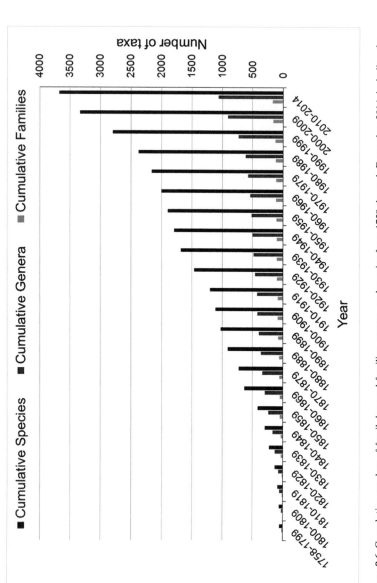

Figure 9.6 Cumulative number of fossil decapod families, genera, and species from 1758 through December 2014, including those that are both extant and extinct.

TABLE 9.1 Selected decapod families, number of Recent and fossil species within each family, and stratigraphic range of each family.

Family	Common name	Recent species	Fossil species	Oldest fossil species	Duration of family
Palinuridae	spiny lobsters	55	56	Anisian (Middle Triassic)	245 my
Scyllaridae	slipper lobsters	85	13	Albian (Early Cretaceous)	112 my
Paguridae	hermit crabs	510	62	Tithonian (Late Jurassic)	150 my
Portunidae	swimming crabs	332	146	Ypresian (Eocene)	55 my
Nephropidae	clawed lobsters	49	90	Berriasian (Early Cretaceous)	145 my
Glypheidae		2	83	Olenekian (Early Triassic)	249 my
Raninidae	frog crabs	39	186	Albian (Early Cretaceous)	112 my
Carpiliidae	Coral crabs, queen crabs, reef crabs (http://www.itis.gov)	4	30	Ypresian (Eocene)	55 my

Discussion and Conclusions

Decapod species in the fossil record are recognized, therefore, based primarily upon morphology. There is a tendency to erect new fossil decapod species from new localities, in most cases meaning a new rock unit or new age for the genus embraced. The problems related to fossil decapod species recognition make it most viable, when working on comprehensive studies such as paleobiogeography and diversity, to work at the genus level with fossil data. This eliminates the issues of singletons, lack of knowledge of range of variation among and between individuals, and geographic isolation. Genus data provide a more robust dataset, necessary in a group with a high percentage of singletons. Working at the genus level requires strong knowledge of the biological and paleontological literature so that morphological characters used by biologists and paleontologists are familiar to and well-known by the researchers. The increasing integration of biologically important morphological characters into the paleontological literature has led to phylogenies for both extant and extinct decapods that are congruent with those that have been generated solely based upon genetic data (Karasawa and Schweitzer 2006; Karasawa et al. 2008, 2011, 2013; Bracken-Grissom et al. 2013). Thus, despite the problems associated with recognition of fossil decapod species, current paleontological methodology seems to be consistent with biological conclusions.

Acknowledgments

W. Allmon, Paleontological Research Institution, Ithaca, NY, and M. Yacobucci, Bowling Green State University, OH, invited the authors to present a talk at the 2011 Annual Meeting of the Geological Society of America that led to this paper. Conversations with Allmon and A. Klompmaker, then at KSU and now at University of Florida at Gainesville, contributed to this manuscript. Sven Tänkner of the NaturMuseum Senckenberg, Frankfurt, Germany, provided the photo of *Astacus astacus*. R. Lemaitre and K. Reed (Crustacea) and M. Florence (Paleobiology), Smithsonian Institution, United States National Museum of Natural History, Washington, DC; A. Kollar, Invertebrate Paleontology, Carnegie Museum of Natural History, Pittsburgh, Pennsylvania; A. Kroh, Geological and Palaeontological Department of the Naturhistorisches Museum Wien; M. Coyne, Geological Survey of Canada, Ottawa, Ontario, Canada; M. Riley, Sedgwick Museum, Cambridge University, United Kingdom; and A. E. Sanders, Charleston Museum, South Carolina, USA, facilitated access to their collections. NSF grant EF-0531670 to the authors funded museum research, examining thousands of specimens, which led to development of the database upon which this manuscript is based. We thank the dozens of curators and collections managers who made our work possible. D. Brandt, Michigan State University, Yacobucci, and an anonymous reviewer provided constructive reviews of the manuscript.

References

Aberhan, M., and W. Kiessling. 2012. Phanerozoic marine biodiversity: a fresh look at data, methods, patterns and processes. In *Earth and Life*, ed. J. A. Talent, 3–22. New York: Springer Science+Business Media.

Ahyong, S. T. 2012. Polychelid lobsters (Decapoda: Polychelida: Polychelidae) collected by the CIDARIS expeditions off Central Queensland, with a summary of Australian and New Zealand distributions. *Memoirs of the Queensland Museum—Nature* 56(1): 1–8.

Anker, A., C. Hurt, and N. Knowlton. 2007. Three transisthmian snapping shrimps (Crustacea: Decapoda: Alpheidae: *Alpheus*) associated with innkeeper worms (Echiura: Thalassematidae) in Panama. *Zootaxa* 1626: 1–23.

Beschin, C., A. Busulini, A. De Angeli, and G. Tessier. 2007. I Decapodi dell'Eocene inferiore di Contrada Gecchelina (Vicenza—Italia settentrionale) (Anomura e Brachiura). *Museo di Archeologia e Scienze Naturali "G. Zannato", Montecchio Maggiore (Vicenza)* 2007: 9–76.

Blow, W. C., and R. C. Manning. 1996. Preliminary descriptions of 25 new decapod crustaceans from the Middle Eocene of the Carolinas, U.S.A. *Tulane Studies in Geology and Paleontology* 29 (1): 1–26, pls. 1–5.

Bracken-Grissom, H. D., M. E. Cannon, P. Cabezas, R. M. Feldmann, C. E. Schweitzer, S. T. Ahyong, D. L. Felder, R. Lemaitre, and K. A. Crandall. 2013. A comprehensive and integrative reconstruction of evolutionary history for Anomura (Crustacea: Decapoda). *BMC Evolutionary Biology*, 13: 128. doi:10.1186/1471-2148-13-128.

Dana, J. D. 1852. Parts I and II, Crustacea. U.S. Exploring Expedition During the Years 1838, 1839, 1840, 1841, 1842, under the Command of Charles Wilkes, U.S.N., 13, 1–1618, 1 map; separate folio atlas with 96 pls. Philadelphia: C. Sherman.

Davie, P. J. F. 2002. Crustacea: Malacostraca: Eucarida (Part 2): Decapoda Anomura, Brachyura. In *Zoological Catalogue of Australia 19.3B*, ed. A. Wells and W. W. K. Houston, 1–641. Melbourne, Australia: CSIRO Publishing.

Davie, P. J. F., and P. K. L. Ng. 2013. A review of *Chiromantes obtusifrons* (Dana, 1851) (Decapoda: Brachyura: Sesarmidae), with descriptions of four new sibling-species from Christmas Island (Indian Ocean), Guam and Taiwan. *Zootaxa* 3609: 1–25.

De Grave, S., N. D. Pentcheff, S. T. Ahyong, T.-Y. Chan, K. A. Crandall, P. C. Dworschak, D. L. Felder, R. M. Feldmann, C. H. J. M. Fransen, L. Y. D. Goulding, R. Lemaitre, M. L. Low, J. W. Martin, P. K. L. Ng, C. E. Schweitzer, S. H. Tan, D. Tshudy, and R. Wetzer. 2009. A classification of Recent and fossil genera of decapod crustaceans. *Raffles Bulletin of Zoology Supplement* 21: 1–109.

De Haan, W. 1833–1850. Crustacea. In *Fauna Japonica sive Descriptio Animalium, quae in Itinere per Japoniam, Jussu et Auspiciis Superiorum, qui summum in India Batava Imperium Tenent, Suscepto, Annis 1823–1830 Collegit, Notis, Observationibus et Adumbrationibus Illustravit*, ed. P. F. von Siebold, i–xvii, i–xxxi, ix–xvi, 1–243, pls. A–J, L–Q, 1–55, circ. tab. 2. Lugduni Batavorum [= Leyden]: J. Müller et Co.

De Saint Laurent, M. 1979. Sur la classification et la phylogénie des Thalassinides: définition de la superfamille des Axioidea, de la sous-famille des Thalassininae et deux genres nouveaux (Crustacea Decapoda). *Comptes Rendus Hebdomadaires des Séances de l'Académie des Sciences, Paris* (D) 288: 1395–1397.

Étallon, A. 1859. Description des Crustacés de la Haute Saône et du Haut-Jura. *Bulletin de la Societé Géologique de France* (2) 16 [1858]: 169–205, pls. 3–6.

Eudes-Deslongchamps, J. A. 1835. Mémoire pour servir à l'histoire naturelle des Crustacés fossils. *Mémoire de la Societé Linnéenne de Normandie* 5: 37–46, pl. 1.

Fabiani, R. 1910. I crostacei terziari del Vicentino. *Bolletin Museo Civico Vicenza* 1: 1–40.

Feldmann, R. M. 1998. Parasitic castration of the crab, *Tumidocarcinus giganteus* Glaessner, from the Miocene of New Zealand: coevolution within the Crustacea. *Journal of Paleontology* 72: 493–498.

Feldmann, R. M., and M. de Saint Laurent. 2002. *Glyphaea forest* n. sp. (Decapoda) from the Cenomanian of Northern Territory, Australia. *Crustaceana* 75 (3–4): 359–373.

Feldmann, R. M., and C. B. McPherson. 1980. Fossil decapod crustaceans of Canada. *Geological Survey of Canada Paper* 79–16: 1–20.

Feldmann, R. M., and C. E. Schweitzer. 2006. Paleobiogeography of Southern Hemisphere decapod Crustacea. *Journal of Paleontology* 80: 83–103.

Feldmann, R. M., and C. E. Schweitzer. 2007. Sexual dimorphism in fossil and extant Raninidae (Decapoda: Brachyura). *Annals of Carnegie Museum* 76: 39–52.

Feldmann, R. M., C. E. Schweitzer, and A. Encinas. 2010. Neogene decapod Crustacea from southern Chile. *Annals of Carnegie Museum* 78: 337–366.

Feldmann, R. M., A. Kolahdouz, B. Biranvand, and G. Schweigert. 2007. A new family, genus, and species of lobster (Decapoda: Achelata) from the Gadvan Formation (Early Cretaceous) of Iran. *Journal of Paleontology* 81: 405–407.

Feldmann, R. M., C. E. Schweitzer, S. Casadío, and M. Griffin. 2011. New Miocene Decapoda (Thalassinidea; Brachyura) from Tierra del Fuego, Argentina: paleobiogeographic implications. *Annals of Carnegie Museum* 79: 91–123.

Feldmann, R. M., C. E. Schweitzer, O. Bennett, O. Franţescu, N. Resar, and A. Trudeau. 2011. Decapod crustaceans from the Eocene of Egypt. *Neues Jahrbuch für Geologie und Paläontologie, Abhandlung* 262: 323–353.

Feldmann, R. M., C. E. Schweitzer, S. Hu, Q. Zhang, C. Zhou, T. Xie, J. Huang, and W. Wen. 2012. Decapoda from the Luoping biota (Middle Triassic) of China. *Journal of Paleontology* 86: 425–441.

Forest, J., and M. de Saint Laurent. 1989. Nouvelle contribution à laconnaissance de *Neoglyphea inopinata* Forest & de Saint Laurent, à propos de la description de la femelle adulte. In *Résultats des Campagnes MUSORSTOM 5*, ed. J. Forest. *Mémoires du Muséum national d'Histoire naturelle, Paris (A)* 144: 75–92.

Fraaije, R. H. B. 2003. Evolution of reef-associated decapod crustaceans through time, with particular reference to the Maastrichtian type area. *Contributions to Zoology* 72(2–3): 119–130.

Franţescu, O. D. 2011. Brachyuran decapods (including five new species and one new genus) from Jurassic (Oxfordian-Kimmeridgian) coral reef limestones from Dobrogea, Romania. *Neues Jahrbuch für Geologie und Paläontologie, Abhandlungen* 259:271–297.

Glaessner, M. F. 1965. Vorkommen fossiler Dekapoden (Crustacea) in Fisch-Schiefern. *Senckenbergiana Lethaia* 46 (a): 111–122.

Glaessner, M. F. 1969. Decapoda. In *Treatise on Invertebrate Paleontology, R (4) (2)*, ed. R. C. Moore, R400–R533, R626–R628. Boulder, CO: Geological Society of America; Lawrence: University of Kansas Press.

Gouws, G., B. A. Stewart, and P. E. Reavell. 2001. A new species of freshwater crab (Decapoda, Potamonautidae) from the swamp forests of Kwazulu-Natal, South Africa: biochemical and morphological evidence. *Crustaceana,* 74: 137–160.

Guérin-Méneville, F. E. 1832. Notice sur quelques modifications à introduire dans les Notopodes de M. Latreille et établissement d'un nouveau genre dans cette tribu. *Annales des Sciences naturelles* 25: 283–289.

Herrick, F. H. 1909. Natural history of the American lobster. *Bulletin of the Bureau of Fisheries* 29: 149–408, pls. 33–47.

Hyžný, M., and J. Schlögl. 2011. An early Miocene deep-water decapod crustacean faunule from the Vienna Basin (Western Carpathians, Slovakia). *Palaeontology* 54: 323–349.

Karasawa, H., and C. E. Schweitzer. 2006. A new classification of the Xanthoidea *sensu lato* (Crustacea: Decapoda: Brachyura) based on phylogenetic analysis and traditional systematics and evaluation of all fossil Xanthoidea *sensu lato*. *Contributions to Zoology* 75 (1/2): 23–73.

Karasawa, H., C. E. Schweitzer, and R. M. Feldmann. 2008. Revision of the Portunoidea Rafinesque, 1815 (Decapoda: Brachyura) with emphasis on the fossil genera and families. *Journal of Crustacean Biology* 28: 82–127.

Karasawa, H., C. E. Schweitzer, and R. M. Feldmann. 2011. Phylogenetic analysis and revised classification of podotrematous Brachyura (Decapoda) including extinct and extant families. *Journal of Crustacean Biology* 31: 523–565.

Karasawa, H., C. E. Schweitzer, and R. M. Feldmann. 2013. Phylogeny and systematics of extant and extinct lobsters. *Journal of Crustacean Biology* 33: 78–123.

Latreille, P. A. 1802–1803. *Histoire naturelle, générale et particulière, des Crustacés et des Insectes* 3: 1–468. Paris: F. Dufart.

Latreille, P. A. 1825. *Entomologie, ou histoire naturelle des Crustacés, des Arachnides et des Insectes. In Genre de Crustacés. Encyclopédie méthodique. Histoire naturelle* 10: 1–832. Paris.

Leach, W. E. 1814. Crustaceology. In *Edinburgh Encyclopaedia, 7*, ed. D. Brewster, 383–437, pl. 221. Edinburgh.

Linnaeus, C. von. 1758. *Systema Naturae per Regna tria Naturae, secundum classes, ordines, genera, species, cum characteribus, differentiis, synonymis, locis* (ed. 10), 1: 1–824. Holmiae [= Stockholm]: Laurentii Salvii.

Manning, R. B., and D. L. Felder. 1991. Revision of the American Callianassidae (Crustacea: Decapoda: Thalassinoidea). *Proceedings of the Biological Society of Washington* 104: 764–792.

Manning, R. B., and L. B. Holthuis. 1989. Two new genera and nine new species of geryonid crabs (Crustacea, Decapoda, Geryonidae). *Proceedings of the Biological Society of Washington* 102: 50–77.

Mantelatto, F. L., R. Robles, C. D. Schubart, and D. L. Felder. 2009. Molecular phylogeny of the genus *Cronius* Stimpson, 1860, with reassignment of *C. tumidulus* and several American species of *Portunus* to the genus *Achelous* De Haan, 1833 (Brachyura: Portunidae). In *Decapod Crustacean Phylogenetics*, ed. J. W. Martin, K. A. Crandall, and D. L. Felder, 567–579. Boca Raton, FL: CRC Press.

Matthews, L. M. 2006. Cryptic biodiversity and phylogeographical patterns in a snapping shrimp species complex. *Molecular Ecology*, 15: 4049–4063.

Milne Edwards, H. 1834–1840. *Histoire naturelle des Crustacés, comprenant l'anatomie, la physiologie, et la classification de ces animaux* 1 [1834]: 1–468; 2 [1837] : 1–532; 3 [1840]: 1–638, Atlas: 1–32, pls. 1–42.

Quenstedt, F. A. 1856–1858. *Der Jura*, 1–842, 100 pls. Tübingen: Verlag der H. Lauppschen Buchhandlung.

Rathbun, M. J. 1935. Fossil Crustacea of the Atlantic and Gulf Coastal Plain. *Geological Society of America, (Special Paper)* 2: i-viii, 1–160.

Schlotheim, E. F. 1822. Nachträge zur Petrefactenkunde. *Beiträge zur näheren Bestimmung der versteinerten und fossilen Krebsarten*. Gotha, Thuringia: Becker.

Schweigert, G. 2001. Dimorphismus bei Krebsen der Gattung *Cyleryon* (Decapoda, Eryonidae) aus dem Oberjura Süddeutschlands. *Stuttgarter Beiträge zur Naturkunde, Serie B, Geologie und Paläontologie* 305: 1–21.

Schweitzer, C. E. 2001. Paleobiogeography of Cretaceous and Tertiary decapod crustaceans of the North Pacific Ocean. *Journal of Paleontology* 75: 808–826.

Schweitzer, C. E. 2003. Utility of proxy characters for classification of fossils: an example from the fossil Xanthoidea (Crustacea: Decapoda: Brachyura). *Journal of Paleontology* 77: 1107–1128.

Schweitzer, C. E., and R. M. Feldmann. 2000a. First notice of the Chirostylidae (Decapoda) in the fossil record and new Tertiary Galatheidae (Decapoda) from the Americas. *Bulletin of the Mizunami Fossil Museum* 27: 147–165.

Schweitzer, C. E., and R. M. Feldmann. 2000b. Reevaluation of the Cancridea Latreille, 1803 (Decapoda: Brachyura) including three new genera and three new species. *Contributions to Zoology* 69: 233–250.

Schweitzer, C. E., and R. M. Feldmann. 2000c. New fossil portunids from Washington, USA, and Argentina and a reevaluation of generic and family relationships within the Portunoidea Rafinesque (Decapoda: Brachyura). *Journal of Paleontology* 74: 636–653.

Schweitzer, C. E., and R. M. Feldmann. 2001a. New Cretaceous and Tertiary decapod crustaceans from western North America. *Bulletin of the Mizunami Fossil Museum* 28: 173–210.

Schweitzer, C. E., and R. M. Feldmann. 2001b. Differentiation of fossil Hexapodidae Miers (Decapoda: Brachyura) from similar forms. *Journal of Paleontology* 75(2): 330–345.

Schweitzer, C. E., and R. M. Feldmann. 2009. Revision of the Prosopinae sensu Glaessner, 1969 (Crustacea: Decapoda: Brachyura) including 4 new families and 4 new genera. *Annalen des Naturhistorischen Museums in Wien* (A) 110: 55–121.

Schweitzer, C. E., and R. M. Feldmann. 2010a. The oldest Brachyura (Decapoda: Homolodromioidea, Glaessneropsoidea) (Jurassic: Pliensbachian, Bajocian, Bathonian) known to date. *Journal of Crustacean Biology* 30: 251–256.

Schweitzer, C. E., and R. M. Feldmann. 2010b. Earliest known Porcellanidae

(Decapoda: Anomura: Galatheoidea) (Jurassic: Tithonian). *Neues Jahrbuch für Geologie und Paläontologie, Abhandlungen*, 258: 243–248.

Schweitzer, C. E., R. M. Feldmann, A. Encinas, and M. Suárez. 2006. New Cretaceous and Eocene Callianassoidea (Thalassinidea, Decapoda) from Algarrobo, Chile. *Journal of Crustacean Biology* 26: 73–81.

Schweitzer, C. E., R. M. Feldmann, G. Gonzáles-Barba, and F. J. Vega. 2002. New crabs from the Eocene and Oligocene of Baja California, México and an assessment of the evolutionary and paleobiogeographic implications of Mexican fossil decapods. *Paleontological Society Memoir* 59 (supplement to *Journal of Paleontology* vol. 76): 43 pp.

Schweitzer, C. E., M. Iturralde-Vinent, J. L. Hetler, and J. Velez-Juarbe. 2006. Oligocene and Miocene decapods (Thalassinidea and Brachyura) from the Caribbean. *Annals of Carnegie Museum* 75: 111–136.

Schweitzer, C. E., R. M. Feldmann, A. Garassino, H. Karasawa, and G. Schweigert. 2010. *Systematic list of fossil decapod crustacean species. Crustaceana Monographs*, 10, 1–222. Leiden: Brill.

Schweitzer, C. E., R. M. Feldmann, J. Fam, W. A. Hessin, S. W. Hetrick, T. G. Nyborg, and R. L. M. Ross. 2003. *Cretaceous and Eocene decapod crustaceans from southern Vancouver Island, British Columbia, Canada*: 1–66. Ottawa, Ontario: NRC Research Press.

Schweitzer Hopkins, C., and R. M. Feldmann. 1997. Sexual dimorphism in fossil and extant species of *Callianopsis* de Saint Laurent. *Journal of Crustacean Biology* 17(2): 236–252.

Stenzel, H. B. 1945. Decapod crustaceans from the Cretaceous of Texas. *University of Texas Publication* 4401: 401–477.

Stigall, A. L. 2012. Speciation collapse and invasive species dynamics during the Late Devonian "Mass Extinction." *GSA Today* 22(1): 4–9.

Stimpson, W. 1862. Notes on North American Crustacea, in the Museum of the Smithsonian Institution. No. II. *Annals of the Lyceum of Natural History of New York* 7 [1862]: 176–246, pls. 2, 5 [read April 1860]. [Pages 49–118, pls. 2, 3 on separate.]

Thoma, B. P., C. D. Schubart, and D. L. Felder. 2009. Molecular phylogeny of Western Atlantic representatives of the genus *Hexapanopeus* (Decapoda: Brachyura: Panopeidae).In *Decapod Crustacean Phylogenetics*, ed., J. W. Martin, K. A. Crandall, and D. L. Felder, 551–565. Boca Raton, FL: CRC Press, Taylor and Francis Group.

Tucker, A. B. 1998. Systematics of the Raninidae (Crustacea: Decapoda: Brachyura), with accounts of three new genera and two new species. *Proceedings of the Biological Society of Washington* 111: 320–371.

Van Bakel, B. W. M., D. Guinot, P. Artal, R. H. B. Fraaije, and J. W. M. Jagt. 2012. A revision of the Palaeocorystoidea and the phylogeny of raninoidian crabs (Crustacea, Decapoda, Brachyura, Podotremata). *Zootaxa* 3215: 1–216.

Van Straelen, V. 1925. Contribution à l'étude des Crustacés Décapodes de la période jurassique. *Mémoires d'Académie Royale de Belgique*, (Science), [collected in number 4, series 2 = (2) 4] 7: 1–462, pls. 1–10.

Weller, J. M. 1961. The species problem. *Journal of Paleontology*, 35: 1181–1192.

Williams, A. B. 1984. *Shrimps, lobsters, and crabs of the Atlantic Coast of the Eastern United States*, 1–550.Washington, DC: Smithsonian Institution Press.

Yazdi, M., A. Bahrami, and F. J. Vega. 2010. Additions to Albian (Cretaceous) Crustacea from Iran. *Boletín de la Sociedad Geológica Mexicana* 62 (2): 207–211.

CHAPTER TEN

Fossil Species as Data: A Perspective from Echinoderms

William I. Ausich

Introduction

Our intuition is that species-level fossil data are problematic for studying the tempo and mode of evolutionary history. Paleontologists working in the 19th and early 20th centuries operated under a paradigm for species concepts that allowed very little intraspecific variability, thus leading to considerable oversplitting at the species level. The resolution of species-level data is further degraded by time averaging, the incomplete fossil record, poor and uneven sampling, among other things. All of these factors conspire to yield an incomplete fossil record, especially at the species level, and result in the use of genus or higher-level taxonomic data for most taxic paleobiology (e.g., Sepkoski, 1981, 2002). The resolution of species-level data is certainly poor globally for answering many questions, but this should not be assumed in all cases.

In this chapter, the fidelity of species-level data is considered in order to evaluate the usefulness of species for answering evolutionary and paleobiological questions. Although species-level data are problematic for many studies, significant information is recorded by fossil species. A fossil species represents a unique morphological entity in time and space, despite varying degrees of resolution for species as a whole.

The Incomplete Fossil Record

Although we have a staggering knowledge of prehistoric life, the fossil record is incomplete. Our comprehensive understanding of fossil species on a global basis is terrible; this is unavoidable. The fossil record is composed primarily of successful organisms with hard parts whose abundance and geographic distribution was sufficiently high to have persisted through the preservation milieu and become part of the fossil record. Soft-bodied organisms, lightly skeletonized organisms, organisms with a multipart skeleton, organisms that lived in environments where erosion was the dominant process, etc. are all significantly underrepresented. Further, paleontological data are subject due to a variety of factors, such as having only certain paleoenvironments well represented (e.g., for the Paleozoic marine record data from epicontinental seas dominates the fossil record) (e.g., Smith et al., 2001), and fossil data have not been sampled evenly (Raup, 1976; Peters and Foote, 2001, 2002; Smith, 2001, and others). These and other sampling biases have led to the common use of post-hoc sampling evaluation, such as rarefaction and resampling statistics (e.g., Alroy et al, 2001; Peters and Foote, 2001, 2002).

Although species-level data are of dubious completeness for many types of studies, there are circumstances where this is not the case. This is especially true in oversampled regions, such as the eastern United States, that have been heavily collected for more than 150 years. All but the most rare taxa may be well known, with intrapopulation variability well constrained. Clearly, new species remain undescribed and species-level data require modern systematic treatment; but in a region, a specific clade may be well sampled and understood. Mississippian crinoids of the eastern United States are an example of a well-documented fauna. The first Mississippian crinoid species from the eastern United States was described in the 1840s. By 1899, 70 percent of presently known species names from this region were described; 90 percent of all Mississippian species from eastern North America were described by 1980 (Ausich, 2009) (fig. 10.1). Many of these names are no longer valid, and relatively few new taxa have been discovered in recent decades. Arguably, we can consider Mississippian crinoids from the eastern United States to be well sampled.

In special circumstances, even recent collecting efforts can be evaluated for completeness. Ausich and Copper (2010) monographed Late Ordovician through Early Silurian crinoids from Anticosti Island, Quebec,

Naming History of Lower Mississippian Crinoid Species

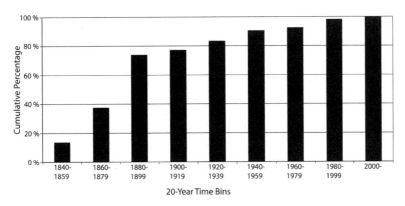

Figure 10.1 Cumulative percentage of described (nominal) North American, Lower Mississippian crinoid species in 20-year bins (from Ausich, 2009).

Canada. Logistically, Anticosti Island is a relatively involved locale to visit, all major collections are known, and expeditions designed to collect echinoderms had never been attempted until the work leading to Ausich and Copper (2010). Based on the collecting history of Anticosti crinoids, comparing number of taxa, number of specimens, and an estimate of relative time spent collecting in each formation, Ausich (2010) argued that Anticosti Island crinoid data were sufficiently well sampled to provide meaningful data for analysis of local evolutionary and paleobiologic patterns during the Late Ordovician and Early Silurian. Whereas sample histories cannot be estimated in most circumstances, this example illustrates that the completeness and fidelity of species-level data may not be hopeless in all cases.

The Species Question

Settling on a single agreed-on definition for a species has eluded biologists. Benton and Pearson (2001), Hey (2001), Coyne and Orr (2004), and de Queiroz (2007) listed more than 20 species definitions that vary as a consequence of the data used for definition or the questions being asked. Two basic concepts are most applicable to fossils: the evolutionary species concept (Simpson, 1951; Wiley, 1978; and Mayden, 1997) and the diagnosable species concept (Nelson and Platnick, 1981; Cracraft, 1983; Nixon and Wheeler, 1990) (see de Queiroz, 2007). Despite all of the issues that degrade the species-level fossil record, species using the diagnosable concept are defined by a unique set of discrete characters. Whether this is, in fact, a unique set of characters should be validated by modern systematic treatment. In this definition, species distinction and defining characters are testable hypotheses, because species diagnostic characters are commonly an array of quantitative and qualitative characters that can be analyzed statistically and continually tested as new data become available. Thus, a described species can be considered to be on a peak of the morphologic landscape that is distinct from other peaks. The punctuated equilibrium speciation concept (Eldredge and Gould, 1972) predicts stable morphology through a species' duration, which should render it diagnosable. Further, the fact that specimens have been proposed as hybrids (Ausich and Meyer, 1994; Goodfield and Gould, 1996) underscores the distinctive position in morphospace of well-diagnosed species.

The diagnosable species concept may become problematic if finely resolved temporal sampling yields more gradual morphological transitions between species (Gingerich, 1974) or punctuated gradualism (Malmgren et al., 1983). However, because of the nature of the fossil record, morphological gradations are relatively uncommon in many groups.

Regardless of the incompleteness of the fossil record, a well-defined species represents a unique position in a time and morphospace. A species defined with a time-averaged sample simply represents a unique position in morphospace averaged through several ecological time slices, arguably eliminating noise in understanding the morphology of species. Consequently, species-level characters are valid data for examining patterns in the fossil record that are dependent on morphological characters, such as phylogenetic, disparity, or biostratigraphic studies. The species fossil record is incomplete; but well-defined species can be regarded as exemplars for many types of analyses.

Reliability of Historic Species Concepts

Kammer and Ausich (e.g., 2006) have been engaged in a research program to evaluate species-level evolutionary patterns in North American Mississippian crinoids across the Osagean-Meramecian (early Viséan–middle Viséan) boundary, which was during the macroevolutionary transition between the Middle and Late Paleozoic crinoid evolutionary faunas (CEF)

(Baumiller, 1993; Ausich et al., 1994). This involved revision of species-level systematics of late Osagean crinoids in the Eastern Interior Basin and the surrounding area (Missouri, Iowa, Illinois, Indiana, Kentucky, Tennessee) (e.g., Ausich and Kammer, 1990, 1991; Kammer and Ausich, 1992, 1993; Ausich et al., 1994; Kammer et al., 1997, 1998). The first study of these crinoid faunas was by Hall (1858, 1859, 1860, 1861a, 1861b), who described the fauna in the Mississippian type region along the Mississippi River valley. In this initial contemplation of these crinoids, Hall, figuratively, if not literally, spread crinoids out onto a table and grouped like morphologies into species. To our surprise, Hall basically recognized the same late Osagean species that we recognized after our comprehensive revisions. A few new species have been recognized since Hall's work; but, again, he identified the basic species concepts in use today. Subsequent workers (e.g., Miller and Gurley, 1893, 1894a, 1894b, 1895a, 1895b, 1896a, 1896b, 1896c, 1897) were responsible for the considerable oversplitting of species. In this series of papers, Miller and Gurley named more than 234 new Devonian and Mississippian camerate crinoid species, of which more than 65% are now junior synonyms. Further, modern species-level systematics must be completed for the remainder of the Miller and Gurley Devonian through middle Osagean taxa.

The source of oversplitting of late Osagean crinoid species by subsequent workers typically resulted from a very restricted concept for the variability within a species (Lane, 1963), different stratigraphic positions, different geographic areas, or some combination. For example, fossil crinoids preserved in the yellowish crinoidal limestones of the Burlington Limestone of Iowa just "look different" from those preserved in the dark gray siltstones of the Edwardsville Formation of Indiana. However, in many instances the crinoids that "look different" have identical species-diagnostic morphological characters (Kammer and Ausich, 1992).

In summary, historic species-level systematics are of varying quality by today's standards, and modern systematic consideration is needed in all cases. However, in well-studied regions, high-resolution species-level data can be obtained.

The Crinoid Fossil Record and Ecological-Time Data

Even if species are defined by discrete characters, fossil data are generally considered time averaged; thus, perhaps, limiting the interpretive power of fossil species data, especially before the origination of aggressive bioeroders that greatly limited the residence time of skeletal debris on the sea floor. Further, very few fossils are preserved in their exact living position. Thus, with the extreme rarity of autochthonous fossil preservations, it is commonly assumed that most fossil occurrences represent allochthonous assemblages. Therefore, it can be argued that time-averaged allochthonous assemblages are sufficiently degraded so that most fossil assemblages have little utility for interpreting ecological-time processes.

However, most fossil occurrences are neither authochthonous nor allochthonous. Most shallow-marine fossil assemblages are parautochthonous, i.e., transported but preserved in the facies in which it lived (e.g., Kidwell et al., 1986). This contention is supported by mapping of the Great Bahamas Bank (Newell et al., 1959; Purdy, 1963) (Bathurst, 1976, figs. 129, 130). Newell et al. (1959) and Purdy (1963) mapped habitats, communities, and lithofacies (sediment) across the Great Bahamas Bank. By and large, the distribution of these three parameters were coincident across the bank, demonstrating that organisms are typically preserved in the facies in which they lived. Of course, organisms may be transported to exotic facies by storms, turbidity currents, and shoreline processes; but the rock record typically preserves clear evidence for significant transportation with bedding characteristics, sedimentary structures, and other petrographic attributes.

Further, time averaging is not always an issue even in parautochthonous deposits. Crinoids have a multielement endoskeleton that begins disarticulation within a few days after death (Meyer, 1971; Liddell, 1975; Donovan, 1991; Baumiller and Ausich, 1992; Ausich, 2001). Dead crinoid individuals on the sea floor are subject to physical and biological disturbances and will begin to disarticulate rapidly, with disarticulation resistance differing among clades (Meyer et al., 1989; Thomka et al., 2011). Preservation of complete individuals can only occur where individuals were rapidly and permanently buried. The primary burial process was undoubtedly tempestite deposition (Taylor and Brett, 1996; Donovan, 1991; Ausich, 2001). Thus, the primary record of crinoid crown preservation was one of instantaneous burial; one of sea floor "ecological snapshots" of ancient communities with little or no time averaging (fig. 10.2).

This unique style of Lagerstätten occurred in all organisms with multielement skeletons, such as arthropods, vertebrates, and plants. Complete specimens preserved in such Lagerstätten may provide unparalleled resolution of morphological variability within and among populations, provide meaningful relative abundance data, etc. This results in species-level data

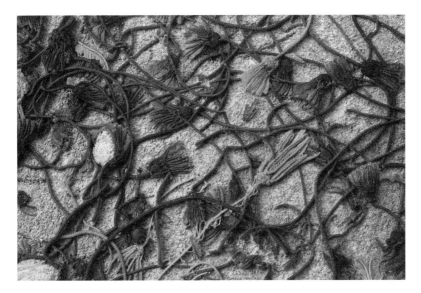

Figure 10.2 Ecological snapshot of crinoid assemblage from the Lower Mississippian Maynes Creek Formation, Legrande, Iowa (specimen in collection of Beloit College).

that are sufficient to ask ecological-time questions, such as inter- and intrapopulation morphology and dynamics and paleoecological questions. These high-resolution data are restricted to complete or nearly complete individuals. Preservation of individual skeletal elements (e.g., crinoid columnals, trilobite sclerites, vertebrate teeth, or leaves) was subject to the same transportation and time-averaging complications of most fossils.

Conclusions

Species-level data in the fossil record are of variable quality and completeness. It cannot be used uncritically for paleobiological interpretations. However, species defined by the diagnosable species concept with modern systematic scrutiny are unique morphologies in time and space. Therefore, these species may be used as exemplars for questions on phylogeny, disparity, biostratigraphy, and paleoecology. Because most fossil occurrences are parautochthonous in nature, species-level data for paleoecological analysis includes both time-averaged information and/or, in many assemblages, ecological-snapshot data of complete multielement organisms, such as echinoderms, arthropods, and vertebrates.

References

Alroy, J., C. R. Marshall, R. K. Bambach, K. Bezuska, M. Foote, F. T. Fürsich, T. A. Hansen, S. M. Holland, L. C. Ivany, D. Jablonski, D. K. Jacobs, D. C. Jones, M. A. Kosnick, S. Lidgard, S. Low, D. M. Raup, K. Roy, J. J. Sepkoski, Jr., M. G. Sommers, P. J. Wagner, and A. Webber. 2001. Effects of sampling standardization on estimates of Phanerozoic marine diversification. Proceedings of the National Academy of Sciences, 98:6261–6266. Ausich, W. I. 2001. Echinoderm taphonomy, pp. 171–227. In J. Lawrence and M. Jangoux (eds.), Echinoderm Studies, vol. 6 (2001). Balkema Press, Rotterdam

Ausich, W. I. 2009. These are not the crinoids your Granddaddy knew. MAPS Digest, 32(1):4–19.

Ausich, W. I. 2010. Post-hoc sampling analysis of crinoid collections from Anticosti Island, Quebec, Canada. Memoirs of the Association of Australasian Paleontologists, 39:12–25.

Ausich, W. I., and P. Copper, P. 2010. The Crinoidea of Anticosti Island, Québec (Late Ordovician to Early Silurian). Palaeontographica Canadiana, 29, 157 pp.

Ausich, W. I., and T. W. Kammer. 1990. Systematics and phylogeny of the late Osagean and Meramecian crinoids Platycrinites and Eucladocrinus from the Mississippian stratotype region. Journal of Paleontology, 64:759–778.

Ausich, W. I., and T. W. Kammer. 1991. Late Osagean and Meramecian Actinocrinites from the Mississippian stratotype region (Echinodermata: Crinoidea). Journal of Paleontology, 65:485–499.

Ausich, W. I., T. W. Kammer, and T. K. Baumiller. 1994. Demise of the Middle Paleozoic crinoid fauna: a single extinction event or rapid faunal turnover? Paleobiology, 20:345–361.

Ausich, W. I., and D. L. Meyer. 1994. Hybrid crinoids in the fossil record (Early Mississippian, Phylum Echinodermata). Paleobiology, 20:362–367.

Bathurst, R. G. C. 1976. Carbonate Sediments and Their Diagenesis (2nd ed.). Elsevier, Amsterdam, 658.

Baumiller, T. K. 1993. Survivorship analysis of Paleozoic Crinoidea: effect of filter morphology on evolutionary rates. Paleobiology, 19:304–321.

Baumiller, T. K., and W. I. Ausich. 1992. The broken-stick model as a null hypothesis for fossil crinoid stalk taphonomy and as a guide for distribution of connective tissue in fossils. Paleobiology, 18:288–298.

Benton, M. J., and P. N. Pearson. 2001. Speciation in the fossil record. Trends in Ecology and Evolution, 16:405–411.

Coyne, J. A., and H. A. Orr. 2004. Speciation. Sinauer Associates, Sunderland, Massachusetts, 545 p.

Cracraft, J. 1983. Species concepts and speciation analysis. Journal of Current Ornithology, 1:159–187.

de Queiroz, K. 2007. Species concepts and species delineation. Systematic Biology, 56:879–886.

Donovan, S. K. 1991. The taphonomy of echinoderms: calcareous multi-element skeletons in the marine environment, pp. 241–269. *In* S. K. Donovan (ed.), The Processes of Fossilization. Belhaven Press, London.

Eldredge, N., and S. J. Gould. 1972. Punctuated equilibrium: an alternative hypothesis to phyletic gradualism, pp. 82–115. *In* T. J. M. Schopf (ed.), Models in Paleobiology. Freeman, Cooper, and Company, San Francisco.

Gingerich, P. D. 1974. Stratigraphic record of early Eocene *Hyopsodus* and the geometry of mammalian phylogeny. American Journal of Science, 276:1–28.

Goodfield, G .A., and S. J. Gould. 1996. Paleontology and chronology of two evolutionary transitions by hybridization in the Bahamian land snail *Cerion*. Science, 274(5294):1894–1897.

Hall, J. 1858. Palaeontology of Iowa, pp. 473–724. *In* J. Hall and J. D. Whitney, Report of the Geological Survey of the state of Iowa: Embracing the results of investigations made during portions of the years 1855, 56 &: 57, v. 1, part II; Palaeontology, p. 473–724, 29 pl., index to Part II separately paginated, 3 pp.

Hall, J. 1859. Contributions to the palaeontology of Iowa, being descriptions of new species of Crinoidea and other fossils. Supplement to vol. I, part II, of the Geological Report of Iowa, 92 pp.

Hall, J. 1860. Contributions to the palaeontology of Iowa: being descriptions of new species of Crinoidea and other fossils. Iowa Geological Survey supplement to 1(2) of Geological Report of Iowa, 1859a, 4 pp.

Hall, J. 1861a. Descriptions of new species of Crinoidea from the Carboniferous rocks of the Mississippi Valley. Journal of the Boston Society of Natural History, 3:261–328.

Hall, J. 1861b. Descriptions of new species of crinoidea; from investigations of the Iowa Geological Survey, Preliminary notice. C. van Benthuysen, Albany, New York, 18 pp.

Hey, J. 2001. The mind of the species problem. Trends in Ecology and Evolution, 18:326–329

Kammer, T. W., and W. I. Ausich. 1993. Advanced cladid crinoids from the middle Mississippian of the East-Central United States: intermediate-grade calyces. Journal of Paleontology 67:614–639.

Kammer, T. W., and W. I. Ausich. 1992. Advanced cladid crinoids from the middle Mississippian of the East-Central United States: primitive-grade calyces. Journal of Paleontology, 66:461–480.

Kammer, T. W., and W. I. Ausich. 2006. The "Age of Crinoids": a Mississippian biodiversity spike coincident with widespread carbonate ramps. Palaios, 21:236–248.

Kammer, T. W., T. K Baumiller, and W. I. Ausich. 1997. Species longevity as a function of niche breadth: evidence from fossil crinoids. Geology, 25: 219–222.

Kammer, T. W., T. K. Baumiller, and W. I. Ausich. 1998. Evolutionary significance of differential species longevity in Osagean-Meramecian (Mississippian) crinoid clades. Paleobiology, 24:155–176.

Kidwell, S. M., F. T. Fürsich, and T. Aigner. 1986. Conceptual framework for the analysis and classification of fossil concentrations. Palaios, 1:228–238.

Lane, N. G. 1963. Meristic variation in the dorsal cup of monobathrid camerate crinoids. Journal of Paleontology, 37:917–930.

Liddell, W. E. 1975. Recent crinoid taphonomy. Geological Society of America Abstracts with Programs, 7:1169.

Malmgren, B. A., W. A. Berggren, and G. P. Lohmann. 1983. Evidence for punctuated gradualism in the Late Neogene *Globorotalia tumida* lineage of planktonic foraminifera. Paleobiology, 9:377–389.

Mayden, R. L. 1997. A hierarchy of species concepts: the denouement in the saga of the species problem, pp. 381–424. *In* M. F. Claridge, H. A. Dawan, and M. R. Wilson (eds.), Species: The Units of Biodiversity. Chapman and Hall, London.

Meyer, D. L. 1971. Post-mortem disintegration of Recent crinoids and ophiuroids under natural conditions. Geological Society of America Abstracts with Programs, 3:645–646.

Miller, S. A., and W. F. E. Gurley. 1893. Description of some new species of invertebrates from the Palaeozoic rocks of Illinois and adjacent states. Illinois State Museum Bulletin, 3, 81 pp.

Miller, S. A., and W. F. E. Gurley. 1894a. Upper Devonian and Niagara crinoids. Illinois State Museum Bulletin, 4, 37 pp.

Miller, S. A., and W. F. E. Gurley. 1894b. New genera and species of Echinodermata. Illinois State Museum Bulletin, 5, 53 pp.

Miller, S. A., and W. F. E. Gurley. 1895a. Description of new species of Palaeozoic Echinodermata. Illinois State Museum Bulletin, 6, 62 pp.

Miller, S. A., and W. F. E. Gurley. 1895b. New and interesting species of Palaeozoic fossils. Illinois State Museum Bulletin, 7, 89 pp.

Miller, S. A., and W. F. E. Gurley. 1896a. Description of new and remarkable fossils from the Palaeozoic rocks of the Mississippi Valley. Illinois State Museum Bulletin, 8, 65 pp.

Miller, S. A., and W. F. E. Gurley. 1896b. New species of crinoids from Illinois and other states. Illinois State Museum Bulletin, 9, 66 pp.

Miller, S. A., and W. F. E. Gurley. 1896c. New species of Echinodermata and a new crustacean from the Palaeozoic rocks. Illinois State Museum Bulletin 10, 91 pp.

Miller, S. A., and W. F. E. Gurley. 1897. New species of crinoids, cephalopods, and other Palaeozoic fossils. Illinois State Museum Bulletin, 12, 69 pp.

Meyer, D. L., W. I. Ausich, and R. E. Terry. 1989. Comparative taphonomy of echinoderms in carbonate facies: Fort Payne Formation (Lower Mississippian) of Kentucky and Tennessee. Palaios, 4:533–552.

Nelson, G., and N. I. Platnick. 1981. Systematics and Biogeography. Columbia University Press, New York.

Newell, N. D., J. Imbrie, E. G. Purdy, and D. L. Thurber. 1959. Organism communities and bottom facies, Great Bahamas Bank. Bulletin of the American Museum of Natural History, 117(4):177–228.

Nixon, K. C., and Q. D. Wheeler. 1990. An amplification of the phylogenetic species concept. Cladistics, 6:211–223.

Peters, S. E., and M. Foote. 2001. Biodiversity in the Phanerozoic: a reinterpretation. Paleobiology, 27:583–601.

Peters, S. E., and M. Foote. 2002. Determinants of extinction in the fossil record. Nature, 416:420–424.

Purdy, E. G. 1963. Recent calcium carbonate facies on the Great Bahamas Bank, 2, sedimentary facies. Journal of Geology, 71:472–497.

Raup, D. M., 1976. Species diversity in the Phanerozoic: an interpretation. Paleobiology, 2:289–297.

Sepkoski, J. J., Jr. 1981. A factor analysis description of the Phanerozoic marine fossil record. Paleobiology, 7:36–53.

Sepkoski, J. J., Jr. 2002. A compendium of fossil marine animal genera. Bulletins of American Paleontology, 363, 560 pp.

Simpson, G. G. 1951. The species concept. Evolution, 5:285–298.

Smith, A. B. 2001. Large-scale heterogeneity of the fossil record: implications for Phanerozoic biodiversity studies. Philosophical Transactions of the Royal Society of London B356:351–367.

Smith, A. B., A. S. Gale, and N. E. A. Monks. 2001. Sea-level change and rock-record bias in the Cretaceous: a problem for extinction and biodiversity studies. Paleobiology, 27:21–253.

Taylor, W. L., and C. E. Brett. 1996. Taphonomy and paleoecology of echinoderm *Lagerstätten* from the Silurian (Wenlockian) Rochester Shale. Palaios, 11:118–140.

Thomka, J. R. R. D. Lewis, D. Mosher, R. K. Pabian, and P. F. Holterhoff. 2011. Genus-level taphonomic variation within cladid crinoids from the Upper Pennsylvanian Barnsdall Formation, Northeastern Oklahoma. Palaios, 26:377–389.

Wiley, E. O. 1978. The evolutionary species concept reconsidered. Systematic Zoology, 27:17–26.

CHAPTER ELEVEN

Species and the Fossil Record of Fishes

William E. Bemis

In short, we shall have to treat species in the same manner as those naturalists treat genera, who admit that genera are merely artificial combinations made for convenience. This may not be a cheering prospect; but we shall at least be freed from the vain search for the undiscovered and undiscoverable essence of the term species.—C. Darwin, 1859: 485

Though I feel that this is probably the most practical species concept (it's basically what goes on now), it does have that unsatisfying "I know it when I see it" quality. Who would have guessed that species and pornography could be identified using the same, Supreme Court–sanctioned method?—S. Werning, 2013

Introduction

This paper briefly explores concepts of species of "fishes" in the fossil record.[1] For an evolutionary biologist also interested in systematics, it is impossible to study any fossil species without careful study of and reference to extant species. Thus, this paper is informed by anatomical comparisons to extant fishes as well as their nomenclatural history, as exemplified by the *Catalog of Fishes* (Eschmeyer, 1998a, 2015), with the goal being synthesis of neontological and paleontological perspectives. The enormous literature on species concepts, speciation, and systematic philosophy includes contributions specifically focused on fishes, such as papers in Ruffing et al.

---

1. In this paper, I use "fishes" in its nonmonophyletic sense, i.e., I do not discuss examples of tetrapods. Such usage is conventionally flagged by enclosing the term in quotation marks to indicate nonmonophyly, but this is cumbersome and distracting, so from this point forward the word fishes will be used without quotation marks.

(2002) and Harrington and Near (2012), as well as a recent general treatment by Wilkins (2009) and extended discussions in Wiley and Lieberman (2011). But in this paper, I am chiefly concerned with practicality, for in my view, species names in the fossil record of fishes are primarily tools for discovery and organized study of paleodiversity and for communicating that information to others. Darwin (1859: 485) considered that species names, like generic names, are primarily about convenience, and convenience is important whether you are studying extant or extinct organisms.

Although it is nearly always implied rather than stated, systematic paleoichthyologists recognize species in the fossil record based on morphology and a species concept similar to one stated by Gareth Nelson and Norman Platnick (1981: 11):

> To a very large extent, this is the species concept actually used in practice: those samples that a biologist can distinguish (diagnose), are called species.

The emphasis on differential diagnosis is a pre-Darwinian concept, employed, for example, by early students of fossil fishes such as Sedgwick and Murchison (1828: 143–144) to distinguish species of Devonian sarcopterygian fishes in the genera †*Dipterus* and †*Osteolepis* in deposits from northern Scotland. Only under the rarest circumstances—extraordinary preservation, a large sample size, and extremely narrow stratigraphic range—might we be justified in thinking that fishes known only from the fossil record operated in ways similar to extant biological species, i.e., that they were members of populations that actually or potentially interbred (e.g., Dobzhansky, 1935; Mayr, 1942).

Naming new species of fossil fishes serves two related but different purposes: (1) species names provide practical constructs for understanding and communicating information about paleodiversity by linking particular specimens to particular localities and stratigraphic data; and (2) species names provide tools for phylogenetic and biogeographic studies. In regard to the second of these purposes, many paleoichthyologists choose to emphasize generic- rather than specific-level comparisons as discussed at the end of this chapter in an example from Forey et al. (2003). This no doubt reflects uncertainties inherent in developing reliable differential diagnoses at the specific level. Nevertheless, paleoichthyologists continue to describe new species of fossil fishes. This is largely because, in the absence of a better system, species names provide convenient tags for specimens and localities and ways to clearly communicate that information to others.

In the 19th century, it was common to name new species of fossil fishes based solely on differences in locality or stratigraphic range. Because there are relatively few paleoichthyologists in any given generation of scientists, we continue to wrestle with the challenges that such practices created. For example, Henry Shaler Williams (1881) named a new fossil lungfish †*Dipterus ithacensis* from the Chemung Group and associated Ithaca Shale (Devonian) of Ithaca, New York. His paper is an abstract written in the third person, apparently by someone who attended the conference proceedings (e.g., the first sentence reads: "The author described several small fish bones from the Devonian rocks at Ithaca, N. Y."). Williams was then a professor of paleontology at Cornell University; he subsequently served on the faculty at Yale (American National Biography Online, 2014). His material of †*D. ithacensis* was not figured in the 1881 paper, no types were designated, and searches in 2014 of the collections at the Paleontological Research Institution in Ithaca and the Division of Vertebrate Paleontology at the Yale Peabody Museum failed to turn up any of Williams' specimens. Dipnoan tooth plates are distinctive and easily recognized, and Williams (1887) later described and figured other species of †*Dipterus* from the northeastern United States, so he clearly had experience with the group, although he does not mention †*D. ithacensis* in his 1887 work. Given the uncertainties surrounding Williams' 1881 description of †*D. ithacensis*, however, we cannot even be sure that his specimens were dipnoans, and the meager details in the paper certainly do not differentially diagnose †*D. ithacensis* from any other species of †*Dipterus*. About 25 years later, Eastman (1907: 163), writing on species of lungfishes in the genus †*Dipterus* from the Chemung group (Devonian), stated:

> Precise determination of *Dipterus* teeth from the Chemung proper of New York and Pennsylvania is a matter of some difficulty, owing to imperfection of type material upon which the various "species" are founded, and an insufficient series of specimens for illustrating the range of variation common to both sets of dental plates, upper and lower.

Perhaps Williams (1881) named †*D. ithacensis* with the expectation that further collecting would recover enough material to resolve any questions, but in this case, an additional century of collecting has not furthered our understanding. The name †*D. ithacensis* remains in our literature despite the fact that it is certainly a *nomen dubium* (the concept of *nomen dubium* and its applications to paleoichthyology are discussed later in this chapter).

Paleoichthyologists currently lack the kinds of tools that have helped propel modern systematic study of living fishes, particularly a comprehensive catalog of the genera and species of fossil fishes. For example, we do not have a realistic idea about how many extant species of fishes are represented in the fossil record, and there is no easy way to even estimate this number from bibliographic tools and databases currently available. Much would need to be done to develop anything approaching the *Catalog of Fishes* (Eschmeyer, 1998a, 2014), but a similarly scholarly catalog for fossil fishes could become an outstanding tool to promote future research in systematic paleoichthyology by bringing together the many species names and scattered literature of our field. Together with Joseph T. Gregory, William N. Eschmeyer began an effort in the 1908s that documented about 3,600 available generic names and 5,200 available species names for fossil fishes (W. N. Eschmeyer, personal communication, April 2014), but as a very rough guess, this is perhaps only half of the total known named diversity of fossil fishes.

General Comments on the Fossil Record of Fishes

The record of fossil fishes is vast, both in terms of individual fossils and named species, partly because of their great stratigraphic range and from their nature as usually small, often abundant aquatic vertebrates that live in environments where they are prone to fossilization. Some species of fossil fishes are known from remarkably complete, exceptionally preserved whole-body specimens, which may preserve evidence of soft tissues and are suited to detailed anatomical study and description. For example, fishes are common members of faunas from several well-known conservation Lagerstätten. These include the Late Devonian Escuminac Formation of Miguasha Québec (Schultze and Cloutier, 1996; Cloutier, 2013), the Cretaceous Santana Formation in northeastern Brazil (Maisey, 1991), and the Eocene Green River Formation of southwestern Wyoming (Grande, 2013); many examples of fishes from marine Lagerstätten are summarized in Bottjer et al. (2003). An outstanding example of exceptional preservation is †*Eusthenopteron foordi*, an osteolepiform from Miguasha known from whole-body, three-dimensional specimens. Because osteolepiforms are important for understanding sarcopterygian phylogeny, many researchers have investigated †*E. foordi* from Miguasha, most notably Erik Jarvik, who published 30 papers on its anatomy (Cloutier, 2013; for a summary of those research findings, see Jarvik, 1980). Other cases of exceptional preservation include coral reef fishes from the Eocene Monte Bolca in Italy that even preserve color patterns (e.g., Bellwood, 1996). Muscles and other soft tissues are known in some articulated Devonian sharks such as †*Cladoselache* from the Cleveland Shale (Dean, 1902) or †placoderms from the Devonian Gogo Formation such as †*Eastmanosteus* (Trinajstic et al., 2007).

Other fossil fishes are known only from incomplete or disarticulated postmortem materials or hard parts such as teeth shed naturally during life. For example, many species of Mesozoic and Cenozoic sharks are named based only on teeth (for examples, see Cappetta, 2012). A similar situation applies to fossil species named only on the basis of otoliths (Nolf, 1985, 2013). Otoliths can be well-mineralized and can form beautiful fossils, but it is relatively rare to find a whole articulated fish fossil with intact and visible otoliths, a condition described as in situ, because otoliths form inside the membranous labyrinths of the inner ear and are not attached to bones of the skull. As a result, many fossil taxa named on the basis of isolated otoliths are unlikely to ever be matched up with whole specimens. A surprisingly large number of taxa have been named based on otoliths:

> 2666 otolith-based fossil fish species have presently been described or named. They include 1391 valid species; 464 species considered to be synonyms of already described fossil species, or of Recent species; 811 species here grouped under "obsolete names." (Nolf, 2013: 4)

A few species of fossil fishes are astonishingly abundant in the record, such as the state fossil of Wyoming, †*Knightia eocaena*. This is perhaps the most commonly collected whole-bodied fossil fish in the world, with hundreds of thousands of more-or-less complete fossilized skeletons known. However, most species of fossil fishes are known from only a few partially complete or fragmentary specimens.

In many cases, only a single incomplete fossil specimen from a locality is known, but it may have such importance for the record and for interpreting phylogenetic history that it becomes well known. An example is the Late Cretaceous paddlefish †*Paleopsephurus wilsoni*, known only from a single partial skull and poorly preserved caudal region discovered in association with a †hadrosaurid dinosaur (†*Edmontosaurus annectens* UMMP 20000; currently on display at the University of Michigan Museum of Natural History; MacAlpin, 1941a, b, 1947; Grande and Bemis,

1991; Grande et al., 2002; P. D. Gingerich, personal communication, May 2014). Materials from other localities have been referred to †*P. wilsoni*, but they lack any diagnostic features of †*P. wilsoni* and were considered †*incertae sedis* C by Grande and Bemis (1991: 85–86, 97–98). No other diagnosable specimens of †*P. wilsoni* have been described in the intervening years. But †*P. wilsoni* is uniquely important in understanding the phylogenetic and biogeographic history of paddlefishes (Polyodontidae), for despite the incompleteness of the known specimen, it preserves a useful mixture of characters for interpreting the phylogeny of the entire family (Grande et al., 2002: fig. 19). †*Paleopsephurus wilsoni* may have been a biological species in the same sense that the single extant species of North American paddlefish, *Polyodon spathula*, is a biological species, but it is not particularly insightful or helpful to speculate about this. For paleoichthyologists, the name †*P. wilsoni* functions primarily as a phylogenetic placeholder for a particular mixture of characters.

Comparative osteology informs systematic studies of both living and fossil species of fishes. Because of the inherent complexity of the skeletal system of fishes—for example, *Amia calva* has more than 680 discrete skeletal elements not including fin rays, gill arch toothplates, scales, teeth, or otoliths (Grande and Bemis, 1998: ix)—fish systematists can draw upon an unusually large potential set of character data from the skeleton. This makes it possible to make detailed side-by-side comparisons of the skeleton of fossil and living species and to integrate extant and fossil taxa into phylogenetic systematic studies, as exemplified by Grande (2010). Comparative osteology remains a cornerstone of taxonomic and phylogenetic studies of fishes, but usually when starting a new study of fossil fishes belonging to an extant genus or family, it is necessary to restudy and describe the osteology of the extant members because the available descriptions and literature are inadequate for the study of fossils. Soft tissue characters including coloration are very important in the description of living species of fishes, and molecular characters increasingly predominate in phylogenetic studies that only consider extant taxa. That said, most extant groups of fishes have a fossil record that can inform phylogenetic study. For example, Sparidae, a family of about 120 extant species of abundant near-shore marine fishes, has an important Eocene record, including some extraordinarily well-preserved specimens from Monte Bolca, Italy, that inform phylogenetic study of the family (Day, 2003). Another important role for osteological and phylogenetic studies of fossil fishes is to calibrate divergence times for molecular phylogenetic studies.

Such calibrations are only as good as our understanding of the interrelationships of the fossils, which necessarily involves comparative osteological study. Funding priorities in systematic ichthyology today may emphasize molecular phylogenetic research, but given the extensive fossil record of fishes, it is only a matter of time before the comparative study of osteological characters regains attention for its fundamental role in integrating the study of extant and fossil fishes.

Species Concepts in Vertebrate Paleontology

From the beginning of the Modern Synthesis in the 1930s and 1940s, vertebrate paleontologists have wrestled with concepts of species in the fossil record. In addition to Simpson's landmark book, *Tempo and Mode in Evolution* (Simpson, 1944), papers in three early symposia stand out. First is a series of papers in *American Naturalist* under the general title *Supra-Specific Variation in Nature and in Classification* (e.g., Kinsey, 1937; Simpson, 1937; and Gregory, 1937). Papers read at a joint meeting of the American Society of Ichthyologists and Herpetologists and the American Society of Mammalogists yielded a volume edited by Charles Bogert (1943), *Criteria for Vertebrate Subspecies, Species, and Genera*. About a decade later, Sylvester-Bradley (1956) edited a volume for the Systematics Association entitled *The Species Concept in Paleontology*, which includes Haldane's (1956) famous two-page paper in which he decries the whole idea of species concepts and characterizes disputes about the validity of species as being "primarily a linguistic rather than biological dispute." Also in that volume is a paper by George (1956) on biospecies, chronospecies, and morphospecies.

An early leader in thinking about the meaning of species in the fossil record of vertebrates, George Gaylord Simpson (1937:267) had a practical and biologically based approach to the study of fossils:

> If two species do not differ appreciably in morphology, it is certainly true that they are closely related, and an error in supposing them to be one species is not an essential falsification of the general record.

Here, Simpson makes the case for not oversplitting when naming new species in the fossil record. This is a good idea given the inherent limitations of the fossil record of fishes.

In a subsequent paper, "Criteria for genera, species, and subspecies in zoology and paleozoology," Simpson (1943: 146–147) stated that "the species in nature is something different from the species in classification." This quotation is echoed and amplified by Stanley (1979: 8): "There is no question that fossil entities recognized as species are not strictly comparable to species of the living world." Clearly, by 1943, Simpson already regarded species in the fossil record as simultaneously a nomenclatural concept and a biological concept. Despite the proliferation of species concepts in seven intervening decades, this duality of nomenclature and biology for all practical purposes captures the range of possibilities that paleoichthyologists consider when they use the word species.

Nomenclature and Taxonomy of Extant and Fossil Species of Fishes

The 4th edition of the International Code of Zoological Nomenclature (designated hereafter as the Code; International Commission on Zoological Nomenclature, 1999, 2014) applies equally to extant and fossil species of fishes. For example, if a paleoichthyologist names a new genus and species based on fossil material and a living species is subsequently found that belongs within that genus, then the genus name of the fossil species applies to the extant species. There were precursors to the Code. For example, the 1842 code of nomenclature (Eschmeyer, 1998b: 2847) predated the 1881 meeting that ultimately resulted in the Code; it identified some key concepts later incorporated into the Code, such as the principle of priority. The Code continues to evolve as new means for publication of new species names and revisions of existing species develop (for more on the history and future of the Code, see Ride and Younès, 1986; Eschmeyer, 1998b; Ride, 1999; International Commission on Zoological Nomenclature, 1999, 2014).

Key concepts concern available and valid names. Provisions conferring availability of names are described in Article 10 of the Code (International Commission on Zoological Nomenclature, 1999, 2014), which are succinctly summarized by Eschmeyer et al. (2010: 20): an available name is "a scientific name that meets the criteria of publication, authorship and other restrictions of the International Code of Zoological Nomenclature." Once available, a species name remains so unless the Commission rules otherwise (International Commission on Zoological Nomenclature, 1999,

2014: Article 10.6). In contrast, a valid name is determined by the principle of priority. "Article 23.1 Statement of the Principle of Priority: The valid name of a taxon is the oldest available name applied to it, unless that name has been invalidated or another name is given precedence by any provision of the Code or by any ruling of the Commission." In the words of Eschmeyer et al. (2010: 20), a valid name is "a species, subspecies or genus that is considered to be a legitimate and recognizable 'good' taxon." There are far many more available names than there are valid names. For example, as of 10 March 2014 there were 57,907 available names for extant fishes, with 5,104 valid genera, and 33,059 valid species (Eschmeyer, 2014). One year later (20 February 2015), there were 58,298 available names, 5,128 valid genera, and 33,377 valid species (Eschmeyer, 2015).

A nomenclatural concept unfortunately central to paleoichthyology is *nomen dubium*, which means "a name of unknown or doubtful application" (International Commission on Zoological Nomenclature, 1999, 2014). Most systematists, including Grande and Bemis (1998: 19), interpret "doubtful application" to mean inadequate original diagnosis or inadequate material to differentially distinguish it from other species. Untold numbers of species that are *nomina dubia* exist in paleoichthyology and pose challenges for rigorous revisionary systematic studies because of the amount of material that must be studied and, essentially, rejected as uninformative for morphological or phylogenetic studies. As a particularly striking example, Hilton and Grande (2006) studied the three nominal species of pre-Pleistocene sturgeons in the genus *Acipenser* and many specimens from North America that had been referred to the genus *Acipenser*. Much sturgeon material spanning the Late Cretaceous to Pliocene is known, but nearly all of it is fragmentary, and Hilton and Grande (2006: 676) concluded that *all* of the named and referred materials "are *nomina dubia* and regarded as Acipenseridae indeterminate genus and species." Rather than creating a new *nomen dubium* for incomplete undiagnosable material, an author can use open nomenclatural flags such as cf. (the abbreviation for *confer*, L., compare to) or aff. (the abbreviation for *affinis*, L., closely related) to indicate that the reader should compare the material to an existing valid species (such flags are not yet addressed by the Code, although they should be; for more on the general topic of "open nomenclature" in paleontology, see Bengston, 1988). For example, Grande et al. (2000) described a specimen from the Late Paleocene Paskapoo Formation from South Central Alberta as †*Amia cf. pattersoni*, a species from the Early Eocene Green River Formation of Wyoming, stating that it is "very closely related to, or conspecific with, *Amia pattersoni* Grande and Bemis, 1998." They opted to leave this material "cf." until additional specimens can be found to clarify this.

As Grande and Bemis (1998: 18) noted: "Comprehensive comparative phylogenetic studies incorporating fossil taxa would be considerably easier if the terms *incertae sedis* or indeterminate genus and species were used more frequently instead of creating more names of doubtful application." Neontological ichthyologists now only rarely need to use *incertae sedis* because molecular studies are increasingly helping to resolve phylogenetic relationships of enigmatic extant taxa previously considered *incertae sedis* such as pygmy sunfishes (*Elassoma*; see Near et al., 2012a). A perhaps even more intriguing example concerns the rare monotypic deep-sea fish *Stylephorus chordatus*. For most of the last century, *Stylephorus* was considered a close relative of ribbonfishes and oarfishes based on the supposedly unique ability of members of this group to project the premaxilla and maxilla as a unit during feeding. But whole mitogenomic analyses by Miya et al. (2007) suggested that *Stylephorus* is the sister taxon to Gadiformes (cods and allies), a position supported by analyses of multiple nuclear genes (Near et al. 2012b). Were *Elassoma* and *Stylephorus* known only as fossils, we would have been unable to correctly place them phylogenetically based only on morphological characters.

Paleoichthyologists commonly employ some conventions related to species-group names not yet incorporated into the Code. Among these, the dagger symbol (†) is particularly useful. Grande and Bemis (1991, 1998: 22) offer an explicit definition: "All taxa with daggers (†) preceding names are exclusively fossil (i.e., extinct)." Among other things, this convention reminds us that information about the fossil taxon is necessarily less complete than it would be for an extant taxon and it helps convey this meaning in comprehensive studies of extant and fossil taxa. It would be a good idea to include this concept in future editions of the Code. Use of quotation marks around a species name to flag it as a *nomen dubium* also is not specified or addressed by the Code, but it can be a very useful convention within the text of a large study to help a reader quickly understand that a name is problematic (see Grande and Bemis, 1998: 19).

Linnaeus (1758, 1766) described many extant species of fishes, and many of his type specimens are available at the Linnean Society of London (Wheeler, 1985). An interesting peculiarity of these types, and perhaps the reason that they have survived for 250 years, is that they are preserved as dry specimens—because Linnaeus treated them like plants and pressed them flat for mounting on botanical paper. Working with these type specimens today is a lot like working with fossil fishes because internal anatomical features are missing.

Since Linnaeus, systematic ichthyology has enjoyed a remarkably high level of scholarship and attention to detail concerning the diversity and taxonomy of fishes. We benefit in particular from the ongoing multi-decadal effort led by William Eschmeyer to comprehensively catalog species-level names, which resulted in the printed edition of the *Catalog of Fishes* (Eschmeyer, 1998a) and its continuing electronic updates, hosted by the California Academy of Sciences (Eschmeyer, 2015). More than 2,600 journals and 32,000 individual articles provide the reference base for the *Catalog of Fishes*. It is this carefully studied and comprehensive literature base that makes the *Catalog of Fishes* such a valuable resource, for it makes it easy to learn who described what genus or species. It also allows users to quickly generate authoritative synonymies. Recently, van der Laan et al. (2014) traced the authorship of names for the extant families of fishes in a catalog of family-group names. Because of the *Catalog of Fishes* and the catalog of family-group names, we have far more ready access to information about the taxonomy and history of species discovery of fishes than we do for any other group of vertebrates or, in fact, for any other large group of animals. For example, using data from the *Catalog of Fishes*, Eschmeyer et al. (2010) described patterns in the discovery of the diversity of marine fishes. Among many other case histories, they noted that, beginning in 1846, Pieter Bleeker described 1373 species of marine fishes; of these, 571 are today considered valid, yielding a 41.59% "success rate" in describing valid species (Eschmeyer et al., 2010: 24, tab. 3; the highest current success rate for any student of living marine fishes is John (Jack) Randall, who has described more than 700 species with a 96.45% success rate).

Of course, the numbers of available and valid species of extant fishes change all the time: about 424 new species were described in 2013 and at least 390 in 2014 (Eschmeyer, 2015). Moving target or not, we have in the *Catalog of Fishes* a remarkable scholarly framework and tool for thinking about the diversity and evolution of extant species as well as the history of species discovery. The same cannot be said for fossil fishes.

Paleoichthyologists still rely on historical works, such as Woodward's four-volume (1889, 1891, 1895, 1901) *Catalogue of the Fossil Fishes in the British Museum (Natural History)*, which catalogued one of the most important collections of fossil fishes. Type catalogs are available for some other collections of fossil fishes, such as the fossil fishes in the Field Museum of Natural History (Bruner, 1992) and others listed in Cleevely (1983: 26–37). The *Bibliography of Fossil Vertebrates* (2014) indexes published literature on fossil vertebrates from 1509 to 1993 (all years except 1969 to 1980 are available to search online), but it does not delve into necessary details about the publications that named new genera and species. New approaches in biodiversity informatics (e.g., Patterson et al., 2010; http://gni.globalnames.org/) can automatically track any string of alphanumeric characters used to refer to a taxon. This approach builds upon digital Name Repositories, such as the *Catalog of Fishes*, for its search capabilities, but in the absence of a dedcated, comprehensive, and scholarly *Catalog of Fossil Fishes* that has been carefully interpreted using the Code, the utility of biodiversity informatics for paleoichthyology will be limited. Popular websites include Fishbase (Froese and Pauly, 2014) and Wikipedia pages, such as the "List of prehistoric bony fish genera," described as "an attempt to create a comprehensive listing of all genera from the fossil record that have ever been considered to be bony fish." This list is, at best, a tertiary document based largely on the list of genera of fossil marine fishes by Sepkoski (2002) and it is not grounded in citations to the original descriptions of taxa. Thus as of now, we have nothing approaching a useful, let alone comprehensive, global catalog of the genera let alone the species of fossil fishes. This would be a challenging but worthy project for the future.

Although species discovery and description of extant and fossil fishes continues, comprehensive revisionary systematic studies at the generic and higher levels are very important. For example, restudying taxa from Eocene Fossil Lake known on the basis of 19th-century descriptions eventually yielded several new genera and species and much new anatomical and phylogenetic information on global species-level diversity of paddlefishes (Grande and Bemis, 1991; Grande et al., 2002), amiids (Grande and Bemis, 1998), hiodontids (Hilton and Grande, 2008), and lepisosteids (Grande, 2010). A common theme of modern revisionary systematic research is the ability to apply time-consuming methods, such as detailed mechanical preparation with needles or acid transfer preparation (Toombs and Rixon, 1959). Fossils prepared using such techniques often yield remarkable new morphological details that can be helpful at the species level as well as for phylogenetic studies. Such preparation and revisionary work is very time consuming and expensive to complete and publish, and it is perhaps less appreciated today than it should be, with the result that few young paleoichthyologists can pursue such approaches. CT scanning offers different benefits, the major ones being that it is far easier and much faster than either mechanical or acid-transfer preparation. But despite the improved resolution of newer instruments (e.g., Bemis and Bemis, 2015; Bemis et al., 2015; Moyer et al., 2015), CT scans will never fully replace carefully prepared fossil fishes.

Biological Species of Extant and Fossil Fishes

Of more than 33,300 valid extant species of fishes (Eschmeyer, 2015), most were described based on preserved specimens prior to the wide availability of molecular phylogenetic techniques, which, as an approximate baseline, occurred sometime in the middle 1980s. This means that the original species descriptions were based solely on morphological study. In most cases, original descriptions of extant fishes focus on external features, such as proportions, fin ray counts, pigment patterns and coloration, with skeletal and molecular characters only becoming known as a product of subsequent revisionary study. *We continue to differentially diagnose extant species of fishes based primarily on morphology and do not know to what extent most of these named species actually operate in nature as biological species.*

Concepts related to genetic differentiation of species continue to play, at best, only ancillary roles in original species descriptions. An interesting exception concerns the living Indonesian Coelacanth, *Latimeria menadoensis*, which was initially differentially diagnosed from the African Coelacanth, *L. chalumnae*, based on mitochondrial markers putatively supported by nine differences in meristic and morphological features (Pouyaud et al., 1999). The reliability and value of these nine morphological differences is difficult to assess because only a single specimen of *L. menadoensis* was available for study at that time. The two coelacanth species occur in allopatry, on opposite sides of the Indian Ocean more than 8,000 km apart, so some genetic differentiation would be expected. As pointed out by Holder et al. (1999: 12616), the ranges of variation in four of the meristic and morphological features proposed to distinguish the two species overlap, but they conclude that "nonetheless, *L. chalumnae* and *L. menadoensis* appear to be separate species based on divergence of mitochondrial DNA." We may presume that the two species of *Latimeria* operate as biological species that do not naturally interbreed.

We understand very little about biological species in the fossil record of fishes. It is possible in some cases and localities to recover large numbers of essentially complete fossilized skeletons of a single species in mass-mortality assemblages. As already noted, the small herring †*Knightia eocaena* is perhaps the most commonly collected species of fossil fish. It occurs in the Middle Early Eocene Fossil Butte Member of southwestern Wyoming. Grande (2013: 177) estimated that more than 600,000 specimens of †*K. eocaena* were collected there between 1870 and 2010. He attributed their abundance in the record in part to high fecundity, which is typical of extant Clupeidae. He also noted that extant herrings are notoriously sensitive to environmental changes in water temperature and dissolved oxygen and are subject to mass die-offs and that this may help explain how so many individual fossil fishes came to be in one place. In the case of †*Knightia*, it is possible to secure as many specimens as necessary to compare with other species in the genus. Thus, we can closely compare †*K. eocaena* with its less abundant congener from the same locality and strata, †*K. alta*, to understand diagnostic anatomical differences such as the much deeper body profile of †*K. alta*. The differences are reliably present and easily seen. Thus, we may have reasonable confidence that these two sympatric species of †*Knightia* operated as biological species in the same sense that two extant species of *Alosa*, the American shad (*A. sapidissima*) and blueback herring (*A. aestivalis*) operate as good biological species that spawn in the same riverine environments.

The best example of fossil fishes as a tool to explore evolutionary questions about speciation concerns †semionotid fishes from freshwater rift valley lakes in Newark Supergroup deposits of the Late Triassic and Early Jurassic of North America studied by Amy R. McCune (McCune et al., 1984; McCune, 1986, 1987a,b, 1990, 1996, 2004). McCune examined more than 2,000 specimens of †*Semionotus* from 45 named localities in the Newark Supergroup. Worldwide, more than 50 species of †semionotids have been described from Mesozoic marine and freshwater deposits, but it is in the Newark Supergroup lakes that they achieve their greatest species diversity, with about 40 species known from these lake basins. As McCune (1996: 34) notes, "Some of these species are probably not valid, but there are also a number of species not yet described." The species of †*Semionotus* from the Newark Supergroup lakes range in shape from slender and fusiform to deep-bodied. There are also important differences in the detailed anatomy of dorsal ridge scales (DRSs) between the occiput and the first dorsal fin: in some species, DRSs have a simple shape; in others, they may bear short or thin spines, have globular or robust shapes, or even be concave. One of the peculiarities of the Newark Supergroup lakes that makes them especially suited for asking evolutionary questions is their extraordinary cyclicity: new lakes repeatedly formed in the rift valleys, reached great depths, and then dried up (McCune, 2004: fig. 18.3). Lake formation, persistence, and death occurred on a 22,000-year basis known as a Van Houten cycle: at least 60 distinct Van Houten cycles are known. During times when a lake was very deep, mixing of surface and bottom waters did not occur, with the result that fish carcasses fell to the bottom into anoxic mud, where they were undisturbed by benthic organisms. The fine layers of mud and calcium carbonate deposited around carcasses demarcate annual increments, so it is possible to track morphological changes in †semionotids over time. In a particularly striking example, the first species with concave DRSs shows up early in the Early Jurassic P4 cycle in the Towaco Formation of New Jersey; based on the microstratigraphy, a second species with a concave DRS shows up 90 years later. Two Van Houten cycles, or about 44,000 years, later there are seven species of †Semionotus in the Boonton Formation of New Jersey that have concave DRSs. McCune (2004) interprets this as a radiation of species of †Semionotus that can be tracked through the record with remarkable precision and compared to lacustrine radiations of extant fishes, such as cichlids in rift valley lakes of Africa.

Morphological Species of Extant and Fossil Fishes

It can be difficult to differentially diagnose extant species of fishes based only on morphology. For example, Auguste Henri André Duméril (1869, 1870) named 19 new species of sturgeons in the genus *Acipenser* based on materials he studied at the Muséum National d'Histoire Naturelle in Paris. The problem is that, based on subsequent study, all but one of his new species of *Acipenser* proved to be junior synonyms of species named by Linnaeus (1758) and Mitchill (1815). Duméril had a 5.3% success rate in naming new species of *Acipenser*. But we can understand some of his challenges and why he made mistakes. First, the species of *Acipenser* are notoriously variable (Bemis et al., 1997; Hilton and Bemis, 1999, 2012; Hilton and Grande, 2006; Hilton et al., 2011). It remains difficult today to identify many of the 17 extant valid species of *Acipenser*, particularly small specimens with no locality data. Some species of Acipenseridae naturally hybridize in nature, a further complication. So nearly 250 years after Linnaeus (1758) named two valid species of *Acipenser*, the genus still proves inherently hard to study. Secondly, in the middle of the 19th century before the availability of formaldehyde as a preservative or freezers to hold specimens until they could be examined, it was very difficult to preserve large fishes for study. This is particularly the case for species that can grow as large as the Atlantic sturgeon, *A. oxyrinchus*, which historically reached lengths of 4.27 m (Eric J. Hilton, personal communication, May 2014) or the sturgeon *A. sturio*, for which there is an authoritative length record of 6 m. The usual method of preserving such large specimens in the 19th century was to gut and stuff them with straw before drying.[2] Drying results in so much distortion that specimens can be very difficult to compare. In a sense, Duméril dealt with his specimens of *Acipenser* in much the same way that a paleoichthyologist might deal with poorly preserved specimens from different localities that he or she had never visited. In both cases, restraint in creating new names is the best practice.

A counterexample concerns cryptic living species of bonefishes in the Linnean genus *Albula*. Because bonefishes live in shallow marine and brackish waters, they are familiar subtropical and tropical fishes. Well-preserved fossil bonefishes are known from Upper Cretaceous and Paleocene deposits (e.g., Mayrinck et al., 2010). Together with tarpons (Megalopidae) and ladyfishes (Elopidae), bonefishes lie near the base of the radiation of teleosts, and have attracted much interest from ichthyologists and paleoichthyologists concerned with the origin and diversification of teleosts (e.g., de Figueiredo et al., 2012). As many as 23 nominal species of *Albula* were named, but, as pointed out by Whitehead (1986), by 1940 all of them had been synonymized with *Albula vulpes* (Linnaeus 1758). This global distribution was interpreted as a result of a very long planktonic larval phase, which seemingly could explain how such a shorefish could reach coasts around the world isolated by deep ocean basins. In the 1980s, however, electrophoretic studies suggested the presence of two cryptic species of *A. vulpes* in Hawaiian waters. This was confirmed by the discovery of meristic differences in vertebral numbers, although the cryptic species from Hawaii are extremely similar to *Albula vulpes* in other morphological characteristics. Later, Colburn et al. (2001) partially sequenced the mtDNA marker Cytochrome b and discovered that there

2. Even today, collections rarely preserve large fishes intact in fluid, preferring to either skeletonize them or preserve only portions of the specimen in fluid.

were not just three species of *Albula* but potentially eight. Some 19th-century species names have been resurrected, and three new species have been named in the last decade (Hidaka et al., 2008, Pfeiler et al., 2011, and Kwun and Kim, 2011). As of 2014, there were 11 valid species of *Albula*, distinguished by molecular markers and very slight morphological differences of the sort that would be extremely difficult to detect in fossils.

Many important species of fossil fishes have not been restudied since their naming and incomplete original descriptions, often more than a century ago. It was common practice in the 19th century to name different species from the same or nearby localities based on single specimens even if the material available did not permit differential diagnoses compared to more complete materials. This is one reason why there are so many *nomina dubia* in paleoichthyology. An example comes from the work of Joseph Leidy (1873a: 98; 1873b: 185–189), who named two new genera and three species of bowfins (Amiidae) from two localities in southwestern Wyoming based on isolated individual vertebrae. Later, many more complete specimens of bowfins were discovered in the region and some were referred to Leidy's species by, for example, Boreske (1974). But because Leidy's species could not be differentially diagnosed, it was impossible to unequivocally associate such referred taxa with Leidy's specimens. Grande and Bemis (1998: 19) considered that †"*Protamia uintaensis*" (based on ~10 isolated vertebrae) †"*Protamia media*" (based on 1 vertebra), †"*Protamia gracilis*" (based on 1 vertebra) and †"*Hypamia elegans*" (based on 1 vertebra) were all *nomina dubia*. We resolved this problem by redescribing the referred taxa under new generic and species names.

A related problem concerns specimens of different sizes collected at the same location that have been assigned different species names. This is the case for three species that Egerton (1858: 883) named in the genus †*Chondrosteus* from Lower Liassic deposits of Lyme Regis, England: †*C. acipenseroides*, †*C. crassior*, and †*C. pachyurus*. These sturgeon-like fishes have long been thought to be at or near the base of Acipenseriformes, and thus are potentially important to interpreting evolutionary relationships of living and fossil sturgeons, paddlefishes, and allies. Many specimens of †*Chondrosteus* are beautifully preserved, including some nearly complete whole-body specimens, disarticulated skulls that reveal key osteological details, and caudal regions preserving details of the axial skeleton and fin rays. There were no new studies of these specimens between Woodward's (1895) catalogue and a detailed osteological and phylogenetic study by Eric J. Hilton and Peter L. Forey. Based on more than 70 specimens, Hilton and Forey (2009: 431) synonymized all of Egerton's species as †*Chondrosteus acipenseroides*, accepting earlier synonymies for †*C. acipenseroides* and †*C. crassior* (e.g., Woodward, 1895: 27) and concluding that the differences between †*C. pachyurus* and †*C. acipenseroides* are an artifact of ontogenetic stage and not diagnostic at the species level.

Sometimes the opposite problem occurs, as reported in recent and fascinating studies by Donald J. Stewart (2013a,b) on extant species of *Arapaima*, a genus of freshwater fishes from South America. Strikingly colorful and majestic in public aquaria, *A. gigas* is routinely considered the largest exclusively freshwater fish in the world; it grows in captivity to more than 3m TL. *Arapaima* is also aquacultured and sought as food and game despite the status of *A. gigas* as the only CITES listed freshwater fish from South America (Stewart, 2013b). If the systematics of any extant genus of fishes should be well known, then you might expect it would be *Arapaima*. But this is not the case. Albert Günther (1868) synonymized without explanation the four nominal species recognized by Valenciennes (1847: *A. gigas, A. arapaima, A. mapae*, and *A. agassizii*) as *A. gigas*. There things stayed for 145 years. According to Stewart (2013a,b) this lack of attention may be due in part to the large sizes of adults, which makes it challenging to collect, preserve and store them in collections. Stewart (2013a) resurrected *A. arapaima, A. mapae* and *A. agassizii*, redescribing this last and rarest species based on illustrations in the type description by Spix and Agassiz (1829).[3] Stewart (2013b) described a new fifth species in the genus, *A. leptosoma*. Stewart (2013b: 470) notes that the total number of preserved specimens in world collections known to be from the type locality of *A. leptosoma* "can be counted on one hand," going on to state, "Our understanding of the taxonomy and distributional ecology of *Arapaima* is substantially hindered by that paucity of study materials." In many respects, the challenges of making systematic studies of the species of *Arapaima* resemble taxonomic and practical challenges typical of paleoichthyology.

A common reason for naming new species in the fossil record of fishes is to distinguish materials from the same locality. For example, fishes from the Cenomanian of Namoura, Lebanon, are beautifully preserved in a lithographic limestone, ideally suited for preparation using the acid transfer method (Forey et al., 2003). Fossils prepared in this way are suitable for finely detailed descriptions and differential diagnoses at the species level.

---

3. The type and only known specimen was at the Zoologische Staatssammlung in Munich, Germany; it was destroyed during the Second World War.

Thus, †*Serrilepis prymnostrigos* and †*S. minor* are two new species in the genus †*Serrilepis* "distinguished chiefly by the condition of the squamation" (Forey et al., 2003: 294); two species in the new genus †*Triplomystus* are †*T. noorae* and †*T. oligoscutatus* are distinguished by "body proportions and meristic characters" (Forey et al., 2003: 275); and †*Armigatus namourensis* and †*A. alticorpus* are distinguished by relative body depth (Forey et al., 2003: 282).

Forey et al. (2003, table 17) also broadly surveyed species and genera of fossil fishes known from eight Cenomanian localities distributed around the Tethys Sea. Several of these localities are thought to represent shallow-water, near-shore environments, whereas others are interpreted as offshore; they range from nearly equatorial to 40° N. These localities are: (1) Hakel, (2) Hajula, and (3) Namoura, Lebanon; (4) English Chalk in Southeastern England; (5) Komen in Slovenia; (6) Jebel Tselfat, Morocco; (7) Jerusalem; and (8) Estrémadure, Portugal. Noting that there is little overlap in species across all of these localities, and only one genus common to the four most speciose localities (localities 1, 4, 5, and 6), Forey et al. (2003: 321) wrote:

> This is undoubtedly influenced a great deal by taxonomic eclecticism. Comparisons are probably more realistic at generic than at specific level. Here we can see that there are considerably more shared genera between Komen, Lebanon and Morocco than any of these localities and England.

This quote emphasizes an important and lasting point about the study of species of fossil fishes: paleoichthyologists tend to focus on genera and their validity and diagnoses far more than on the validity and diagnoses of species. Species names for fossil fishes thus primarily serve as practical tools for cataloging paleodiversity. Species of fossil fishes may yet have much to teach us about paleodiversity and, in some cases, evolutionary processes, but we still need basic tools such as a *Catalog of Fossil Fishes* to understand what has already been described and how best to understand the biological meaning of these taxa. Only then can we begin to assess how best to integrate the study of extant and fossil fishes.

Future Considerations

If, as interpreted here, species names for fossil fishes primarily serve as practical tools for cataloging paleodiversity and secondarily as tools for phylogenetic and biogeographic analyses, then a pressing need for the future is the development of a comprehensive, global, universally accessible digital *Catalog of Fossil Fishes*. It should be designed to be comparable to and compatible with the *Catalog of Fishes* (Eschmeyer, 1998a, 2015) and build directly upon all original species descriptions since 1758. It would need to be hosted by an institution, much as the California Academy of Sciences hosts the Catalog of Fishes, with the clear understanding that such hosting needs to be a long-term commitment. As was done to develop the *Catalog of Fishes*, the first step is to document the genera (Eschmeyer and Bailey, 1990). The goal would be to catalog: (1) all available generic names; (2) all valid generic names; and (3) all type localities, including stratigraphic information. Subsequently, this effort could be extended to document all available species names, all valid species names, and all type localities. Such a multi-decadal project could become an important catalyst for revisionary studies of fossil and living fishes.

In closing, paleontologists and neontologists should not regard most species of fossil fishes as equivalent to biological species of living fishes. Without specific evidence from, for example, the availability of large numbers of specimens collected at the same locality in the same horizon, species names of fossil fishes represent little more than convenient tags for communication about specimens, localities, and particular mixtures of characters.

Acknowledgments

I am grateful to William N. Eschmeyer, Warren Allmon, Dana Friend, Joshua K. Moyer, Stacy Farina, Katherine E. Bemis, Eric Hilton, Amy McCune, and Richard van der Laan for discussions and improvements to the manuscript. Andrea Cerruti and Arthur Egitto assisted with manuscript preparation.

References

American National Biography Online. Article on Henry Shaler Williams. http://www
.anb.org/articles/13/13–02669.html. Electronic version accessed May 7, 2014.
Bellwood, D. R. 1996. The Eocene fishes of Monte Bolca: The earliest coral reef fish assemblage. Coral Reefs 15: 11–19.
Bemis, K. E. and W. E. Bemis. 2015. Functional and developmental morphology of tooth replacement in the Atlantic Wolffish, *Anarhichas lupus* (Teleostei: Zoarcoidei: Anarhichadidae). In: G. Arratia and D. Johnson, eds. Symposium Proceedings: Fishes and Morphology Today. Copeia 103(4): 886–901.

Bemis, W. E., E. K. Findeis, and L. Grande. 1997. Overview of Acipenseriformes. Environmental Biology of Fishes 48: 25–71.

Bemis, W. E., J. K. Moyer, and M. Riccio. 2015. Homology of lateral cusplets in the teeth of lamnid sharks (Lamniformes: Lamnidae). In: G. Arratia and D. Johnson, eds. Symposium Proceedings: Fishes and Morphology Today. Copeia 103(4): 961–972.

Bengston, P. 1988. Open nomenclature. Paleontology 31: 223–227.

Bibliography of Fossil Vertebrates. 2014. http://www.bfvol.org/. Electronic version accessed April 14, 2014.

Bogert, C. M. 1943. Introduction. Pages 107–108 in: C. M. Bogert, ed. *Criteria for Vertebrate Subspecies, Species and Genera.* Annals of the New York Academy of Science 44.

Boreske, J. R. 1974. A review of the North American fossil amiid fishes. Bulletin of the Museum of Comparative Zoology, Harvard University 146:1–87.

Bottjer, D. J., W. Etter, J. W. Hagadorn, and C. M. Tang, eds. 2003. *Exceptional Fossil Preservation: A Unique View on the Evolution of Marine Life.* Columbia University Press, New York.

Bruner, J. C. 1992. A catalogue of type specimens of fossil fishes in the Field Museum of Natural History. Fieldiana Geology, n.s. 23, 1431: 1–54.

Cappetta, H. 2012. *Chondrichthyes: Mesozoic and Cenozoic Elasmobranchii: Teeth.* H.-P. Schultze, ed., *Handbook of Paleoichthyology Volume 3E.* Verlag Dr. Friedrich Pfeil, München.

Cleevely, R. J. 1983. *World Paleontological Collections.* British Museum (Natural History), London. 1–365.

Cloutier, R. 2013. Great Canadian Lagerstätten 4. The Devonian Miguasha biota (Québec): UNESCO World Heritage Site and a time capsule in the early history of vertebrates. Geoscience Canada 40: 149–163.

Colburn, J., R. E. Crabtree, J. B. Shaklee, E. Pfeiler, and B. W. Bowen. 2001. The evolutionary enigma of bonefishes (*Albula* spp.): Cryptic species and ancient separations in a globally distributed shorefish. Evolution 55: 807–820.

Darwin, C. 1859. *On the Origin of Species by Means of Natural Selection, or the Preservation of Favoured Races in the Struggle for Life.* John Murray, London.

Day, J. J. 2003. Evolutionary relationships of the Sparidae (Teleostei: Percoidei): Integrating fossil and Recent data. Transactions of the Royal Society of Edinburgh: Earth Sciences 93: 333–353.

Dean, B. 1902. The preservation of muscle fibres in sharks of the Cleveland shale. American Geologist 30: 273–278.

de Figueiredo, F. J., V. Gallo, and M. E. C. Leal. 2012. Phylogenetic relationships of the elopomorph fish †*Paraelops cearensis* Silva Santos revisited: Evidence from new specimens. Cretaceous Research 37: 148–154.

Dobzhansky, T. 1935. A critique of the species concept in biology. Philosophy of Science 2: 344–355.

Duméril, A. H. A. 1869. Note sur trois poissons de la collection du Muséum un sturgeon, un polyodonte, et un malarmat, accompagnée de quelques considérations générales sur les groupes auxquels ces espèces appartiennent. Nouvelles Archives du Museum d'Histoire Naturelle de Paris 4: 93–116.

Duméril, A. H. A. 1870. *Histoire Naturelle des Poissons, ou Ichthyologie Générale,* vol. 2. Paris. 624 pp.

Eastman, C. R. 1907. *Devonic Fishes of the New York Formations.* New York State Museum, Albany, 1–235.

Egerton, P. M. G. 1858. On *Chondrosteus,* an extinct genus of the Sturionidae, found in the Lias Formation at Lyme Regis. Philosophical Transactions of the Royal Society 1858: 871–885.

Eschmeyer W. N., ed. 1998a. *Catalog of Fishes.* 3 vols. California Academy of Sciences, San Francisco.

Eschmeyer, W. N. 1998b. Appendix A. Genera and species and the International Code of Zoological Nomenclature. Pages 2847–2881 in: W. N. Eschmeyer, ed., *Catalog of Fishes.* California Academy of Sciences, San Francisco.

Eschmeyer, W. N., ed. 2014. *Catalog of Fishes: Genera, Species, References.* (http://research.calacademy.org/research/ichthyology/catalog/fishcatmain.asp). Electronic version accessed April 12, 2014. (This version was edited by W. N. Eschmeyer.)

Eschmeyer, W. N., ed. 2015. *Catalog of Fishes: Genera, Species, References.* (http://research.calacademy.org/research/ichthyology/catalog/fishcatmain.asp). Electronic version accessed February 20, 2015. (This version was edited by W. N. Eschmeyer.)

Eschmeyer, W. N., and R. Bailey. 1990. *Genera of Fishes.* California Academy of Sciences, San Francisco.

Eschmeyer, W. N., R. Fricke, J. D. Fong, and D. A. Polack. 2010. Marine fish diversity: History of knowledge and discovery (Pisces). Zootaxa 2525: 19–50.

Forey, P. L., L. Yi, C. Patterson, and C. E. Davies. 2003. Fossil fishes from the Cenomanian (Upper Cretaceous) of Namoura, Lebanon. Journal of Systematic Palaeontology 1: 227–330.

Froese, R., and D. Pauly, eds. 2014. FishBase. World Wide Web electronic publication. www.fishbase.org. Accessed April 12, 2014.

George, T. N. 1956. Biospecies, chronospecies, and morphospecies. Pages 123–137 in: P. C. Sylvester-Bradley, ed., *The Species Concept in Paleontology.* Systematics Association, London.

Grande, L. 2010. An empirical synthetic pattern study of gars and closely related species (Lepisosteiformes) based mostly on skeletal anatomy: The resurrection of Holostei. American Society of Ichthyologists and Herpetologists, special sublication 7. Allen Press, Lawrence, Kansas.

Grande, L. 2013. *The Lost World of Fossil Lake.* University of Chicago Press, Chicago.

Grande, L., and W. E. Bemis. 1991. Osteology and phylogenetic relationships of fossil and Recent paddlefishes (Polyodontidae) with comments on the interrelationships of Acipenseriformes. Society of Vertebrate Paleontology Memoir 1:1–121; supplement to Journal of Vertebrate Paleontology 11(1).

Grande, L., and W. E. Bemis. 1998. A comprehensive phylogenetic study of amiid fishes (Amiidae) based on comparative skeletal anatomy: An empirical search for interconnected patterns of natural history. Society of Vertebrate Paleontology memoir 4: i–x, 1–690; supplement to Journal of Vertebrate Paleontology 18(1).

Grande, L., F. Jin, Y. Yabumoto, and W. E. Bemis. 2002. †*Protopsephurus liui*, a well-preserved primitive paddlefish (Acipenseriformes: Polyodontidae) from the Lower Cretaceous of China. Journal of Vertebrate Paleontology 22:209–237.

Grande, L., G.-Q. Li, and M. V. H. Wilson. 2000. *Amia cf. pattersoni* from the Paleocene Paskapoo Formation of Alberta. Canadian Journal of Earth Science 37: 31–37.

Gregory, W. K. 1937. Supra-specific variation in nature and in classification: A few examples from mammalian paleontology. American Naturalist 71: 268–276.

Günther, A. 1868. *Catalogue of the Fishes of the British Museum. Catalogue of the Physostomi, containing the Families Heteropygii, Cyprinidae, Gonorhynchidae, Hyodontidae, Osteoglossidae, Clupeidae, Chirocentridae, Alepocephalidae, Notopteridae, Halosauridae, in the Collection of the British Museum v. 7*: i–xx + 1–512. British Museum (Natural History), London.

Haldane, J. B. S. 1956. Can a species concept be justified? Pages 95–96 in: P. C. Sylvester-Bradley, ed., *The Species Concept in Paleontology*. Systematics Association, London.

Harrington, R. C. and T. J. Near. 2012. Phylogenetic and coalescent strategies of species delimitation in snubnose darters (Percidae: Etheostoma). Systematic Biology. 61: 63–79.

Hidaka, K., Y. Iwatsuki, and J. E. Randall. 2008. A review of the Indo-Pacific bonefishes of the *Albula argentea* complex, with a description of a new species. Ichthyological Research 55: 53–64.

Hilton, E. J., and W. E. Bemis. 1999. Skeletal variation in shortnose sturgeon (*Acipenser brevirostrum*) from the Connecticut River: Implications for the study of fossil fishes. Pages 69–94 in: G. Arratia and H.-P. Schultze, eds., *Mesozoic Fishes II: Systematics and the Fossil Record*. Verlag Dr. Friedrich Pfeil, München.

Hilton, E. J., and W. E. Bemis. 2012. External morphology of shortnose sturgeon, *Acipenser brevirostrum* (Acipenseriformes: Acipenseridae) from the Connecticut River, with notes on variation as a natural phenomenon. Pages 243–275 in: B. Kynard, P. Bronzi, and H. Rosenthal, eds., *Life History and Behaviour of Connecticut River Shortnose and Other Sturgeons*. World Sturgeon Conservation Society, special publication 4.

Hilton, E. J., and L. Grande. 2006. Review of the fossil record of sturgeons, family Acipenseridae (Actinopterygii: Acipenseriformes), from North America. Journal of Paleontology 80: 672–683.

Hilton, E. J., and L. Grande. 2008. Fossil mooneyes (Teleostei, Hiodontiformes, Hiodontidae) from the Eocene of western North America, with a reassessment of their taxonomy. Pages 221–251 in: L. Cavin, A. Longbottom, and M. Richter, eds., *Fishes and the Break-Up of Pangea*. Geological Society of London, special publication 295.

Hilton, E. J., L. Grande, and W. E. Bemis. 2011. Skeletal anatomy of the shortnose sturgeon, *Acipenser brevirostrum* Lesueur, 1818, and the systematics of sturgeons (Acipenseriformes, Acipenseridae). Fieldiana Life and Earth Sciences, no. 3: 1–168.

Hilton, E. J., and P. L. Forey. 2009. Redescription of †*Chondrosteus acipenseroides* Egerton, 1858 (Acipenseriformes, Chondrosteidae) from the Lower Lias of Lyme Regis (Dorset, England), with comments on the early evolution of sturgeons and paddlefishes. *Journal of Systematic Palaeontology* 7: 427–453.

Holder, M. T., M. V. Erdmann, T. P. Wilcox, R. L. Caldwell, and D. M. Hillis. 1999. Two living species of coelacanths? Proceedings of the National Academy of Sciences 96: 12616–12620.

International Commission on Zoological Nomenclature. 1999. International Code of Zoological Nomenclature, 4th ed. London.

International Commission on Zoological Nomenclature Online. 2014. (Incorporating Declaration 44, amendments of Article 74.7.3, with effect from 31 December 1999 and the Amendment on e-publication, amendments to Articles 8, 9, 10, 21 and 78, with effect from 1 January 2012). The International Trust for Zoological Nomenclature. c/o The Natural History Museum, London. http://www.nhm.ac.uk/hosted-sites/iczn/code/. Accessed January 27, 2014.

Jarvik, E. 1980. *Basic structure and Evolution of Vertebrates*. Vols. 1 and 2. Academic Press, London.

Kinsey, A. C. 1937. Supra-specific variation in nature and in classification from the view-point of zoology. American Naturalist 71: 206–222.

Kwun, H. J., and J. K. Kim. 2011. A new species of bonefish, *Albula koreana* (Albuliformes: Albulidae) from Korea and Taiwan. Zootaxa 2903: 57–63.

Leidy, J. 1873a. Notice of remains of fishes in the Bridger Tertiary Formation of Wyoming. Proceedings of the Academy of Natural Sciences, Philadelphia 1873:97–99.

Leidy, J. 1873b. Contributions to the extinct vertebrate fauna of the western territories. Report of United States Geological Survey of the Territories. Part 1: 1–358. F. V. Hayden, U.S. Geologist.

Linnaeus, C. 1758. *Systema naturae per regna tria naturae, secundum classes, ordines, genera, species, cum characteribus, differentiis, synonymis, locis, Tomus I*. Editio decima, reformata, Holmiae.

Linnaeus, C. 1766. *Systema Naturae*. Editio duodecima, reformata. Impensis Direct Laurentii Salvii, Holmiae.

MacAlpin, A. 1941a. *Paleopsephurus wilsoni*, a new polyodontid fish from the Upper Cretaceous of Montana, with a discussion of allied fish, living and fossil (abstract). Bulletin of the Geological Society of America 52: 1989.

MacAlpin, A. 1941b. *Paleopsephurus wilsoni*, a new polyodontid fish from the Upper Cretaceous of Montana, with a discussion of allied fish, living and fossil. Unpublished Ph.D. dissertation, University of Michigan.

MacAlpin, A. 1947. *Paleopsephurus wilsoni*, a new polyodontid fish from the Upper Cretaceous of Montana, with a discussion of allied fish, living and fossil. Contributions from the Museum of Paleontology, University of Michigan 6: 167–234.

Maisey, J. G. 1991. *Santana Fossils: An Illustrated Atlas*. T. F. H. Publications, Neptune, New Jersey.

Mayr, E. 1942. *Systematics and the Origin of Species from the Viewpoint of a Zoologist*. Columbia University Press, New York.

Mayrinck, D., P. M. Brito, and O. Otero. 2010. A new albuliform (Teleostei: Elopomorpha) from the Lower Cretaceous Santana Formation, Araripe Basin, northeastern Brazil. Cretaceous Research 31: 227–236.

McCune, A. R. 1986. A revision of *Semionotus*, with redescriptions of valid European species. Palaeontology 29: 213–233.

McCune, A. R. 1987a. Toward the phylogeny of a fossil species flock: Semionotid fishes from a lake deposit in the Early Jurassic Towaco Formation, Newark Basin. Yale Peabody Museum of Natural History Bulletin 43: 1–108.

McCune, A. R. 1987b. Lakes as laboratories of evolution: Endemic fishes and environmental cyclicity. Palaios 2: 446–454.

McCune, A. R. 1990. Evolutionary novelty and atavism in the *Semionotus* Complex: Relaxed selection during colonization of an expanding lake. Evolution 44: 71–85.

McCune, A. R. 1996. Biogeographic and stratigraphic evidence for rapid speciation in semionotid fishes. Paleobiology 22: 34–48.

McCune, A. R. 2004. Diversity and speciation of semionotid fishes in Mesozoic rift lakes. Pages 362–379 in: U. Dieckman, M. Doebli, and J. A. J. Metz, eds., *Adaptive Speciation*. Cambridge University Press.

McCune, A. R., K. S. Thomson, and P. E. Olsen. 1984. Semionotid fishes from the Mesozoic great lakes of North America. Pages 27–44 in: A. Echelle and I. Kornfield, eds., *Evolution of Fish Species Flocks*. University of Maine Press, Orono.

Mitchill, S. L. 1815. The fishes of New York described and arranged. Transactions of the Literary and Philosophical Society of New York volume 1 (article 5) (for 1814): 355–492.

Miya, M., N. I. Holcroft, T. P. Satoh, M. Yamaguchi, M. Nishida, and E. O. Wiley. 2007. Mitochondrial genome and a nuclear gene indicate a novel phylogenetic position of deep-sea tube-eye fish (Stylephoridae). Ichthyological Research 54: 323–332.

Moyer, J. K., M. Riccio, and W. E. Bemis. 2015. Development and microstructure of tooth histotypes in the blue shark, *Prionace glauca* (Carcharhiniformes: Carcharhinidae) and the great white shark, *Carcharodon carcharias* (Lamniformes: Lamnidae). Journal of Morphology. 276: 797–817.

Near, T. J., M. Sandel, K. L. Kuhn, P. J. Unmack, P. C. Wainwright, and W. L. Smith. 2012a. Nuclear gene-inferred phylogenies resolve the relationships of the enigmatic pygmy sunfishes, *Elassoma* (Teleostei: Percomorpha). Molecular Phylogenetics and Evolution. 63: 388–395.

Near, T. J., R. I. Eytan, A. Dornburg, K. L. Kuhn, J. A. Moore, M. P. Davis, P. C. Wainwright, M. Friedman, and W. L. Smith. 2012b. Resolution of ray-finned fish phylogeny and timing of diversification. Proceedings of the National Academy of Sciences 109: 13698–13703.

Nelson, G., and N. Platnick. 1981. *Systematics and Biogeography: Cladistics and Vicariance.* Columbia University Press, New York.

Nolf, D. 1985. *Otolithi Piscium.* In H.-P. Schultze, ed., *Handbook of Paleoichthyology. Volume 10.* Gustav Fischer, Stuttgart.

Nolf, D. 2013. *The Diversity of Fish Otoliths, Past and Present.* E. Steurbaut, R. Bzobohaty, and K. Hoedemakers, eds. Royal Belgian Institute of Natural Sciences, Brussels.

Patterson, D. J., J. Cooper, P. M. Kirk, R. L. Pyle, and D. P. Remsen. 2010. Names are key to the big new biology. Tree 25: 686–691.

Pfeiler, E., A. M. Van Der Heiden, R. S. Ruboyianes, and T. Watts. 2011. *Albula gilberti*, a new species of bonefish (Albuliformes: Albulidae) from the eastern Pacific, and a description of adults of the parapatric *A. esuncula*. Zootaxa 3088: 1–14.

Pouyaud, L., S. Wirjoatmodjo, I. Rachmatika, A. Tjakrawidjaja, R. Hadiaty, and W. Hadie. 1999. Une nouvelle espèce de coelacanthe. Preuves gènètiqueset morphologiques. Comptes Rendus de l'Académie des Sciences—Series III— Sciences de la Vie 322: 261–267.

Ride W. D. L., and T. Younès, eds. 1986. *Biological Nomenclature Today.* International Union of Biological Sciences Monograph Series 2: 1–70.

Ride, W. D. L. 1999. Introduction. Pages XIX–XXIX In International Commission on Zoological Nomenclature, eds., *International Code of Zoological Nomenclature*, 4th ed. Also available at http://www.nhm.ac.uk/hosted-sites/iczn /code/; accessed January 27, 2014.

Ruffing, R. A., P. M. Kocovsky, and J. R. Stauffer, Jr. 2002. An introduction to species concepts and speciation of fishes. Fish and Fisheries, 3: 143–145. doi: 10. 1046/j.1467-2979.2002.00087.x

Schultze, H.-P., and R. Cloutier, eds. 1996. *Devonian Fishes and Plants of Miguasha, Quebec, Canada.* Verlag Dr. Friedrich Pfeil, München.

Sedgwick, A., and R. I. Murchison. 1828. On the structure and relations of the deposits contained between the Primary Rocks and the Oolitic Series in the North of Scotland. Transactions of the Geological Society of London 3: 125–160.

Sepkoski, J. J., Jr. 2002. *A Compendium of Fossil Marine Animal Genera.* Edited by D. Jablonski and M. Foote. Bulletins of American Paleontology 363. Paleontological Research Institution, Ithaca.

Simpson, G. G. 1937. Supra-specific variation in nature and in classification from the view- point of paleontology. American Naturalist 71: 236–267.

Simpson, G. G. 1943. Criteria for genera, species, and subspecies in zoology and paleozoology. Pages 145–178 in: C. M. Bogert, ed., *Criteria for Vertebrate Subspecies, Species, and Genera.* Annals of the New York Academy of Science 44.

Simpson, G. G. 1944. *Tempo and Mode in Evolution.* Columbia University Press, New York.

Spix, J. B. von, and L. Agassiz. 1829. *Selecta Genera et Species Piscium Brasiliensium.* Typis C. Wolf, Monachii (München).

Stanley, S. M. 1979. *Macroevolution: Pattern and Process.* W. H. Freeman and Company, San Francisco.

Stewart, D. J. 2013a. Re-description of *Arapaima agassizii* (Valenciennes), a rare fish from Brazil (Osteoglossomorpha: Osteoglossidae). Copeia 2013: 38–51.

Stewart, D. J. 2013b. A new species of *Arapaima* (Osteoglossomorpha: Osteoglossidae) from the Solimões River, Amazonas State, Brazil. Copeia 2013: 470–476.

Sylvester-Bradley, P. C., ed. 1956. *The Species Concept in Paleontology.* Systematics Association, London.

Toombs, H. A., and A. E. Rixon. 1959. The use of acids in the preparation of vertebrate fossils. Curator 2: 304–312.

Trinajstic, K. C. Marshall, J. Long, and K. Bifield. 2007. Exceptional preservation of nerve and muscle tissues in Late Devonian placoderm fish and their evolutionary implications. Biology Letters 3: 197–200.

Valenciennes, A., in Cuvier, G., and A. Valenciennes. 1847. *Histoire Naturelle des Poissons.* Vol. 19. Bertrand, Paris. (1969 reprint edition, A. Asher and Company, Amsterdam.)

van der Laan, R., W. N. Eschmeyer, and R. Fricke. 2014. Family group names of fishes. Zootaxa 3882: 1–230. http://dx.doi.org/10.11646/zootaxa.3882.1.1

Werning, S. 2013. How species are like pornography: Species concepts and the fossil record.http://blogs.plos.org/paleo/2013/04/01/how-species-are-like-pornography/. Accessed May 7, 2014.

Wiley, E. O., and B. S. Lieberman. 2011. *Phylogenetics: The Theory of Phylogenetic Systematics.* 2nd ed. Wiley-Blackwell, Hoboken, New Jersey.

Wilkins, J. S. 2009. *Species: A History of the Idea.* University of California Press, Berkeley.

Williams, H. S. 1881. Notes on some fish-remains from the Upper Devonian rocks in New York State. Proceedings of the American Association for the Advancement of Science 30: 192–193.

Williams, H. S. 1887. On the fossil faunas of the Upper Devonian: The Genesee section, New York. United States Geological Survey Bulletin 41: 1–123.

Wheeler, A. 1985. The Linnaean fish collection in the Linnean Society of London. Zoological Journal of the Linnean Society 84: 1–76.

Whitehead, P. J. P. 1986. The synonymy of *Albula vulpes* (Linnaeus, 1758) (Teleostei, Albulidae). Cybium 10: 211–230.

Woodward, A. S. 1889. *Catalogue of the Fossil Fishes in the British Museum (Natural History)*. Vol. 1. British Museum (Natural History), London.

Woodward, A. S. 1891. *Catalogue of the Fossil Fishes in the British Museum (Natural History)*. Vol. 2. British Museum (Natural History), London.

Woodward, A. S. 1895. *Catalogue of the Fossil Fishes in the British Museum (Natural History)*. Vol. 3. British Museum (Natural History), London.

Woodward, A. S. 1901. *Catalogue of the Fossil Fishes in the British Museum (Natural History)*. Vol. 4. British Museum (Natural History), London.

CHAPTER TWELVE

The Impact of Invasive Species on Speciation: Lessons from the Fossil Record

Alycia L. Stigall

Introduction

Life on Earth began with a single species, yet approximations of modern species diversity range from 8.7 to 100 million extant species (e.g., May 1988; Mora et al. 2011). Ernst Mayr estimated that "99.99% or more of all evolutionary lineages have become extinct" (2001, 140). If this is reasonably accurate, then well over 87 billion species have existed during the history of our planet. The development of billions of species—via billions of speciation events—during the history of life dictates that speciation, the process by which new species form, is of central importance to the history of life on our planet. Thus, we must understand the processes that both promote and hinder speciation if we seek to understand the past and future of biodiversity on our planet. Better constraining the relationships between earth system events and speciation becomes particularly significant in the context of current global climatic and environmental changes.

Both the fossil record and neontological data demonstrate that speciation rates have varied through time and among clades (e.g., Avise and Walker 1998; Sepkoski 1998; Alroy 2008; Stadler 2011). Well-known terms such as "adaptive radiation" and "living fossils," which reference bursts of rapid speciation and long intervals of restricted speciation within clades, respectively, convey the generality of this rate variation. Some of the variation in speciation rates can be attributed to biologic or internal causes. For example, speciation rate has been demonstrated to be higher among lineages with specialist vs. generalist ecologies (e.g., Vrba 1987, 1992; Kammer et al. 1997) and among species with narrow rather than broad geographic ranges (e.g., Jackson 1974; Rode and Lieberman 2004; Jablonski 2008) because populations with these characteristics may be isolated from each other more easily. External factors, however, can also play significant roles in moderating speciation rate. For example, temporal intervals with tectonic fragmentation often exhibit elevated levels of speciation (e.g., Lieberman 2003, 2012).

In this chapter, I build on the framework of species and speciation concepts developed in the preceding chapters of this volume to examine the processes and mechanisms of speciation elucidated from analyses of invasive species in the fossil record during two invasive regimes, the Late Devonian Biodiversity Crisis and the Late Ordovician Richmondian Invasion. Specifically, the case studies presented in this chapter emphasize the role that external drivers, such as sea level change and the arrival of invasive species within an ecosystem, and internal factors, including niche breadth and dispersal potential, may have in promoting or limiting speciation processes.

Invasive Species and Potential Impacts on Speciation

Discussions of speciation rate are inexorably linked to a particular species concept (see Allmon [chapter 3] and Allmon and Sampson [chapter 4], this volume). For the purpose of this chapter, a species is defined as a group of organisms that maintain genetic continuity by interbreeding among members of the group and that is distinct from other reproductive entities. The genetic integrity of a species persists through geologic time, and thus species have temporal as well as spatial attributes. Employing a genetic rather than character-based definition provides a framework to examine speciation as a process. Within this definition, speciation is the process by which a population becomes genetically isolated from the ancestral population. Speciation, therefore, relates to a unique event that occurred during a discrete interval of geologic time that transpired at a specific location within a specific lineage of organisms. To fully assess speciation, therefore, we must understand the temporal, spatial, and evolutionary attributes of the process.

Reproductive isolation may occur either via shifts in reproductive timing or behavior within the geographic range of the ancestor (sympatric speciation) or via geographic separation of the incipient species from the ancestral population (allopatric speciation) (Mayr 1963). Sympatric speciation is often undetectable in the fossil record because reproductive shifts may not correlate with morphological changes (Benton and Pearson 2001). During allopatric speciation, however, incipient species adapt to environmental conditions different from those of the ancestral range and are thus more likely to have discrete (fossilizable) morphological characters that allow separation of the new species from the ancestral morphology. Consequently, analyses of speciation in the fossil record are largely restricted to analyses of allopatric events (Lieberman 2000, Stigall 2010b).

There are two primary mechanisms of allopatric speciation: (1) *vicariance*, in which the ancestral population becomes passively divided into two or more large subpopulations by the formation of a barrier, each of which diverges to form a new species, and (2) *dispersal*, in which a subpopulation actively moves away from the ancestral range and establishes a geographically isolated population that subsequently diverges (fig. 12.1; Wiley and Mayden 1985; Lieberman 2000). Vicariance and dispersal are characterized by discrete biogeographic patterns relative to the geographic range of daughter species and ancestral species (fig. 12.1). Consequently, it is possible to distinguish vicariance and dispersal events in fossil taxa where evolutionary relationships and biogeographic distributions are known (fig. 12.2; Lieberman 2000).

In modern ecosystems, human impacts, such as habitat fragmentation and the transport of invasive species, have the potential to greatly increase the frequency of vicariance and dispersal, respectively. Unfortunately, many instances of habitat fragmentation result in elimination of local populations and ultimately to species extinctions. Thus, impact of modern habitat fragmentation on speciation processes may be limited. Therefore, I will focus the discussion herein on the potential for introductions of nonnative species to impact speciation processes.

The dramatic increase in human-facilitated species invasions in recent decades has been predicted to have a significant impact on speciation in the modern biota (Levin 2003; Wiens and Graham 2005). Modern species invasions are initiated by the transportation of a (typically) small population into an ecosystem to which this species is alien. Only about 10% of these dispersal events succeed in establishing a viable population in the new location (Lockwood et al. 2007). If these new populations are unable

Vicariance

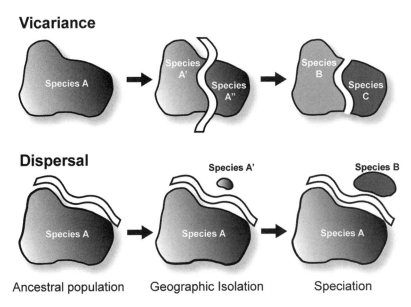

Dispersal

Ancestral population   Geographic Isolation   Speciation

Figure 12.1 Differential geographic patterns of speciation by vicariance and dispersal. During vicariance, the ancestral population (species A) becomes passively divided by a geographic barrier. Incipient species (species A' and A″) form during geographic isolation and later diverge to become new species (species B and C). In speciation by dispersal, a subpopulation of the ancestral species (species A) actively migrates across a geographic barrier to form an incipient species (species A') that later diverges to become a new species (species B). Modified from Stigall (2013).

to exchange genetic material with the parent population, these newly isolated populations may eventually diverge into a new allopatric species. Similar processes facilitated by earth history factors, such as transgressive events, would be identified as speciation by dispersal in the fossil record (see fig. 12.2). Because the fossil record does not record ecological timescales, the processes operating during speciation itself cannot be assessed within the context of this chapter. Only successful episodes of speciation can be ascertained in the fossil record, but analyzing the distribution and factors that promote or hinder successful speciation can provide key insights into the development of biodiversity through time.

Modern invasive species have been documented to alter the evolutionary pathway of native species via competitive exclusion, niche displacement, hybridization, predation, and extinction (Mooney and Cleland 2001). Furthermore, invaders themselves may evolve in response to novel biotic

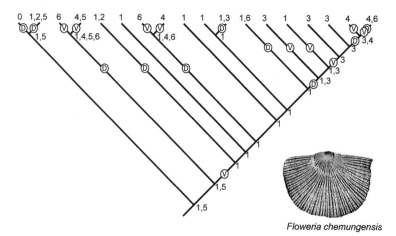

Floweria chemungensis

Figure 12.2 Differentiation of speciation mode on an area cladogram. Here, numerals representing biogeographic areas are mapped onto a cladogram for the brachiopod genus *Floweria*. Biogeographic states at the nodes were optimized using modified Fitch parsimony of Lieberman (2000). Speciation events where daughter species occupy a subset of the ancestral range are interpreted as vicariance events, marked V. Speciation events where daughter species occupy areas additional to the ancestral areas are interpreted as dispersal events, marked D. Geographic areas denoted by: 0 = Europe, 1 = Northern Appalachian Basin, 2 = Southern Appalachian Basin, 3 = Michigan Basin, 4 = Iowa/Illinois Basin, 5 = Missouri, 6 = Western United States. Modified from Stigall Rode (2005).

and abiotic interactions in their new environment (Broennimann et al. 2007; Pearman et al. 2008). The extent to which native or invasive species respond to ecosystem changes via an adaptive response, such as shifting of an ecologic niche, is currently not well constrained (Simberloff et al. 2013). Synthetic reviews of the literature have reported both high levels of niche conservatism (Peterson 2011) and substantial amounts of niche evolution (Pearman et al. 2008), whereas analyses of longer timescales in the deep fossil record have demonstrated that niche response varies between rapid and gradual environmental changes (Brame and Stigall 2014; Stigall 2012b, 2014).

Analyzing the evolutionary impacts of species invasions within the fossil record provides a framework for examining the impact of species introductions on evolutionary timescales. Although humans were not significant transportation vectors throughout geologic time, other processes, such as tectonic rearrangements, continental flexure, and sea level change did facilitate analogous species movements in the geologic past (Rode and Lieberman 2004). In this context, an ancient invasive species is defined as a species that evolved in one biogeographic region and subsequently expanded its geographic ranges into a region from which it was previously excluded due to a geographic barrier (Stigall 2010a). Significant invasion events have occurred throughout the Phanerozoic and include intervals such as the Pliocene Great American Biotic Interchange (Webb 2006) and Trans-Arctic Interchange (Vermeij 1991), the Late Devonian Biodiversity Crisis (Rode and Lieberman 2004; McGhee et al. 2013), the spread of the Late Ordovician Hirnantian Fauna (Rong et al. 2007), and the Late Ordovician Richmondian Invasion (Holland 1997; Stigall 2010b). For many of these, analyses of biodiversity dynamics have focused on patterns of extinction or differential survival of taxa rather than speciation (e.g., Webb 2006). However, examining the speciation dynamics of these and similar intervals in Earth history is critical for understanding the long-term trajectory for modern biodiversity of our planet. In this chapter, I review how rampant species invasions impacted speciation during the Late Devonian Biodiversity Crisis and Late Ordovician Richmondian Invasion and provide a synthetic overview of how such invasion events impact the process of speciation.

Invasive Species and Speciation during the Late Devonian

The Middle Devonian was an interval of high endemicity among biogeographic regions, and the coral reef ecosystem reached its greatest geographic distribution Earth's history (Copper 1994; McGhee 1996). By the end of the Frasnian Stage of the Late Devonian, however, metazoan reefs were largely extinct and a cosmopolitan fauna dominated the epicontinental seas (Johnson 1970). The transition from the highly endemic faunas of the Middle Devonian to the cosmopolitan Late Devonian fauna was facilitated by widespread species invasions (McGhee 1997; Rode and Lieberman 2004) and has been referred to as the Great Devonian Interchange (GDI) by McGhee (1997). The biodiversity loss associated with this transition, referred to as the Frasnian-Famennian Mass Extinction, ranks as one of the largest biodiversity crises in Earth history (Raup and Sepkoski 1982). This event was not technically a mass extinction, however, because the Frasnian *extinction* rate was not statistically higher than the background extinction rate throughout the Phanerozoic (fig. 12.3; Bambach et al. 2004; Alroy 2008). Instead, reduced *speciation* was the primary cause of this decline in biodiversity (Bambach et al. 2004; Stigall 2010a,

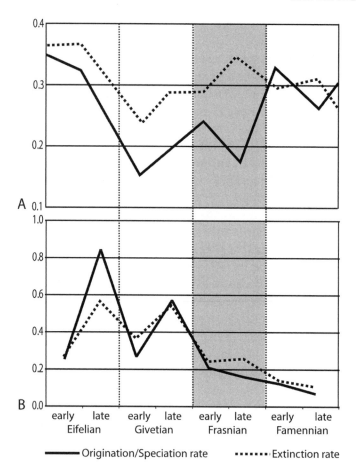

Figure 12.3 Temporal changes in extinction and origination/speciation across the Middle to Late Devonian interval.
The Late Devonian Biodiversity Crisis interval is shaded grey. (A) Proportion of generic extinction or origination per interval. Modified from Bambach et al. (2004). (B) Instantaneous per-capita speciation and extinction rates (of Foote 2000) for all fifty species within two brachiopod genera (*Schizophoria* and *Floweria*), one bivalve genus (*Leiopteria*) combined. Modified from Stigall (2010a). Congruent results from both the generic and species level analyses indicate substantially reduced origination/speciation rates are during the crisis interval, whereas extinction rates are lower during the crisis than the preceding interval.

2012a), a point first appreciated by McGhee (1984, 1988). The concurrence of pervasive species invasions and reduced speciation rates suggests a link may exist between invasions and speciation depression. To test that hypothesis, both the nature of the Middle to Late Devonian species invasions and the mechanisms of speciation reduction must first be examined.

Characterizing Late Devonian species invasions

The prevalence of Frasnian interbasinal species invasions has been noted for many clades including corals, trilobites, stromatoporoids, brachiopods, fish, and conodonts (see review in McGhee 1996). However, invasion timing has not been tightly constrained for many of these clades. Rode and Lieberman (2004) examined patterns of species migration among 341 species of the most common brachiopod and bivalve genera of Laurentia at the temporal resolution of conodont zone. Their analysis included assessing species identification based using discrete morphological characters of specimens in museum collections (cf. Allmon, chapter 3, this volume). Rode and Lieberman (2004) produced a series of GIS-generated range maps for each species through time, from which they identified interbasinal species invasion events.

In this dataset, invasion events were concentrated into discrete intervals that correlated with transgressive events (fig. 12.4), which indicates that sea level change was a prominent factor facilitating dispersal through this interval (Rode and Lieberman 2004). Similarly, Racki (1993) and Brice et al. (1994) identified species migration events on other Laurussian shelves that were coincident with times of relative sea level increases. In Laurentia, sea level increased both episodically and cumulatively from the Givetian to Frasnian (Johnson et al. 1985). These sea level fluctuations, when coupled with the cyclical episodes of orogenesis and tectonic quiescence along the Acadian front that alternately raised and lowered the relief of intracratonic arches, facilitated the breaching of intracratonic arches and provided a pathway for species invasions (Rode and Lieberman 2004, Stigall Rode and Lieberman 2005b).

The species that successfully invaded new tectonic basins shared a suite of characteristics: they occupied broader geographic ranges than native species, they survived the crisis interval at a higher frequency than noninvasive species, and they occupied a broader suite of depositional environments than noninvasive species (Rode and Lieberman 2004). Furthermore, by the end of the Frasnian, the only species (invasive or endemic) that survived into the Famennian were those with large geographic ranges (Stigall Rode and Lieberman 2005a). This combination of features indicates that most Devonian invaders and crisis survivors were species with broad environmental tolerances, or ecological generalists. A positive correlation between geographic range and niche breadth has been recovered in numerous studies (e.g., Jackson 1974; Brown 1984; Thompson et al. 1999; Gaston and Spicer 2001; Fernández and Vrba 2005; but see Williams

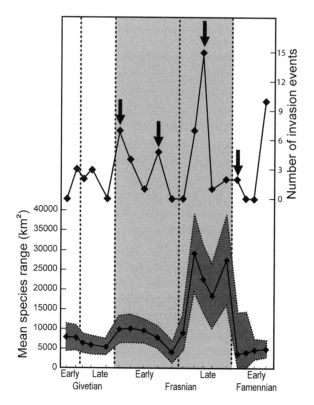

Figure 12.4 Invasion intensity and mean geographic range size for Devonian benthos. Invasion intensity records the number of interbasinal invasion events per conodont zone based on a compendium of species within 29 brachiopod and bivalve genera. Geographic range size calculated for the same set of taxa. Both the mean geographic range size for brachiopod and bivalve species and interbasinal invasion intensity increase during Frasnian crisis interval, shaded, compared with Middle Devonian (Givetian) background levels. Arrows in upper graph indicate transgressive episodes of Johnson et al. (1985). Modified from Rode and Lieberman (2004).

et al. 2006 for a counterexample). Ecological generalists typically exhibit larger geographic ranges than specialist species, which are characterized by highly constrained niches (Mayr 1963; Stanley 1979). The fact that only broadly ranging species survived the Late Devonian Biodiversity Crisis suggests that the influx of invasive species produced a shift in the biotic environment. Specifically, the pervasive establishment of wide-ranging invasive taxa established an ecological environment in which individuals belonging to specialist species could not survive, likely due to competitive interactions (Stigall 2012a). Competition among individuals of incumbent and invasive species is well documented in modern environments and is frequently cited as the key reason for community structure breakdown and reassembly following successful species invasions (Tilman 1994; Sanders et al. 2003; Lockwood et al. 2007; Mata et al. 2013). The differential success of the invasive taxa occurs when the ecological breadth of that species is larger than the incipient species and the invasive population is able to secure a disproportionately larger share of the available resources (e.g., Sanders et al. 2003).

Impact of Late Devonian species invasions on speciation

The geographic mode of speciation within fossil clades can be assessed by examining biogeographic distributions within an evolutionary framework (fig. 12.2). To conduct this type of analysis, geographic ranges are optimized onto the internal nodes of a species-level cladogram and shifts in geographic ranges between ancestor-descendant pairs are analyzed (Lieberman 2000; Wiley and Lieberman 2011). Episodes of vicariance are identified when a descendant occupies only a subset of the ancestral geographic range, whereas episodes of speciation by dispersal are identified when a descendant occupies a region different from or additional to the range of their immediate ancestor (fig. 12.2).

Application of this type of analysis to four clades of Devonian marine invertebrates—an order of phyllocarid crustaceans (Archaeostraca), two genera of rhynchonelliform brachiopods *(Floweria* and *Schizophoria (Schizophoria))*, and a genus of bivalves *(Leptodesma (Leiopteria))*— recovered evidence for vicariance in only 28% of speciation events and evidence for dispersal in 72% of speciation events (Stigall 2012a). This result is nearly opposite that reported from the modern fauna (74% vicariance vs. 26% dispersal; Brooks and McLennan 1991). In addition, speciation by vicariance is more frequent than speciation by dispersal within clades of Cambrian through early Devonian trilobites (Lieberman and Eldredge 1996; Lieberman 1997, 2003; Congreve and Lieberman 2008, 2010) and Ordovician brachiopods (e.g., Wright and Stigall 2013; Bauer and Stigall 2014). Therefore, the decline in Devonian speciation rate can be classified more specifically as a decline in vicariant speciation. Notably, the Late Devonian reduction in speciation by vicariance contributed to a nearly complete loss of all speciation in each of these four lineages (fig. 12.3; Stigall 2010a).

The key to understanding Late Devonian speciation collapse is to determine why vicariant speciation effectively halted during this interval. A crucial requirement for successful vicariant speciation is geographic isolation. However, sustained isolation was problematic in the Devonian seas. The range analyses of Rode and Lieberman (2004) revealed a dramatic expansion in mean geographic range size during the Frasnian, which would have reduced opportunities for isolation. This is particularly true among the lineages that survived the crisis interval; the ecological niche distribution modeling of Stigall Rode and Lieberman (2005a) demonstrated that only taxa expanding their geographic ranges during the crisis interval survived into the Famennian. Both native and invasive lineages would have had few opportunities to sustain geographic isolation required for speciation as expansion of geographic ranges facilitated by sea-level rise prevented effective long-term vicariance from ancestral populations. Rather, incipient species were more likely to be subsumed by geographic extension of the expanding ancestral species than to remain isolated and develop into new species.

The relative increase in geographic range size corresponded to a reduction in the number of narrowly ranging ecological specialist species. Although the endemic biotas of the Middle Devonian included a mix of both specialist and generalist taxa, Late Devonian seas were dominated by ecological generalists including both surviving native species and the new invasive species. Modern invasive species are characterized by broad environmental tolerances (Lockwood et al. 2007). Devonian invaders were likely similar, because ecological tolerances of Devonian invaders must have been sufficiently broad to allow colonization of both the invasion pathway and the new tectonic basin. Studies of modern and Cenozoic invasive species have demonstrated that invader species regularly displace native species through higher resource efficiency (Johansson 2007) or competitive ability (Vermeij 2005). Similar processes operating during the Late Devonian would have contributed to differential extinction of narrowly ranging ecological specialist species. Because specialist taxa, on average, exhibit higher speciation rates and contain more species per lineage than generalist taxa (Vrba1987; Kammer et al. 1997; Ozinga et al. 2012), the relative reduction of specialist lineages in the Devonian ecosystem would have further reduced speciation rates (Stigall 2012a).

In summary, widespread species invasions during the Late Devonian promoted the preferential extinction of specialist species and geographic expansion of generalist species (both native and invasive). This reduced opportunities for the geographic isolation that is required for successful speciation, and thereby produced a negative feedback loop that facilitated the dramatic speciation depression of the Late Devonian.

Invasive Species and Speciation during the Richmondian Invasion

A regional wave of species invasions, termed the Richmondian Invasion, is recorded in the Late Ordovician strata that outcrop in the region surrounding Cincinnati, Ohio. This event comprises several waves of extrabasinal species invasions that recorded in the C4 and C5 depositional sequences (of Holland and Patzkowsky 1996) in the early Richmondian Stage of North American nomenclature (=Katian Stage) (Holland 1997; Holland and Patzkowsky 2007; Meyer and Davis 2009). The Richmondian Invasion caused significant faunal reorganization within the Cincinnatian ecosystem; the stable community gradient of the prior five million years was disrupted and a new community structure emerged after the final set of invasions (Holland and Patzkowsky 2007; Patzkowsky and Holland 2007).

Because the Richmondian Invasion was localized, studies of the evolutionary impact of the Richmondian invaders (e.g., Malizia and Stigall 2011; Brame and Stigall 2014) have focused on a finer geographic scale than the Devonian analyses described in the last section. Nevertheless, similar patterns have emerged between the two events. Notably, speciation was reduced following the invasion and differential biogeographic responses have been observed between native specialist and generalist taxa (Stigall 2010b). Differences in biodiversity patterns also exist between the Late Devonian and Richmondian Invasions. Examining the similarities between the relative impact of Ordovician and Devonian invaders provides a framework to identify common responses to invasive regimes, and exploring the differences may illuminate differential impacts of invasive species based on scale (global vs. regional) of the invasion.

Characterizing Richmondian species invasions

The Richmondian Invasion was a cross-faunal immigration event. Taxa new to the Cincinnati region included species from all trophic groups including tabulate and rugose corals, nautiloid cephalopods, gastropods, bivalves, trilobites, and brachiopods (see Holland 1997 for a list of invader taxa). Particularly notable were introductions of species of rugose corals and rhynchonellid brachiopods. Although some invasive genera (e.g., *Rhynchotrema*) were present in the Cincinnati region during the Mohawkian Stage (Elias

Figure 12.5 Invasive species transportation vectors mapped onto a paleogeographic reconstruction of Laurentia in the Late Ordovician.
The star indicates the Cincinnati region. Current data support invasions from the paleoequa-
tor, and marginal basins near modern Anticosti Island and Oklahoma. A: Transcontinental
Arch, T: Taconic Highlands. Modified from Stigall (2010b).

1983), fewer than 25% of the Richmondian invaders were recurrent (Lam
and Stigall 2015). Thus, the Richmondian Invasion was a true biotic immigration event. Invasive species arrived in several pulses beginning the
C4 sequence and climaxing in the C5 sequence (Holland and Patzkowsky
2007; Holland 2008; Lam and Stigall 2015). Current data indicate that the
invasion was multidirectional (fig. 12.5); some taxa apparently immigrated
from paleoequatorial waters (cf. Elias 1983; Holland 1997), whereas others
immigrated from other temperate epicontinental basins or basins marginal
to Laurentia (cf. Jin 2001; Wright 2012; Wright and Stigall 2013; Bauer and
Stigall 2014; Lam and Stigall 2015).

Like the Devonian example, transgressive events were the most likely pathway by which species movements were facilitated, probably via larval transport assisted by surface current circulation (Stigall 2010b; Lam and Stigall 2015). The pulsed nature of the invasion itself, however, may indicate that the transportation vectors from the various source regions did not operate simultaneously (Lam and Stigall 2015). The invasion has been linked to paleo-oceanographic changes that resulted in warm, low-nutrient waters replacing the former nutrient-rich temperate conditions in the Cincinnati region, thereby facilitating migration of equatorial taxa into the region (Holland 1993; Holland and Patzkowsky 1996). Migration from temperate regions may have been related to this and other oceanographic vectors such as the Intra-Iapetus current and surface currents (see Lam and Stigall 2015).

Stigall (2010b) conducted a GIS-based study of biogeographic patterns of 49 species within 21 genera of rhynchonelliform brachiopods from the C1 through C6 sequences. Species occurrence data were compiled from museum and field collections by inspecting individual specimens. Species identification was based on morphological characteristics using morphological species concept based on original species descriptions from the published literature. Results of range analyses showed that (1) native species with large geographic ranges preferentially survived through the invasion interval, (2) invasive species and native species that persisted into the invasion interval had statistically larger ranges than native species that became extinct, (3) surviving native species had larger geographic ranges than invader species during the invasion interval, (4) surviving native species and invasive species had similarly large ranges after postinvasion community equilibrium was restored (Stigall 2010b). These results indicate that species with broad geographic ranges—whether native or invasive—persisted through the invasive interval of the C4 and C5 sequences and were the dominant species within the postinvasion ecosystem. As described above, species with broad geographic ranges typically also occupy broad ecological niches. Therefore, the Richmondian Invasion follows the same basic pattern as the Great Devonian Interchange: broadly adapted taxa preferentially survived and became more prevalent than narrowly adapted species in the ecosystem during the invasion interval.

Impact of the Richmondian species invasions on speciation

Similar to the Frasnian stage of the Late Devonian, speciation was depressed during the Richmondian Invasion. Only one speciation event can be confidently identified among brachiopod lineages (the most common benthos) of the Cincinnati depositional basin during the C4 sequence (Stigall 2010b). Subsequent phylogenetic biogeographic analyses of four brachiopod genera identified zero speciation events during the invasion interval could be related to vicariance (Wright and Stigall 2013, Bauer and Stigall 2014). Thus reduced speciation, specifically depression of vicariance, characterized the Richmondian Invasion. This is likely due to the preferential extinction of specialist taxa during the invasive interval and the establishment of the same negative feedback loop described above for the Late Devonian (Stigall 2013).

A series of analyses utilizing ecological niche modeling (ENM, see Myers et al. 2015 for overview of methods) were recently conducted that provides an explanation for the loss of vicariance and speciation observed during the Richmondian Invasion. This suite of analyses examined how both native and invasive species adjusted to the invasive regime, specifically whether species maintained the parameters of their ecological niches through time or evolved aspects of their niches and adapted to the changing condition (reviewed in Stigall 2014). Results from the combined set of analyses (i.e., Dudei and Stigall 2010; Malizia and Stigall 2011; Stigall 2011, 2012b; Walls and Stigall 2011; Brame and Stigall 2014; Stigall and Brame 2014) recovered a congruent pattern across native and invasive taxa from four phyla. Species and genera displayed high fidelity habitat tracking (= niche stability in geographic space) during the preinvasion interval but high levels of niche evolution during the invasion and postinvasion intervals (fig. 12.6). During the invasion and postinvasion intervals, geographic niche stability declined indicating that habitat tracking deteriorated. However, taxa exhibited statistically higher similarity of specific niche parameters during and after the invasion.

This combination of high ecological niche similarity with limited habitat tracking indicates that species adjusted to the biotic invasion by reducing the variability of niche parameters (i.e., utilizing a smaller region of their previous niche dimensions) rather than by shifting to a new region of ecospace (Stigall 2011; Brame and Stigall 2014; Stigall 2015). This result is congruent with Holland and Zaffos's (2011) analysis of environmental preference at the genus level. The retreat of surviving taxa into a portion of previously occupied ecospace is also congruent with documented competitive displacement (Tyler and Leighton 2011) and ecospace partitioning (Patzkowsky and Holland 2007) and provides an explanation for the relative success of generalist vs. specialist taxa during the Richmondian Invasion.

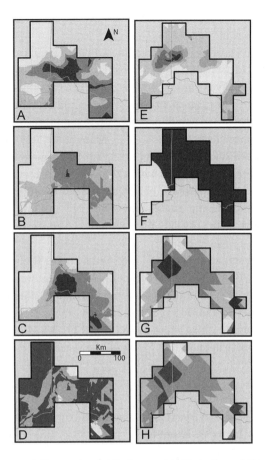

Figure 12.6 Representative species distribution models illustrating relative niche stability patterns in geographic space before and after the Richmondian Invasion. A–D: before the invasion; E–H: after the invasion. *Vinlandostrophia ponderosa:* projected species distribution of *V. ponderosa* in the C3-1 time slice based on the niche model generated from species occurrence and environmental data of the C3-1 time slice (A), predicted distribution of *V. ponderosa* during C3-2 time slice generated by projecting the C3-1 niche model onto the environmental layers of the C3-2 time slice; this map shows the distribution of *V. ponderosa* during C3-2 if the species niche had been completely conserved from timeslice C3-1 (B), predicted distribution of *V. ponderosa* in C3-2 time slice based on the niche model generated from species occurrence and environmental data of the C3-2 time slice (C), overlap of maps B and C; high degree of overlap between the forward projection (B) and the C3-2 distribution model (C) is indicative of niche stability (D). *Leptaena richmondensis:* projected distribution of *L. richmondensis* in the C5-1 time slice based on the niche model generated from species occurrence and environmental data of the C5-1 time slice (E), predicted distribution of *L. richmondensis* during C5-2 time slice generated by projecting the C5-1 niche model onto the environmental layers of the C5-2 time slice; this map shows the distribution of *L. richmondensis* during C5-2 if the species niche had been completely conserved from timeslice C5-1 (F), predicted distribution of *L. richmondensis* in C5-2 time slice based on the niche model generated from species occurrence and environmental data of the C5-2 time slice (G), overlap of maps F and G; high degree of overlap between the forward projection (F) and the C5-2 distribution model (G) exhibits is indicative of niche stability (H). Modified from Malizia and Stigall (2011).

The prevalence of niche contraction would have further reduced opportunities for taxa to shift to new ecospace (Stigall 2011). Because generalist taxa originally occupied broad ecological niches, there was a higher
probability that a portion of that ecospace would remain habitable following the intense interspecific competition that typically develops during
the invasive regime (Sanders et al. 2003; Mata et al. 2013). For specialist
taxa, that probability was lower and extinction was a more likely outcome.
Similarly, any incipient species with a limited population size and niche
breadth would have experienced high competition resulting in a low likelihood for persistence and establishment of a new species. In fact, within-
basin speciation resumed only after community gradients reestablished
the C5 sequence (Stigall 2010b).

The primary impacts of invasive species during the Richmondian Invasion, therefore, were to increase competition which promoted niche evolution among invasive and native generalist taxa, a decline in native specialist taxa, and reorganization of a previously stable ecosystem into one
constrained by competitive interactions (Stigall 2012b). These impacts contributed to an environment that hindered the successful establishment of
new species.

Synthesizing Invasive Species Impacts During the Devonian and Richmondian Invasive Regimes

From the two case studies presented above, several patterns emerge relative to the impact of invasive species on the process of speciation:

- Pervasive establishment of invasive species within an ecosystem produces a
  competitive regime deleterious to the survival of ecological specialists.
- The loss of specialist species reduces the species pool from which new species
  may arise. Because specialists have higher intrinsic speciation rates, the loss of
  such species from the ecosystem results in an overall decline in aggregate speciation rate.
- Native generalist taxa and invaders both undergo niche evolution during community reorganization accomplished primarily by reducing variation within original
  niche parameters.
- The expansion of geographic ranges and prevalence of species with broad environmental tolerances reduces the probability of sustained geographic isolation
  required for allopatric speciation to succeed. Incipient species are more likely

to be subsumed within the ancestral population again than retain geographic isolation long enough to accomplish speciation.

Thus, species invasions engender speciation decline in both direct and indirect ways. Range expansion during the invasive regime directly limits geographic isolation, and consequently genetic exchange can continue unabated among subpopulations, which lowers the probability of successful allopatric speciation. The loss of ecological specialists via competition or other indirect mechanisms reduces the number of lineages from which new species may form and thereby contributes to the reduction in speciation rate. Overall, the prevalence of ecological generalists and reduction in allopatric speciation can produce a depauperate fauna during or after the invasion interval for a sustained amount of time. The duration of these impacts may be related to the magnitude of the invasion; ranging from ~1 million years for regional events like the Richmondian Invasion to several million years for global events like the Great Devonian Interchange.

Notably, although the two case studies document similar patterns and mechanisms of speciation decline, the impact of species invasions on standing diversity differs. The global invasion of the Late Devonian interchange contributed to faunal homogenization and global biodiversity decline. The regional influx of invasive species during the Richmondian Invasion, however, resulted in a more diverse assemblage of taxa within the postinvasion communities (Holland and Patzkowsky 2007). The difference in these outcomes is likely related to the hierarchical scales at which ecosystems were impacted. The Late Devonian invasions persisted over several million years and impacted many depositional basins simultaneously. At this scale, widespread species invasions suppressed speciation for an extensive time. When coupled with the other Earth system events that occurred during the Late Devonian (i.e., widespread basin anoxia, global cooling), this produced a tremendous drop in biodiversity, in terms of both per capita diversity as well as ecological diversity (McGhee et al. 2013). On the other hand, the Richmondian Invasion primarily impacted a single region, and speciation was reduced for only one million years. The more limited nature of the invasion coupled with limited environmental change (i.e., minor climatic warming, oceanographic circulation shifts, and gradual basin infilling) allowed the community structure to reassemble rapidly and the documented ecological partitioning enabled a higher total species count on the seafloor.

Implications for Modern Biodiversity

Invasive species are a major threat to modern ecosystems and cause billions of dollars in economic damage annually (Sala et al. 2000; Pimentel et al. 2005). Due to their significant impact on modern ecosystems and government budgets, the effects of invasive species on modern ecosystems are intensely studied and aspects of their immediate and short-term ecological impacts are well understood (reviewed in Lockwood et al. 2007; Davis 2009). Because the long-term impacts of species invasions are difficult to assess on the ecological timescales available to biologists, studying analogous ancient invasions preserved in the fossil record can help to establish an understanding of the long-term impacts of invasive species on modern biodiversity.

Two case studies examined herein, the Late Devonian Biodiversity Crisis and the Late Ordovician Richmondian Invasion, provide insight into the effect of invasive species on speciation processes and ecosystem structuring. During both intervals, invasive species are characterized by broad ecological tolerances, broad geographic ranges, and higher than average survival potential through the crisis interval. Among the native species, narrowly adapted ecological specialists are more likely to become extinct, while broadly adapted generalist species persisted through the invasion interval by modifying aspects of their ecological niche through niche evolution. In addition, the formation of new species was effectively halted during the invasion intervals due to reduced opportunities for geographic isolation and vicariance.

Modern invasive species have been observed to decimate local community diversity (Blaustein 2001), cause extinction of native taxa via over predation (e.g., Fritts and Rodda 1998), reduce ecosystem complexity (Lockwood et al. 2007), and have other deleterious impacts on local ecosystems. The pressure these invasive taxa apply to speciation, however, cannot be quantified on an ecological (=experimental) timescale, and consequently has not yet received as much attention from conservationists. If the impacts of invasive species of the Late Devonian and Late Ordovician are representative of long-term ecological impacts of invaders, then these impacts should be considered within a conservation paradigm. The implications for specialist taxa are the most significant. Many modern species of this type are already reduced to small population sizes due to habitat loss and are often listed as threatened or endangered on the IUCN red list (http://www.iucnredlist.org/). The additional pressure applied by invasive species is likely to cause additional complications for these lineages. Due to constraints of time, personnel, and finances, it may not be possible to preserve all species alive today, and thus may be prudent to focus efforts on the more broadly adapted of the threatened taxa.

Acknowledgments

Thanks to M. E. Patzkowsky and W. D. Allmon for constructive reviews on an earlier version of this paper and to the editors for the invitation to submit to this edited volume. This research was supported by NSF EAR-0922067 and EF-1206750. This is a contribution to the International Geoscience Programme (IGCP) Project 591- "The Early to Middle Paleozoic Revolution."

References

Alroy, J. 2008. Dynamics of origination and extinction in the marine fossil record. *Proceedings of the National Academy of Sciences, USA*, 105, 11536–11542.

Avise J. C., and Walker, D. 1998. Pleistocene phylogeographic effects on avian populations and the speciation process. *Proceedings of the Biological Society*, 265, 457–463.

Bambach, R. K., Knoll, A. J., and Wang, S. C. 2004. Origination, extinction, and mass depletions of marine diversity. *Paleobiology*, 30, 522–542.

Bauer, J. E., and Stigall, A. L. 2014. Phylogenetic paleobiogeography of Late Ordovician Laurentian brachiopods. *Estonian Journal of Earth Sciences*, 63, 189–194.

Benton, M. J., and Pearson, P. N. 2001. Speciation in the fossil record. *Trends in Ecology and Evolution*, 16, 405–411.

Blaustein, R. J. 2001. Kudzu's invasion into Southern United states life and culture. 55–62. *The Great Reshuffling: Human Dimensions of Invasive Species*. World Conservation Union: Cambridge, UK.

Brame, H. M. R., and Stigall, A. L. 2014. Controls on niche stability in geologic time: congruent responses to biotic and abiotic environmental changes among Cincinnatian (Late Ordovician) marine invertebrates. *Paleobiology* 40, 70–90.

Brice, D., Milhau, B., and Mistiaen, B. 1994. Northern American affinities of Devonian (Givetian–Frasnian) taxa from Boulonnais, north of France: migrations and diachronisms. *Bulletin de la Société géologique de France*, 165, 291–306.

Broennimann, O., Treier, U. A., Müller-Schärer, H., Thuiller, W., Peterson, A. T., and Guisan, A. 2007. Evidence of climatic niche shift during biological invasion. *Ecology Letters*, 10, 701–709.

Brooks, D. R., and McLennan, D. A. 1991. *Phylogeny, Ecology, and Behavior.* University of Chicago Press, Chicago.

Brown, J. H. 1984. On the relationship between abundance and distribution of species. *American Naturalist,* 124, 255–279.

Congreve, C. R., and Lieberman, B. S. 2008. Phylogenetic and biogeographic analysis of Ordovician homalonotid trilobites. *Open Paleontology Journal,* 1, 24–32.

Congreve, C. R., and Lieberman, B. S. 2010. Phylogenetic and biogeographic analysis of deiphonine trilobites. *Journal of Paleontology,* 84, 126–136.

Copper, P. 1994. Ancient reef ecosystem expansion and collapse. *Coral Reefs,* 13, 3–11.

Davis, M. A. 2009. *Invasion Biology.* Oxford University Press, Oxford.

Dudei, N. L., and Stigall, A. L. 2010. Using ecological niche modeling to assess biogeographic and niche response of brachiopod species to the Richmondian Invasion (Late Ordovician) in the Cincinnati Arch. *Palaeogeography, Palaeoclimatology, Palaeoecology,* 296, 28–43.

Elias, R. J. 1983. Middle and Late Ordovician solitary rugose corals of the Cincinnati Arch region. *United States Geological Survey Professional Paper,* 1066-N, N1–N13.

Fernández, M. H., and Vrba, E. S. 2005. Body size, biomic specialization, and range size of large African mammals. *Journal of Biogeography,* 32, 1243–1256.

Foote, M. 2000. Origination and extinction components of taxonomic diversity: general problems. *Paleobiology,* 26 (Sp. 4), 74–102.

Fritts, T. H., and Rodda, G. H. 1998. The role of introduced species in the degradation of an island ecosystem: a case history of Guam. *Annual Review of Ecology and Systematics,* 29, 113–140.

Gaston, K. J., and Spicer, J. I. 2001. The relationship between range size and niche breadth, a test using five species of *Gammarus* (Amphipoda). *Global Ecology and Biogeography,* 10, 179–188.

Holland, S. M. 1993. Sequence stratigraphy of a carbonate-clastic ramp: the Cincinnatian Series (Upper Ordovician) in its type area. *Geological Society of America Bulletin,* 105, 306–322.

Holland, S. M. 1997. Using time/environment analysis to recognize faunal events in the Upper Ordovician of the Cincinnati arch. 309–334. *In* Brett, C. E., Baird, G. C. (eds.), *Paleontological Events: Stratigraphic, Ecological, and Evolutionary Implications.* Columbia University Press, New York.

Holland, S. M. 2008. The type Cincinnatian: an overview. 174–184. *In* McLaughlin, P. I., Brett, C. E., Holland, S. M., Storrs, G. W. (eds.), *Stratigraphic Renaissance in the Cincinnati Arch: Implications for Upper Ordovician Paleontology and Paleoecology.* Cincinnati Museum Center Scientific Contributions 2, Cincinnati.

Holland, S. M., and Patzkowsky, M. E. 1996. Sequence stratigraphy and long-term paleoceanographic change in the Middle and Upper Ordovician of the eastern United States. *Geological Society of America Special Papers*, 306, 117–129.

Holland, S. M., and Patzkowsky, M. E. 2007. Gradient ecology of a biotic invasion: biofacies of the type Cincinnatian Series (Upper Ordovician), Cincinnati, Ohio Region, USA.*Palaios*, 22, 392–407.

Holland, S. M., and Zaffos, A. 2011. Niche conservatism along an onshore-offshore gradient. *Paleobiology*, 37, 270–286.

Jablonski, D. 2008. Species sorting and selection. *Annual Review of Ecology and Evolution*, 39, 501–524.

Jackson, J. B. C. 1974. Biogeographic consequences of eurytopy and stenotopy among marine bivalves and their evolutionary significance. *American Naturalist*, 108, 541–560.

Jin, J. 2001. Evolution and extinction of the North America *Hiscobeccus* brachiopod fauna during the Late Ordovician. *Canadian Journal of Earth Science*, 38, 143–151.

Johansson, J. 2007. Evolutionary responses to environmental changes: How does competition affect adaptation? *Evolution*, 62, 421–435.

Johnson, J. G. 1970. The Taghanic onlap and the end of North American Devonian provinciality. *Geological Society of America Bulletin*, 81, 2077–2106.

Johnson, J. G., Klapper, G., and Sandberg, C. A. 1985. Devonian eustatic fluctuations in Euramerica. *Geological Society of America Bulletin*, 96, 567–587.

Kammer, T. W., Baumiller, T. K., and Ausich, W. I. 1997. Species longevity as a function of niche breadth: evidence from fossil crinoids. *Geology*, 25, 219–222.

Lam, A. R., and Stigall, A. L. 2015. Pathways and mechanisms of Late Ordovician (Cincinnatian) faunal migrations of Laurentia and Baltica. *Estonian Journal of Earth Sciences*, 64, 62–67.

Levin, D. A. 2003. Ecological speciation: lessons from invasive species. *Systematic Botany*, 28, 643–650.

Lieberman, B. S. 1997. Early Cambrian paleogeography and tectonic history: a biogeographic approach. *Geology*, 25, 1039–1042.

Lieberman, B. S. 2000. *Paleobiogeography: Using Fossils to Study global Change, Plate Tectonics, and Evolution*. Kluwer Acad./Plenum Publ., New York.

Lieberman, B. S. 2003. Biogeography of the Trilobita during the Cambrian radiation: deducing geological processes from trilobite evolution. *Special Papers in Palaeontology*, 70, 59–72.

Lieberman, B. S. 2012. Adaptive radiations in the context of macroevolutionary theory: a paleontological perspective. *Evolutionary Biology*, 39, 181–191.

Lieberman, B. S., and Eldredge, N. 1996. Trilobite biogeography in the Middle Devonian: geological processes and analytical methods. *Paleobiology*, 22, 66–79.

Lockwood, J., Hoopes, M., and Marchetti, M. 2007. *Invasion Ecology*. Wiley-Blackwell Publishing, Singapore.

Malizia, R. W., and Stigall, A. L. 2011. Niche stability in Late Ordovician articulated brachiopod species before, during, and after the Richmondian Invasion. *Palaeogeography, Palaeoclimatology, Palaeoecology*, 311, 154–170.

Mayr, E. 1963. *Animal Species and Evolution*. Belknap, Cambridge.

Mayr, E. 2001. *What Evolution Is*. Basic Books, New York.

Mata, T. M., Haddad, N. M., and Holyoak, M. 2013. How invader traits interact with resident communities and resource availability to determine invasion success. *Oikos*, 122, 149–160.

May, R. M. 1988. How many species are there on Earth? *Science*, 241, 1441–1449.

McGhee, G. R., Jr. 1984. Tempo and mode of the Frasnian-Famennian biotic crisis. *Abstracts with Programs, Geological Society of America*, 16, 49.

McGhee, G. R., Jr. 1988. The Late Devonian extinction event: evidence for abrupt ecosystem collapse. *Paleobiology*, 14, 250–257.

McGhee, G. R., Jr. 1996. *The Late Devonian Mass Extinction: The Frasnian/Famennian Crisis*. Columbia University Press, New York.

McGhee, G. R., Jr. 1997. Late Devonian bioevents in the Appalachian Sea: immigration, extinction, and species replacements. 493–508. *In* Brett, C. E., Baird, G. C. (eds.), *Paleontological Events: Stratigraphic, Ecological, and Evolutionary Implications*. Columbia University Press, New York.

McGhee, G. R., Jr., Clapham, M., Sheehan, P. M., Bottjer, D. J., and Droser, M. L. 2013. A new ecological-severity ranking of major Phanerozoic biodiversity crises. *Palaeogeography, Palaeoclimatology, Palaeoecology*, 370, 260–270.

Meyer, D. L., and Davis, R. A. 2009. *A Sea Without Fish: Life in the Ordovician Sea of the Cincinnati Region*. Indiana University Press, Bloomington.

Mooney, H. A., and Cleland, E. E. 2001. The evolutionary impact of invasive species. *Proceedings of the National Academy of Sciences, USA*, 98, 5446–5451.

Mora, C., Tittensor, D. M., Adl, S., Simpson, G. B., and Worm, B. 2011. How many species are there on Earth and in the ocean? *PLoS Biology*, 9(8), e1001127.

Myers, C. E., Stigall, A. L., and Lieberman, B. S. 2015. PaleoENM: applying ecological niche modeling to the fossil record. *Paleobiology*, 41, 226–244.

Ozinga, W. A., Colles, A., Bartish, I. V., Hennion, F., Hennekens, S. M., Pavione, S., Poschold, P., Herman, M., Schaninée, J. H. J., and Prinzing, A. 2012. Specialists leave fewer descendants within a region than generalists. *Global Ecology and Biogeography*, 22, 213–222.

Patzkowsky, M. E., and Holland, S. M. 2007. Diversity partitioning of a Late Ordovician marine biotic invasion: controls on diversity in regional ecosystems. *Paleobiology*, 33, 295–309.

Pearman, P. B., Guisan, A., Broennimann, O., and Randin, C. F. 2008. Niche dynamics in space and time. *Trends in Ecology and Evolution*, 23, 149–158

Peterson, A. T. 2011. Ecological niche conservatism: a time-structured review of evidence. *Journal of Biogeography*, 28, 817–827.

Pimentel, D., Zuniga, R., and Morrison, D. 2005. Update on the environmental and economic costs associated with alien-invasive species in the United States. *Ecological Economics,* 52, 273–288.

Racki, G. 1993. Brachiopod assemblages in the Devonian Kowala Formation of the Holy Cross Mountains. *Acta Palaeontologica Polonica,* 37, 297– 357.

Raup, D. M., and Sepkoski, J. J., Jr. 1982. Mass extinctions in the marine fossil record. *Science,* 215, 1501–1503.

Rode, A. L., and Lieberman, B. S. 2004. Using GIS to unlock the interactions between biogeography, environment, and evolution in Middle and Late Devonian brachiopods and bivalves. *Palaeogeography, Palaeoclimatology, Palaeoecology,* 211, 345–359.

Rong, J.-Y., Chen, X., and Harper, D. A. T. 2007. The latest Ordovician Hirnantian fauna (Brachiopoda) in space and time. *Lethaia,* 35, 231–249.

Sala, O. E., Chapin, F. S., III, Armesto, J. J., Berlow, E., Bloomfield, J., Dirzo, R., Huber-Sanwald, E., Huenneke, L. F., Jackson, R. B., Kinzig, A., Leemans, R., Lodge, D. M., Mooney, H. A., Oesterheld, M., Poff, N. L., Sykes, M. T., Walker, B. H., Walker, M., and Wall, D. H. 2000. Global biodiversity scenarios for the year 2100. *Science,* 287, 1770–1774.

Sanders, N. J., Gotelli, N. J., Heller, N. E., and Gordon, D. M. 2003. Community disassembly by an invasive species. *Proceedings of the National Academy of Sciences* 100, 2474–2477.

Sepkoski, J. J., Jr. 1998. Rates of speciation in the fossil record. *Philosophical Transactions of the Royal Society of London B,* 353, 315–326.

Simberloff, D., Martin. J.-L., Genovesi, P., Maris, V., Wardle, D. A., Aronson, J., Courchamp, F., Galil, B., García-Berthou, E., Pascal, M., Pyšek, P., Sousa, R., Tabacchia, E., and Vilà, M. 2013. Impact of biological invasions: what's what and the way forward. *Trends in Ecology and Evolution,* 28, 58–66.

Stadler, T. 2011. Mammalian phylogeny reveals recent diversification rate shifts. *Proceedings of the National Academy of Sciences, USA,* 108, 6187–6192.

Stanley, S. M. 1979. *Macroevolution: Pattern and Process.* W. H. Freeman and Co., San Francisco.

Stigall, A. L. 2010a. Speciation decline during the Late Devonian Biodiversity Crisis related to species invasions. *PLoS ONE,* 5(12), e15584.

Stigall, A. L. 2010b. Using GIS to assess the biogeographic impact of species invasions on native brachiopods during the Richmondian Invasion in the Type-Cincinnatian (Late Ordovician, Cincinnati region). *Palaeontologia Electronica* 13, 5A. http://palaeo-electronica.org/2010_1/207/index.html.

Stigall, A. L. 2011. Application of niche modeling to analyse biogeographic patterns in Paleozoic brachiopods: evaluating niche stability in deep time. *Memoirs of the Association of Australian Palaeontologists,* 41, 229–255.

Stigall, A. L. 2012a. Speciation collapse and invasive species dynamics during the Late Devonian "Mass Extinction." *GSA Today,* 22(1), 4–10.

Stigall, A. L. 2012b. Using ecological niche modeling to evaluate niche stability in deep time. *Journal of Biogeography*, 39, 772–781.

Stigall, A. L. 2013. Analyzing links between biogeography, niche stability, and speciation: the impact of complex feedbacks on macroevolutionary patterns. *Palaeontology*, 56, 1225–1238.

Stigall, A. L. 2014. When and how do species achieve niche stability over long time scales? *Ecography*, 37, 1123–1132.

Stigall, A. L. 2015. Expanding the role of biogeography and niche evolution in macroevolutionary theory. 301–327. *In* Serrelli, E., Gontier, N. (eds), *Macroevolution: Explanation, Interpretation, and Evidence*. Springer, Interdisciplinary Evolution Research series.

Stigall, A. L., and Brame, H.-M. R. 2014. Relating environmental change and species stability in Late Ordovician seas. *GFF*, 136, 249–253.

Stigall Rode, A. L. 2005. Systematic revision of the Devonian brachiopods *Schizophoria (Schizophoria)* and *"Schuchertella"* from North America. *Journal of Systematic Palaeontology*, 3, 133–167.

Stigall Rode, A. L., and Lieberman, B. S. 2005a. Using environmental niche modelling to study the Late Devonian biodiversity crisis. 93–180. *In* Over, D. J., Morrow, J. R., Wignall, P. B. (eds.). *Understanding Late Devonian and Permian-Triassic Biotic and Climatic Events: Towards an Integrated Approach*. Elsevier, Amsterdam.

Stigall Rode, A. L., and Lieberman, B. S. 2005b. Paleobiogeographic patterns in the Middle and Late Devonian emphasizing Laurentia. *Palaeogeography, Palaeoclimatology, Palaeoecology*, 222, 272–284.

Thompson, K., Gaston, K. J., and Band, S. R. 1999. Range size, dispersal and niche breadth in the herbaceous flora of central England. *Journal of Ecology*, 87, 150–155.

Tilman, D. 1994. Competition and biodiversity in spatially structured habitats. *Ecology*, 75, 2–16.

Tyler, C. L., and Leighton, L. R. 2011. Detecting competition in the fossil record: support for character displacement among Ordovician brachiopods. *Palaeogeography, Palaeoclimatology, Palaeoecology*, 307, 205–217.

Vermeij, G. J. 1991. Anatomy of an invasion: the trans-Arctic interchange. *Paleobiology*, 17, 281–307.

Vermeij, G .J. 2005. Invasion as expectation: a historical fact of life. 315–339. *In* Sax, D. F., Stachowicz, J. J., Gaines, S. D. (eds.), *Species Invasions: Insights into Ecology, Evolution, and Biogeography*. Sinauer Associates, Inc., Sunderland, MA.

Vrba, E. S. 1987. Ecology in relation to speciation rates: some case histories of Miocene-Recent mammal clades. *Evolutionary Ecology*, 1, 283–300.

Vrba, E. S. 1992. Mammals as a key to evolutionary theory. *Journal of Mammalogy*, 73, 1–28.

Walls, B. J., and Stigall, A. L. 2011. Analyzing niche stability and biogeography of

Late Ordovician brachiopod species using ecological niche modeling. *Palaeogeography, Palaeoclimatology, Palaeoecology*, 299, 15–29.

Webb, S. D. 2006. The Great American Biotic Interchange: patterns and processes. *Annals of the Missouri Botanical Garden*, 93, 245–257.

Wiens, J. J., and Graham, C. H. 2005. Niche conservatism: integrating evolution, ecology, and conservation biology. *Annual Review of Ecology, Evolution, and Systematics*, 36, 519–539.

Wiley, E. O., and Mayden, R. L. 1985. Species and speciation in phylogenetic systematics, with examples from North American fish fauna. *Annals of the Missouri Botanical Garden*, 72, 596–635.

Wiley, E. O., and Lieberman, B. S. 2011. *Phylogenetics, 2nd edition*. Wiley and Sons, New York.

Williams, Y. M., Williams, S. E., Alford, R. A., Waycott, M., and Johnson, C. N. 2006. Niche breadth and geographical range: ecological compensation for geographical rarity in rainforest frogs. *Biology Letters*, 2, 532–535.

Wright, D. F. 2012. Macroevolution and paleobiogeography of Middle to Late Ordovician brachiopods: a phylogenetic biogeographic approach. MS thesis, Ohio University.

Wright, D. F., and Stigall, A. L. 2013. Geologic drivers of Late Ordovician faunal change in Laurentia: investigating links between tectonics, speciation, and biotic invasions. *PLoS ONE*, 8, e68353.

CHAPTER THIRTEEN

Fossil Species Lineages and Their Defining Traits: Taxonomic "Usefulness" and Evolutionary Modes

Melanie J. Hopkins and Scott Lidgard

Introduction

Within a decade after punctuated equilibrium was first proposed (Eldredge and Gould 1972), the theory had come under considerable criticism. While most of the uproar was about the veracity of its tenets, or perceived and sometimes proclaimed threats to neo-Darwinian theories (Princehouse 2009, Sepkoski 2009), a few critics also alleged a potential tautology in its formulation. Their fundamental concern was the conflation of the identification of stasis and speciation events with the identification of the fossil species themselves (Levinton and Simon 1980, Levinton 1983, Rieppel 1984; see also Stamos 2002). Levinton and Simon (1980) argued that because, in paleontology, "characters which evolve sporadically (i.e., remain for long periods unaltered and then undergo rapid change) are ideal species-defining characters" (136), it was inevitable that paleontologists saw punctuations in the fossil record and associated them with speciation events.

A few years prior to Levinton and Simon's critique, Gingerich (1976) had hinted at this tautology when he discussed previous systematic work on *Hyopsodus*, an Eocene hoofed mammal. Species had been formerly defined based on a typological approach using a well-preserved skeletal trait. In the jaw of *Hyopsodus*, the length of one molar had been categorized as "small," "medium," or "large," leading workers to postulate separate lineages persisting unchanged through time, with new lineages appearing abruptly. In contrast, Gingerich demonstrated that the distribution of variation in molar length within samples of fossil populations shifted rather steadily over time, implying that such distinctions within and between stratigraphically successive populations were arbitrary. He argued that the pattern was more consistent with gradual rates of divergence than with episodes of rapid, relatively abrupt divergence. Similarly, Schopf (1981) claimed that paleontologists, faced with the incompleteness of the fossil record, tended to assign morphological variation to previously described species rather than erecting numerous "new" species as observed patterns of variation might warrant. He argued that this practice "covertly encourages the view that stasis, rather than incipient speciation, is the norm" (159; see also Scott 1976). Later in the same paper, Schopf noted (1) that the process of preservation limited the amount of material available to the paleontologist for species recognition; (2) that long-lived abundant (and presumably more stable) species are more likely to be sampled than short-lived rare species; and most relevant to this paper, (3) that any interpretation of stasis or nonstasis depended on the taxonomic philosophy of the systematist, citing Raup and Crick's (1981) reexamination of Brinkmann's (1928, 1929) study of morphological change in the Jurassic ammonite *Kosmoceras*. Schopf (1982) later expanded on these same concerns.

Eldredge (1995) apparently interpreted Levinton and Simon's argument as a claim that species could not be recognized in the fossil record. Having put forward that one of the accomplishments of the theory of punctuated equilibrium was reconciliation between the biological species concept and patterns of stasis and transformation in the fossil record, he may have assumed that Levinton and Simon were arguing more specifically that reproductive isolation could not be demonstrated in the fossil record. Thus, he countered that Recent species are typically identified by morphology as well, and that concerns regarding over- or underestimating species diversity because of polymorphism or sibling species, respectively, applied equally to biologists as to paleontologists. Gould responded indirectly to Levinton and Simon's argument by specifying criteria for identifying which punctuations were also likely to be speciation events, thereby defying the charge that paleontologists necessarily associated all punctuations with speciation. In particular, he suggested that cladogenesis must be demonstrated by observing the survival of the ancestral species following the origination of the descendant species (Gould 1982, 2002). Other punctuations were

inconclusive or might represent rapid anagenesis (or branching speciation where the ancestor did not survive). However, Gould did not address
Levinton and Simon's assertion that species in the fossil record exhibit stasis by definition.

This assertion is scrutinized, however, in a current philosophical critique (Turner 2011). Turner argues that because species definitions and
the criteria that discriminate species are not necessarily the same (rather,
morphology is the only available test of species membership in the fossil
record), the tautology arises only when paleontologists both use the morphological species concept *and* claim that species do not undergo much
morphological change during their existence. However, he acknowledges
that if, in practice, paleontologists never classify two morphologically distinct fossil specimens from different time periods as the same species, they
will never end up recognizing cumulative change within species, and thus
they leave themselves with no way of disconfirming stasis.

In this paper, we address these concerns empirically. The data used in
this study have been gathered from paleontological studies published during the last 50 years, each of which documents morphological evolution
at or near the "species level" through successive stratigraphic intervals
(Hopkins and Lidgard 2012). For the purposes of this study, a "species" is
a suite of samples of specimens, all of which have been given one name at
the discretion of the original author(s) (e.g., *Globorotalia truncatulinoides*,
Lazarus et al. 1995). A "lineage" is a suite of samples of specimens that
have been subdivided based on morphological variation that corresponds
with differences in stratigraphic position (and thus temporal arrangement)
and given several names, but are understood to be related to one another
evolutionarily (e.g., *Miniochoerus chadronensis-affinis-gracilis*, Prothero
and Heaton 1996). Species as primary units in systematics are largely a
categorization with definitional import, as witnessed by the plethora of
species concepts and conflicts among concepts when applied to living or
fossil domains of study (for example, Allmon [chapter 3], Miller [chap-
ter 2], and Hageman [chapter 5], this volume). This is not our principal focus. Rather, we are concerned with the evidential role of fossil species and
species lineages—and characters selected to represent them—in discriminating among different manifestations of evolutionary tempo and mode
(Hunt 2012). To this end, we apply a consistent analytical protocol in order
to estimate the mode of evolution of individual, quantitatively described
size and shape traits for such fossil species or lineages (see Methods).
Then, without recourse to these results, we assess the taxonomic utility of

each trait with respect to discriminating its relevant species from other similar species. Our assessment is based directly on the taxonomic description used by each study's author(s) or on the systematic literature corresponding to each species as identified and analyzed in the study. Here, theoretical commitments to a particular species concept and use of a given trait in prior classificatory practices that are normative among students of a particular group of organisms may *both* come into play. Our procedures attempt to be agnostic respecting these two points. Because the overall dataset includes representatives from across eukaryotes and throughout the fossil record, our results should be broadly applicable and cannot be considered unique to one taxonomic group, geographical region, or period of time. By comparing the results of our analysis of the evolutionary modes shown by individual traits with the qualitative assessment of "usefulness" of these traits in taxonomic/phylogenetic species categorization, we are asking: how often do the traits used to delimit fossil species overlap with the traits under analysis as morphological proxies for temporal patterns of species evolution? More pointedly, does the taxonomic "usefulness" of a trait correspond to the mode of evolution it expresses—or the perceived mode of evolution of the species it helps to define?

Single Traits, Species Boundaries, and Evolutionary Modes

For organisms with even moderately complex morphologies, there are inevitably more traits that *could be* chosen to delimit species and measure evolutionary change than actually *are* chosen to serve in proxy roles for these purposes. What may be surprising is the large extent to which the patterns of single traits have dominated empirical fossil tests of evolutionary mode—canonical patterns of stasis, gradualism, and random walks (Hunt 2007, 2012)—in the wake of punctuated equilibrium 40 years ago. In a recent synthetic analysis of evolutionary modes among 635 traits across 153 fossil species lineages, Hopkins and Lidgard (2012) reported that only 17% were generated using multivariate methods. Most of these latter traits come from studies published in the last 20 years, with some notable early exceptions (e.g., Lohmann and Malmgren 1983). In addition, 78% of the univariate size traits are simple length measurements and 77% of the univariate shape traits (excluding meristic traits) are ratios of length measurements. To be clear, these findings derive only from those studies we selected because they recorded trait data from a stratigraphic series of fossil species populations, hereafter referred to as sequences (see Methods). Of the 635 total traits, most were measured and analyzed as single parts rather than as character complexes or in a multivariate context, even where sets of several discrete traits were initially obtained from the same species or lineage. Thus a large portion of the empirical basis for our current understanding of the relative frequencies of different evolutionary modes in the fossil record is related to patterns of variation of single traits, either within coeval populations or through temporal sequences of populations.

The role of single traits in these results takes on added significance as one considers genealogical continuity, even as paleontologists interpret distances or gaps between entities in morphospace, and hence nominate fossil species. To the degree that this continuity must be present, one must also grant that species have fuzzy boundaries (Rieppel 2007, 2008; Harrison and Larson 2014), on a range of timescales and independent of how different systematists choose particular traits and employ them to individuate taxa. Even more importantly, different traits measured from the same monospecific series of fossil population samples or from a series of chronospecies often show different modes or rates of evolution (fig. 13.1), and may conflict with the overall pattern of morphological change (Hopkins and Lidgard 2012, Wagner 2012). Some workers (Raup and Crick 1981, Levinton 1983, Cheetham 1987) have explicitly wondered whether single traits adequately represent species-level change. Notably, within the bryozoan genus *Metrarabdotos*, the minority of traits that depart from the static patterns seen in overall species morphology are relatively unimportant in distinguishing species, particularly ancestors and descendants (Cheetham 1987). Similarly, Smith and Paul (1985) noted that characters that served to distinguish the echinoid *Discoides subucula* from other congeners did not show any detectable change through time within the species.

As noted above, paleontologists rely on morphological variation to determine species. However, concepts differ in how that variation is used to delimit species, depending on the epistemic goal of the study. For example, studies based in cladistic parsimony look for changes in otherwise invariant traits, studies using numerical phenetics may claim measureable traits are all equivalent, and biostratigraphic studies may seek out (directionally) variable traits for finer temporal resolution. Even within these categories, some workers divide specimens more finely than others ("splitters" vs. "lumpers"). For the purposes of this study, we have accepted the species delimitations in the original studies that contributed to the dataset, which in total may comprise different species concepts. Rather than focusing on whether the "species" described in each of the sequences analyzed is a

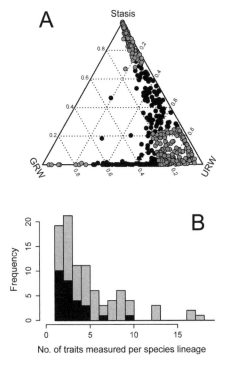

Figure 13.1 Results from Hopkins and Lidgard (2012).
(A) Ternary diagram showing AICc weights for each model for each analyzed trait (see Methods). Strongly supported traits shown in gray. (B) Stacked histogram showing distribution of species lineages where all measured univariate traits show same evolutionary mode (black) and where evolutionary mode varies across measured traits (gray).

"true" species under any particular species concept, we examine how the practice of delimiting species influences the choice and evidential role of traits used in analyzing trends in the fossil record. Because of this, our results may inform trait choice for species-level studies of tempo and mode of evolution, but likely not trait choice for taxonomy and systematics, as this is usually specific to different groups, and also limited by which organismal parts are regularly preserved. For these latter subjects, we refer the reader to other papers in this volume (e.g., Allmon, chapter 3; Ausich, chapter 10; Budd and Pandolfi, chapter 7; Schweitzer and Feldman, chapter 9).

If traits were largely selected because they had prior taxonomic importance, then we expect to see an overlap between the traits chosen for analyses of evolutionary mode and the traits used to define the species. However, such expectations (and interpretation) depend critically on whether a sequence is made up of a single species or of a lineage, as defined above. Typically taxonomically useful traits show more interspecific variation than intraspecific variation. Thus within single species, we expect taxonomically useful traits to be more frequently characterized by stasis. Conversely, within lineages, we expect taxonomically useful traits to be more frequently characterized by directional change or random walk.

Methods

Data collection

We began with the results from a previous study investigating the relative frequency of directional change, unbiased random walk, and stasis in the fossil record (Hunt 2007). This dataset consisted of model selection results for 251 sequences from 32 publications, where a sequence is a temporal series of samples belonging to a fossil species or lineage as defined in the introduction to this chapter. The data come almost entirely from sequences of samples taken from single localities or cores. We excluded 9 sequences because they did not meet our selection criteria (see below). We added an additional 393 sequences from 61 references (Hopkins and Lidgard 2012). Model selection (see below) requires trait means, variances, sample sizes, and relative stratigraphic position or age. We included sequences if there were at least 6 temporal samples with 5 or more specimens; of these, we excluded sequences if the analysis was at the genus level and samples were not clearly monospecific (e.g., Sheldon 1987, Grey et al. 2008) or if one or more of the above four parameters went unreported (typically sample size or variance, e.g., Cheetham 1986, Sorhannus et al. 1988, Cheetham et al. 2007), including situations where only the order of samples was reported (e.g., Reyment et al. 1977). The dataset is made up of both micro- and macrofossils from marine shelf, deep sea, open ocean (pelagic), terrestrial, and lacustrine environments, including brachiopods, bryozoans, echinoderms, mammals, fish, mollusks, conodonts, ostracods, trilobites, planktonic and benthic foraminiferans, diatoms, and radiolarians. Sequences range from 0.01 My to 35.6 My long with a median duration of 2.8 My.

Model selection

In order to determine whether a sequence is best characterized by directional change, a random walk, or stasis, we used the model selection criteria developed by Hunt (2006). Here directional change is represented by a random walk with a mean and variance step size, where the step size is the difference in trait means between successive samples and the step sizes are assumed to be normally distributed. The magnitude of the mean step size determines the magnitude of the shift in trait means in that direction. An unbiased random walk is simply a special case where the mean step size is 0. In both cases the variance of the step size determines the volatility of the random walk. Stasis is modeled using two parameters: the optimum phenotype (θ) and the magnitude of fluctuations around the mean (ω) (Sheets and Mitchell 2001, Hunt 2006). Here the trait values across samples are expected to be normally distributed and the expected step size for each transition is the difference between the ancestral trait value and the optimum; steps will tend to be negative if the trait value in the ancestral population is greater than θ and steps will tend to be positive if the trait value of the ancestral population is less than θ. Thus, in contrast to the random walk models, the expected step size is not constant but varies as a function of the trait value of the ancestral population. In addition, the mean step size need not be 0 for change in the trait value to be constrained by θ. Under a random walk model, divergence is expected to occur with time even if the average step size is 0.

The mode of evolution that best characterizes each sequence was determined using the corrected Akaike information criterion, which takes into account both the log likelihood (L)[1] of each model given the data and the number of free parameters (K) in the model, and includes a correction for finite sample size:

$$AICc = -2L + 2K + (2K[K+1])/(N-K-1),$$

1. The likelihood of a hypothesis, given some outcome, is a central concept of statistical inference. Likelihood is proportional to probability but allows different hypotheses to be compared to one another in relation to a particular outcome, whereas probability deals with a single hypothesis and many possible outcomes. For example, given the hypothesis that the probability of having a male child is 0.5, the probability of having any combination of boys and girls may be calculated and compared. In contrast, if a family is observed to have 2 boys and one girl, the likelihood of the hypothesis can be calculated and compared to other hypotheses (such as the probability of having a male child being equal to 0.3). See A.W.F. Edwards, *Likelihood, Expanded Edition* (Baltimore, MD: Johns Hopkins University Press, 1992).

where N is the number of observations. The approximate probability that each of J models is the best candidate (i.e, the Akaike weight, w_i) is

$$w_i = \exp\left(-\frac{\Delta_i}{2}\right)\Big/ \sum_{j=1}^{J} \exp\left(-\frac{\Delta_j}{2}\right),$$

where $\Delta_i = AIC_i - \min(AIC)$. The sum of the Akaike weights across the set of models is equal to 1 (Burnham and Anderson 2002, Hunt 2006). AICc results are provided in Hopkins and Lidgard (2012) and appendix A. Analyses were carried out using the paleoTS package for R (see Hunt 2006, 2008).

In many cases, two or more models may have very similar AICc weights. Thus even though one of the models is best supported, other models cannot be ruled out confidently. Because of this, we also evaluated separately the subset of sequences for which the best supported evolutionary mode was also much more strongly supported than the other models. We considered a model sufficiently more strongly supported if its AICc weight was 2.7 times the size of the second best supported model. This criterion corresponds to the "rule of 2," where a model is considered well supported compared to the other models if the difference in likelihood values is greater than 2 (Burnham and Anderson 2002). In other words, if $L_1 / L_2 > 2$, then $w_1 / w_2 > 2.7$. However, this is an arbitrary cutoff and should not be interpreted as a level of statistical significance. We chose not to use the stricter criterion of an AICc weight of 0.89 (the likelihood criterion of rejecting a hypothesis when an outcome is eight times [or more] less probable [Wagner et al. 2006]) because this reduced the dataset by over 70% (Hopkins and Lidgard 2012).

Assigning taxonomic importance

In determining whether a trait was useful for classification, the primary reference we used was the paper that presented the temporal sequence data (hereafter referred to as the "primary source"). For cases where the taxonomic affinity of specimens and/or the status of subspecies and species within genera are disputed, we gave the opinion stated in the primary source priority over other opinions because the data provided was based on these designations. For example, even if other workers had split a set of specimens into two groups based on the distribution of variation in a trait, if the primary source lumped specimens together into one "species" regardless of this distribution—i.e., did not use that trait to separate congeners—then that trait would be coded as "not useful" for classification.

In the majority of cases, the taxonomy is not currently disputed. Often, the primary source included discussion on the merits of different traits for taxonomy and thus provided the necessary information to code the analyzed traits. Otherwise, we consulted papers cited within the primary source. Finally, in cases where taxonomy was not discussed nor references provided in the primary source, we turned to previous papers written by the same authors, to contemporary papers written by other authors on the same taxa, or lastly to the original descriptions of the species.

If traits were included in the diagnosis of a taxon or in a discussion of how the taxon differs from congeners, they were considered "useful" for classification. Traits that were not discussed but clearly correlate with useful traits (either logically or empirically) were also considered "useful" for classification. For example, Fermont (1982) used the thickness of the test to distinguish between species of the foraminiferan *Discocyclina*. Since the thickness correlates with the size of the embryon, size-dependent traits such as the degree to which the protoconch is enclosed by the deuteroconch and the number of periembryonic chambers were also considered useful for classification. Traits that were dismissed by workers because they are environmentally sensitive were considered "not useful" in classification, as were traits useful at higher taxonomic levels but shared by congeners. Traits that were not mentioned in any of the consulted literature and did not obviously correlate with useful traits were also coded as "not useful."

With the exception of biometric studies that explicitly use the results from multivariate analysis to group specimens, the interpretation of multivariate traits (i.e., traits defined by some combination of single traits, such as ordination scores) as useful or not useful for classification is less straightforward than for univariate traits. Nevertheless, it is possible to categorize such traits when it is known how multivariate axes describe the morphology. For analyses based on traditional morphometric traits (such as length measurements), the loadings of each trait on different ordination axes may be used to infer which axis would have taxonomic value. Higher-order ordination axes on which taxonomically useful traits load heavily may themselves be considered taxonomically useful when workers infer species difference from larger distances between specimens or clusters in a multivariate space, due to separation along these axes. For analyses based on landmark data, deformation plots of the ordination axes can be used to infer the morphological variation described by that axis and compare it to the variation used to designate the species in the first place. Finally, any traits derived from canonical variates analysis (CVA) were excluded because the results of a CVA depend on predefined groups. While it could be argued that the CV axes are simply a concise description of the traits that were used for the original specimen designations or could be used to identify *additional* traits that separate groups, specimens must already be assigned to a species prior to the analysis, and thus the CV axes themselves cannot logically be used as discriminating characters.

We were able to code 92% of the 635 sequences (appendix A). Justification and references for all coding assignments are provided in appendix B. However, because of the difficulty with coding multivariate traits, we also analyze the subset of sequences where only single traits were measured.

Results and Discussion

Overall, there is a strong relationship between the taxonomic usability of a trait and the mode of evolution that best characterizes change in that trait: the majority of traits that show directional change or a random walk are taxonomically useful while a slight majority of traits that show stasis are not useful or unused (table 13.1, G-test = 24.40, $p < 0.001$). Recall also that in the vast majority of primary studies, taxonomically useful traits had been established *prior to* the morphological trend analyses themselves. However, this relationship between trait usability and evolutionary mode is driven primarily by traits measured from lineages rather than from species (table 13.2, G-test for lineages = 43.170, $p < 0.001$; G-test for species = 1.149, $p = 0.563$). This is true for subsets of the dataset as well, including (1) traits where the best supported model was also strongly supported (G-test for lineages = 20.287, $p < 0.001$; G-test for species = 2.831, $p = 0.243$), (2) shape traits (G-test for lineages = 25.348, $p < 0.001$; G-test for species = 3.096, $p = 0.213$), (3) size traits (G-test for lineages = 19.070, $p < 0.001$; G-test for species = 0.229, $p = 0.892$), and (4) single (i.e., not derived from multivariate analysis) traits (G-test for lineages = 39.412, $p < 0.001$; G-test for species = 2.524, $p = 0.283$) (table 13.2). Thus, traits useful for taxonomic categorization of species do influence the evidential role that they play in discriminating a particular mode of evolution for lineages but not for species. In particular, within lineages, traits that show directional change are also those that have been taxonomically useful and those that show stasis are not useful, whereas within species, useful and not useful traits are almost equally distributed among modes of evolution.

None of the temporal sequences that document a trait best characterized by directional change are longer than 10 million years (My). This is

TABLE 13.1 **Summary of traits measured from all sequences. The majority of traits showing directional change (GRW) and unbiased random walk (URW) are also taxonomically useful while a slight majority of traits showing stasis are not useful.**

	Counts		
Taxonomic usability	GRW	URW	Stasis
not useful	9	107	131
useful	25	200	111

TABLE 13.2 **Summary of traits measured from all lineages and all species as well as subsets of the entire dataset. Within lineages, the majority of traits showing directional change (GRW) or unbiased random walk (URW) are also taxonomically useful while the majority of traits showing stasis are not useful. Within species, modes of evolution are more evenly distributed among useful and nonuseful traits.**

			Counts		
Dataset		Taxonomic usability	GRW	URW	Stasis
all	lineage	not useful	3	46	55
		useful	22	125	32
	species	not useful	6	61	76
		useful	3	65	79
only strongly supported	lineage	not useful	1	23	43
		useful	5	63	27
	species	not useful	0	43	65
		useful	2	45	63
only shape traits	lineage	not useful	2	16	28
		useful	14	71	20
	species	not useful	4	25	40
		useful	1	36	45
only size traits	lineage	not useful	1	30	27
		useful	8	61	12
	species	not useful	2	36	36
		useful	2	29	34
only single traits	lineage	not useful	3	44	54
		useful	21	120	31
	species	not useful	6	58	55
		useful	3	58	72

consistent with the view that directional change can occur over shorter geological timescales, and may be documented only infrequently because temporal resolution is not fine enough to capture the change in the fossil record (Hunt 2010). Some paleontological time series do demonstrate rapid transitions (e.g., Hunt et al. 2008). Transitions so rapid that they are unlikely to be captured by normal taphonomic processes in the fossil record have been shown for living populations (Hendry and Kinnison 1999). However, like Hunt (2007), we found no relationship between duration and mode of evolution (G-test = 6.9521, p = 0.138 using the 33rd and 67th percentiles to divide up the distribution) or temporal resolution (mean spacing between samples) and mode of evolution (G-test = 3.257, p = 0.516). Interestingly, the median duration for the traits showing directional change (4 My) is higher than that for either traits showing a random walk (2.4 My) or stasis (3 My). It is possible to account for this by noting that sequences where traits are measured within a lineage are often (but not necessarily) longer than those where traits are measured over species (fig. 13.2), and most traits showing directional change were measured using taxonomically useful traits within lineages. Nonetheless, because many traits measured over short temporal durations show either a random walk or stasis, it appears that species-lineage directional change *sustained* over millions of years is indeed rare.

Some studies that analyzed traits showing directional change targeted taxonomically useful traits from intervals of the rock record where transitions between species were occurring. For example, Kucera and Widmark (2000) explicitly chose to measure univariate characters that were used to discriminate end-members species within a sequence, and variable loadings indicate that the first eigenshape captures the variation within these characters. Other studies attempted to select at least some traits that had not been used for taxonomy (e.g., Kelley 1983). In fact, over half of the studies used a mixture of both useful and nonuseful traits in their analyses of morphological change through time (fig. 13.3A). In addition, studies that analyzed trends within lineages relied much more frequently on a mixture of traits than studies that analyzed trends within species (Fig. 13.3B, C), though possibly because the median number of traits measured per study is higher for lineages (8) than species (4), and thus there was more "opportunity" to select a mixture.

Very few of the studies analyzing within-lineage patterns included enough sampling to divide the sequence into segments represented by only one species. However, data from the radiolarian *Striatojaponocapsa*

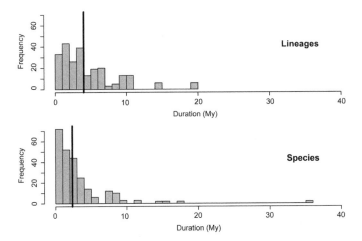

Figure 13.2 Histogram of temporal durations of sequences analyzed from lineages (upper panel) and from species (lower panel).
Black vertical line shows median duration.

plicarum-synconexa lineage (Hatakeda et al. 2007) allowed us to compare the relationship between mode of evolution and taxonomic usefulness at the species level and lineage level within the same sequence. Hatakeda et al. used 6 length measurements, 2 length-length ratios, and 1 meristic trait in *Striatojaponocapsa* from Japan. At the Hoshakuji locality, they sampled 12 horizons, the lower 4 containing *Striatojaponocapsa plicarum* and the upper 8 containing its putative descendent *Striatojaponocapsa synconexa* (fig. 13.4). Within this sequence, 5 traits were best characterized by an unbiased random walk, 3 by stasis, and 1 by directional change (Hopkins and Lidgard 2012). The trait that showed directional change, the width of the basal appendage, is also taxonomically useful (Appendix A, B). However, if only the upper eight samples containing *S. synconexa* are analyzed, change in the width of the basal appendage is best characterized by stasis (AIC weights, GRW: 0.045; URW: 0.247; Stasis: 0.708). Thus in this study where stratigraphic resolution is sufficiently high, our hypothesized pattern—that taxonomically useful traits typically show directional change when measured over lineages and stasis when measured over species—is indeed demonstrated.

Yet this example belies our actual results, which show no strong relationship between mode of evolution and taxonomic usefulness in traits measured within *species*. One reason we may not have found the hypothesized

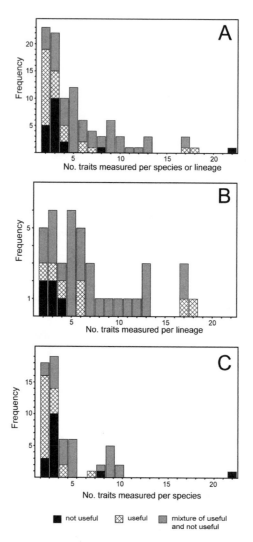

Figure 13.3 Stacked histogram of number of traits measured per study for all studies (A), just lineages (B), and just species (C).

Studies shown in black are those where all traits measured were considered not useful for taxonomy; studies shown with hatch marks are those where all traits measured had taxonomic utility; studies shown in grey measured a mixture of useful and not useful traits.

pattern in species is our method for determining taxonomic utility. We considered traits to be useful if they had been used in species-level taxonomy, specifically if they had been used to discriminate between congeners. However, traits recognized for definition of higher-level taxonomic ranks, such as those used to define the genus or family, may be expected to show stasis within the species that belong to that genus or family. The relative proportion of useful to nonuseful traits showing stasis in species may thus be biased by this focus at the species level: traits that are static at the genus level will obviously be static at the species level, even though these same traits may not be mentioned in species-level determinations. Similarly, the relationship seen between taxonomic usefulness and mode of evolution

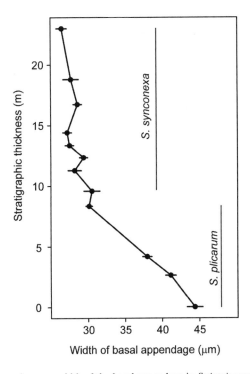

Figure 13.4 Change in mean width of the basal appendage in *Striatojaponocapsa plicarum-synconexa* from a single stratigraphic section (Hoshakuji, Japan; Hatakeda et al. 2007). Change in mean width of the basal appendage is a taxonomically useful character in this genus (see appendix B). The entire sequence is best characterized by directional change (Hopkins and Lidgard 2012), but the trend is best characterized by stasis if only the upper 8 samples containing *S. synconexa* are analyzed (see text).

within lineages may be driven more strongly by differences between useful and nonuseful traits showing directional change and unbiased random walk compared to differences among traits showing stasis. Finally, when traits were not discussed in systematic descriptions, we conservatively interpreted these as "not useful" for taxonomy. However, some of these traits may actually be quite useful. One way to test this would be to identify a subsample of the data for which the systematics and phylogenetics has been recently revised, and compare both the assignments of individual traits as well as the effect on the overall results if taxonomic usefulness is inferred from phylogenetic analysis rather than systematic descriptions (S. Carlson, pers. comm., 2013). With regard to the above concerns, we borrow a recommendation from Hageman (chapter 5, this volume), that "selection of morphologic features . . . should be given careful consideration at the outset, so as to maximize the utility of a given study."

One aspect that remains puzzling concerns the number and distribution of traits that show a random walk: we cannot explain why so many traits showing an unbiased random walk are also taxonomically useful, particularly within lineages. It is possible that this reflects choices made by taxonomists working with varying degrees of geographic and stratigraphic completeness. In addition, the temporal sequences analyzed in this database were sampled largely from single localities. Perhaps at the level of species, taxonomically useful traits show some distribution of variation that is more or less invariant through time but fluctuates locally (e.g., Lieberman et al. 1995). We suggest that one important avenue for future studies will be further exploration of this and other possible explanations.

Overall, our results suggest that the temporal patterns that paleontologists have documented in the fossil record have been influenced by taxonomic practice, insofar as some of our expectations were realized. Nonetheless, the results remain open to some degree of interpretation: traits that have been used for taxonomy do not always show the expected mode of evolution. Given that modes (and thereby rates) of morphological evolution often vary among morphological traits within species lineages (Hopkins and Lidgard 2012), one putative solution is to measure more traits. But simply adding more traits does not by itself guarantee a solution to distinguishing the "real" taxon, phylogenetic tree, or evolutionary mode. Systematists and taxonomists, whether paleontologists working on skeletal morphology or biologists working with molecular genetic sequences, accept the largely untested assumption that adding more data will cause their main result, a hierarchical tree, to converge statistically. Yet the biological

foundation for expected convergence is contested and varies across inference methods and organisms (Wagner 2000, Delsuc et al. 2005, Doolittle
and Bapteste 2007). Moreover, we found that problems in discerning evolutionary mode for fossil species and lineages persist even as the number
of measured traits is increased dramatically (Hopkins and Lidgard 2012).
Another potential problem is the exhaustion of morphological character
states among fossil taxa (Wagner 2000). As fossil clades diversify, homoplasy increases and the ability to distinguish congruent homoplasy from
congruent homology in the structure of phylogenies becomes ever more
problematic. One implication of this work is that even though single traits
are used to characterize species and also to describe evolutionary change
within species, traits (or trait changes) within lineages are not completely
independent from one another. The modular nature of organization within
organisms makes it possible for lineages to evolve in a mosaic fashion (e.g.,
Hopkins and Lidgard 2012). Thus, quantitative assessments of modularity
(e.g., Gerber and Hopkins 2011, Webster and Zelditch 2011) may help identify traits that can more appropriately be treated as independent in discerning the boundaries of fossil species and in phylogenetic and morphological
rate studies.

    By no means do these results imply that *none* of the documented patterns
record real morphological change over time. In fact, for paleontologists to
have divided up species lineages into series of nominate species implies that
"enough" morphological change has occurred to warrant a new name, even
if that decision is informed by the taxonomic experience of the investiga-
tor (as has been the case historically). Frequently, these decisions are also
informed by patterns of variation, particularly geographic variation, in extant species as well as by patterns of standing variation among co-occurring
congeners in the fossil record. Further, in cases where divisions were largely
for biostratigraphic purposes, this change was widespread enough to make
the division useful for correlation across localities.

    Finally, it is clear that prior theoretical commitments may influence
attempts at unbiased empirical studies. If species are defined based on
geographically and temporally invariant characters and these are the only
characters measured, then it will appear (perhaps erroneously) that stasis
dominates in those species. For research on rates and modes of morphological evolution in the fossil record to move forward, we must determine
how different methods of using morphology to recognize species (or morphospecies, or unique biological units) in the fossil record commit us to
seeing particular patterns of evolutionary change.

Summary

Examining the premises needed for logical inference, one concern regarding the theory of punctuated equilibrium is that the use of the morphological species concept in paleontology has led to the conflation of species delimitation with the recognition of stasis and speciation. If this indeed has had a biasing effect on our view of evolution in the fossil record, we can make predictions about the relationship between mode of evolution and taxonomic usefulness of morphological traits. Specifically, we would expect that taxonomically useful traits more frequently show stasis when analyzed within species, but show other modes of evolution, such as directional change, when analyzed within lineages. Our expectation was born out for lineages but not for species, possibly because the focus for determining taxonomic usefulness was at the species level alone. As has been shown in other studies, sustained species-lineage directional change (longer than 10 My) is rare in the fossil record. In addition, traits showing directional change are almost all taxonomically useful and analyzed within lineages. This result suggests that these studies have focused specifically on times of putative species transitions, which may occur relatively quickly on geologic timescales and be captured only rarely at the resolution typically afforded by the fossil record. Despite the fact that studies frequently analyzed a mixture of both taxonomically useful and nonuseful traits, these results indicate that the modes of evolution that paleontologists (when considered as a group) have inferred from the fossil record have been influenced to a greater or lesser degree by taxonomic practice.

Acknowledgments

We are grateful to G. Hunt for making his 2007 AICc results available and for his help and advice using the paleoTS package. Thank you to D. Lazarus, K. Kim, M. Kucera, B. Lauridsen, W. Theriot, E. Erba, and D. Anderson for providing additional data, to P. Yacobucci, S. Carlson, R. Feldman, and an anonymous reviewer for helpful reviews, and to P. Yacobucci and W. Allmon for inviting us to contribute to this volume. This project was supported by a John Caldwell Meeker Postdoctoral Fellowship (Department of Geology, Field Museum of Natural History, Chicago) to MJH, and by a VolkswagenStiftung grant to W. Kiessling, which provided postdoctoral support for MJH while at the Museum für Naturkunde Berlin and the Friedrich-Alexander-Universität Erlangen-Nürnberg.

Appendices

Appendix A (description of traits, AICc results for actual fossil sequences, and assignment of taxonomic usefulness) and appendix B (justification for coding assignments) are available at: [url for appendices here].

References

Brinkmann, R. 1928. Statistisch-phylogenetische Untersuchungen an Ammoniten. *Verhandlungen des 5. Internationalen Kongresses für Vererbungswissenschaft*: 496–513.

Brinkmann, R. 1929. Statistisch-biostratigraphische Untersuchungen an Mittel-jurassischen Ammoniten über Artbegriff und Stammesentwicklung. *Abhandlungen der Gesellschaft der Wissenschaften zu Göttingen, Klasse für Mathematik und Physik, N. F.* 13:1–249.

Burnham, Kenneth P., and David R. Anderson. 2002. *Model Selection and Multimodel Inference, Second Edition.* New York: Springer-Verlag, Inc.

Cheetham, Alan H. 1986. Tempo of Evolution in a Neogene Bryozoan: Rates of Morphologic Change within and across Species Boundaries. *Paleobiology* 12:190–202.

Cheetham, Alan H. 1987. Tempo of Evolution in a Neogene Bryozoan: Are Trends in Single Morphological Characters Misleading? *Paleobiology* 13:286–96.

Cheetham, Alan H., Joann Sanner, and Jeremy B. C. Jackson. 2007. *Metrarabdotos* and Related Genera (Bryozoa: Cheilostomata) in the Late Paleogene and Neogene of Tropical America. *Journal of Paleontology* 81:1–96.

Delsuc, Frédéric, Henner Brinkmann, and Hervé Philippe. 2005. Phylogenomics and the Reconstruction of the Tree of Life. *Nature Reviews Genetics* 6:361–76.

Doolittle, W. Ford, and Eric Bapteste. 2007. Pattern Pluralism and the Tree of Life Hypothesis. *Proceedings of the National Academy of Sciences USA* 104:2043–49.

Eldredge, Niles. 1995. Species, Speciation, and the Context of Adaptive Change in Evolution. In *New Approaches to Speciation in the Fossil Record*, edited by D. H. Erwin and R. L. Anstey, 39–63. New York: Columbia University Press.

Eldredge, Niles, and Stephen J. Gould. 1972. Punctuated Equilibria: An Alternative to Phyletic Gradualism. In *Models in Paleobiology*, edited by T. J. M. Schopf, 82–115. San Francisco: Freeman, Cooper, & Co.

Fermont, W. J. J. 1982. Discocyclinidae from Ein Avedat (Israel). *Utrecht Micropaleontological Bulletins* 27:1–152.

Gerber, Sylvain, and Melanie J. Hopkins. 2011. Mosaic Heterochrony and Evolutionary Modularity: The Trilobite Genus *Zacanthopsis* as a Case Study. *Evolution* 65: 3241–52.

Gingerich, Philip D. 1976. Paleontology and Phylogeny: Patterns of Evolution at the Species Level in Early Tertiary Mammals. *American Journal of Science* 276:1–28.

Gould, Stephen J. 1982. The Meaning of Punctuated Equilibrium and Its Role in

Validating a Hierarchical Approach to Macroevolution. In *Perspectives on Evolution*, edited by R. Milkman, 83–104. Sunderland, MA: Sinauer Associates.

Gould, Stephen J. 2002. *The Structure of Evolutionary Theory*. Cambridge, MA: Belknap Press, Harvard University Press.

Grey, Melissa, James W. Haggart, and Paul L. Smith. 2008. Variation in Evolutionary Patterns across the Geographic Range of a Fossil Bivalve. *Science* 322: 1238–41.

Harrison, Richard G., and Erica L. Larson. 2014. Hybridization, Introgression, and the Nature of Species Boundaries. *Journal of Heredity* 105 (S1):795–809.

Hatakeda, Kentaro, Noritoshi Suzuki, and Atsushi Matsuoka. 2007. Quantitative Morphological Analyses and Evolutionary History of the Middle Jurassic Polycystine Radiolarian Genus *Striatojaponocapsa* Kozur. *Marine Micropaleontology* 63:39–56.

Hendry, Andrew P., and Michael T. Kinnison. 1999. The Pace of Modern Life: Measuring Rates of Contemporary Microevolution. *Evolution* 53:1637–53.

Hopkins, Melanie J., and Scott Lidgard. 2012. Evolutionary Mode Routinely Varies among Morphological Traits within Fossil Species Lineages. *Proceedings of the National Academy of Sciences USA* 109:20520–25.

Hunt, Gene. 2006. Fitting and Comparing Models of Phyletic Evolution: Random Walks and Beyond. *Paleobiology* 32:578–602.

Hunt, Gene. 2007. The Relative Importance of Directional Change, Random Walks, and Stasis in the Evolution of Fossil Lineages. *Proceedings of the National Academy of Sciences USA* 104:18404–08.

Hunt, Gene. 2008. Evolutionary Patterns within Fossil Lineages: Model-Based Assessment of Modes, Rates, Punctuations, and Process. In *From Evolution to Geobiology: Research Questions Driving Paleontology at the Start of a New Century*, edited by R. K. Bambach and P. H. Kelley, 117–31. Paleontological Society Papers, volume 14.

Hunt, Gene. 2010. Evolution in Fossil Lineages: Paleontology and *the Origin of Species*. *American Naturalist* 176 (suppl. 1):S61–S76. doi:10.1086/657057.

Hunt, Gene. 2012. Measuring Rates of Phenotypic Evolution and the Inseparability of Tempo and Mode. *Paleobiology* 38:351–73.

Hunt, Gene, Michael A. Bell, and Matthew P. Travis. 2008. Evolution toward a New Adaptive Optimum: Phenotypic Evolution in a Fossil Stickleback Lineage. *Evolution* 62:700–710.

Kelley, Patricia H. 1983. Evolutionary Patterns of Eight Chesapeake Group Molluscs: Evidence for the Model of Punctuated Equilibria. *Journal of Paleontology* 57:581–98.

Kucera, Michal, and Joen G. V. Widmark. 2000. Gradual Morphological Evolution in a Late Cretaceous Deep-Sea Benthic Foraminiferan, *Parkiella*. *Historical Biology* 14:205–28.

Lazarus, David, Heinz Hilbrecht, Cinzia Spencer-Cervato, and Hans Thierstein.

1995. Sympatric Speciation and Phyletic Change in *Globorotalia truncatulinoides*. *Paleobiology* 21:28–51.

Levinton, Jeffrey S. 1983. Stasis in Progress: The Empirical Basis of Macroevolution. *Annual Review of Ecology and Systematics* 14:103–37.

Levinton, Jeffrey S., and Chris M. Simon. 1980. A Critique of the Punctuated Equilibria Model and Implications for the Detection of Speciation in the Fossil Record. *Systematic Zoology* 29:130–42.

Lieberman, Bruce S., Carl E. Brett, and Niles Eldredge. 1995. A Study of Stasis and Change in Two Species Lineages from the Middle Devonian of New York State. *Paleobiology* 21:15–27.

Lohmann, George P., and Björn A. Malmgren. 1983. Equatorward Migration of *Globorotalia Truncatulinoides* Ecophenotypes through the Late Pleistocene: Gradual Evolution or Ocean Change? *Paleobiology* 9:414–21.

Princehouse, Patricia. 2009. Punctuated Equilibria and Speciation: What Does It Mean to Be a Darwinian? In *The Paleobiological Revolution: Essays on the Growth of Modern Paleontology*, edited by David Sepkoski and M. Ruse, 149–75. Chicago: University of Chicago Press.

Prothero, Donald R., and Timothy H. Heaton. 1996. Faunal Stability during the Early Oligocene Climatic Crash. *Palaeogeography Palaeoclimatology Palaeoecology* 127:257–83.

Raup, David M., and Rex E. Crick. 1981. Evolution of Single Characters in the Jurassic Ammonite *Kosmoceras*. *Paleobiology* 7:200–15.

Reyment, Richard A., Itaru Hayami, and Giles Carbonnel. 1977. Variation of Discrete Morphological Characters in *Cytheridea* (Crustacea: Ostracoda). *Bulletin of the Geological Institutions of the University of Uppsala*, n.s. 7:23–36.

Rieppel, Olivier. 1984. Atomism, Transformism, and the Fossil Record. *Zoological Journal of the Linnean Society* 82:17–32.

Rieppel, Olivier. 2007. Species: Kinds of Individuals or Individuals of a Kind. *Cladistics* 23:373–84.

Rieppel, Olivier. 2008. Origins, Taxa, Names, and Meanings. *Cladistics* 24:598–610.

Schopf, Thomas J. M. 1981. Punctuated Equilibrium and Evolutionary Stasis. *Paleobiology* 7:156–66.

Schopf, Thomas J. M. 1982. A Critical Assessment of Punctuated Equilibria. I. Duration of Taxa. *Evolution* 36:1144–57.

Scott, George H. 1976. Foraminiferal Biostratigraphy and Evolutionary Models. *Systematic Zoology* 25:78–80.

Sepkoski, David. 2009. "Radical" or "Conservative"? The Origin and Early Reception of Punctuated Equilibrium. In *The Paleobiological Revolution: Essays on the Growth of Modern Paleontology*, edited by David Sepkoski and M. Ruse, 301–25. Chicago: University of Chicago Press.

Sheets, H. David, and Charles E. Mitchell. 2001. Why the Null Matters: Statistical Tests, Random Walks, and Evolution. *Genetica* 112–113:105–25.

Sheldon, Peter R. 1987. Parallel Gradualistic Evolution of Ordovician Trilobites. *Nature* 330:561–63.

Smith, Andrew B., and Christopher R. C. Paul. 1985. Variation in the Irregular Echinoid *Discoides* during the Early Cenomanian. In *Evolutionary Case Histories from the Fossil Record*, edited by J. C. W. Cope and P. W. Skelton, 29–37. London: Palaeontological Association, Special Papers in Palaeontology.

Sorhannus, Ulf, Eugene J. Fenster, Lloyd H. Burckle, and Antoni Hoffman. 1988. Cladogenetic and Anagenetic Changes in the Morphology of *Rhizosolenia Praebergonii* Mukhina. *Historical Biology* 1:185–206.

Stamos, David N. 2002. Species, Languages, and the Horizontal/Vertical Distinction. *Biology and Philosophy* 17:171–98.

Turner, Derek. 2011. *Paleontology: A Philosophical Introduction*. Cambridge: Cambridge University Press.

Wagner, Peter J. 2000. Exhaustion of Morphologic Character States among Fossil Taxa. *Evolution* 54:365–86.

Wagner, Peter J. 2012. Modelling Rate Distributions Using Character Compatibility: Implications for Morphological Evolution among Fossil Invertebrates. *Biological Letters* 8:143–46.

Wagner, Peter J., Matthew A. Kosnik, and Scott Lidgard. 2006. Abundance Distributions Imply Elevated Complexity of Post-Paleozoic Marine Ecosystems. *Science* 314:1289–92.

Webster, Mark, and Miriam L. Zelditch. 2011. Modularity of a Cambrian Ptychoparioid Trilobite Cranidium. *Evolution and Development* 13:96–109.

CHAPTER FOURTEEN

Geographic Clines, Chronoclines, and the Fossil Record: Implications for Speciation Theory

Donald R. Prothero, Valerie J. Syverson, Kristina R. Raymond,
Meena Madan, Sarah Molina, Ashley Fragomeni, Sylvana DeSantis,
Anastasiya Sutyagina, and Gina L. Gage

Introduction

When modern evolutionary biology textbooks discuss speciation theory (e.g., Endler, 1977; Ridley, 2004; Levinton, 2001; Strickberger, 2007; Futuyma, 2009), they nearly always focus on the idea that the splitting and divergence of populations along geographic gradients (*geographic clines*) is an important mechanism to produce new species. They frequently point to these geographic clines of subspecies that are partially but not completely genetically separated as incipient species, or a single species in the process of splitting into many species, as postulated by the allopatric speciation model of Mayr (1942). Examples such as the famous *Ensatina* salamanders of California (Stebbins, 1949; but see Wake, 1997) or the "ring species" of *Larus* gulls arranged around the Arctic Circle (Mayr, 1942; but see Liebers et al., 2004) are found in many college-level evolutionary biology texts (e.g., Ridley, 2004; Levinton, 2001; Strickberger, 2007; Futuyma, 2009), and frequently used again and again as examples of species that are in the process of splitting up to form new species.

For many decades, paleontologists have applied the concepts of populations arranged along a gradient of variation to the fossil record as well, and argued that gradual continual change in fossil lineages through time were *chronoclines*, similar in many ways to geographic clines in a single time plane (Eldredge and Gould, 1972). Prior to 1972, there were frequent debates in the scientific literature about how to subdivide these morphological continua changing through time, and whether such definitions of species as slices of a continuum were arbitrary and unnatural. This debate became so pervasive that it had its own label: the "species problem in paleontology" (Prothero, 2013). In some instances, paleontologists actually rejoiced in the incompleteness and gaps of the fossil record, since it provided breaks in continua that were otherwise indivisible without arbitrary criteria. Simpson (1951) went so far as to introduce the concept of "evolutionary species," defined by branching points and splitting, so as to avoid the dilemma of arbitrarily splitting up continuous chronoclines in defining species.

The introduction of the punctuated equilibrium model of speciation (Eldredge and Gould, 1972; Gould and Eldredge, 1977) radically changed the terms of the debate. If, as Eldredge and Gould argued, gradual anagenetic change was truly rare in the fossil record, and punctuation and stasis were the rule, then the longstanding debate over dividing continuous chronoclines was largely a moot issue. In ensuing years, the paleontological community has reached consensus (Jackson and Cheetham, 1999; Geary, 2009; Princehouse, 2009; Ruse and Sepkoski, 2009; Hallam, 2009) that most metazoan species show stasis and geologically rapid change through time, and that well-established examples of long-term anagenetic change are extremely rare. (Anagenetic change is common among protistans, but these organisms are largely clonal and not bound by the criteria of metazoan sexual reproduction—Lazarus, 1983). Thus, the discussion about the nature of speciation has shifted to other mechanisms, such as those that cause genetic isolation and speciation, especially mechanisms that view species as stable entities at a hierarchical level above that of its component populations ("species selection" or "species sorting"), with its own inherent properties (rates of speciation, extinction, etc.) that are not necessarily properties of the individual populations (see discussion in Prothero, 2013).

However, the evolutionary biology textbooks and neontological research community seem not to have factored these challenges into account. The idea that the fragmentation of geographic clines is the stuff of speciation, and that anagenetic evolution of chronoclines is still the important mode of paleontological speciation, is still common in the evolutionary biology textbooks listed above, despite their almost universal rejection in the paleontological community. This raises a question: can we examine some modern clines and see whether the fossil record bears out the idea that changes across geography are the same as changes through geologic time?

Methods

This paper summarizes a recent data set compiled by Prothero and coauthors over the past 8 years. These data are already published in numerous places with free online access (Syverson and Prothero, 2010; DeSantis et al., 2011; Fragomeni and Prothero, 2011; Molina and Prothero, 2011; Madan et al., 2011; Prothero and Raymond, 2011; Raymond and Prothero, 2011; Prothero et al., 2012; Prothero, 2012; Madan et al., 2015). All of our statistical data summaries and tests are available in these papers (mostly in open-access online journals) for those who wish to look at the quality of the data in detail.

To test our hypothesis about whether chronoclines during climate change are relevant to the issue of speciation, we looked at all the abundantly preserved larger birds and mammals from the tar pits at Rancho La Brea, which record the last 35,000 years of climate change from the last glacial to the current interglacial cycle. These fossils also come from an area with an excellent climatic record (Warter, 1976; Coltrain et al., 2004; Ward et al., 2005), with the fossils dated by numerous carbon-14 analyses (Marcus and Berger, 1984; O'Keefe et al,. 2009). For example, Heusser's (1998) study on pollen recovered from well-dated deep-sea cores just offshore showed that southern California went through intervals of extreme climatic and environmental changes over the past 59,000 years. These transitions suggest a climate and landscape much different from the one today. According to Heusser (1998), the region changed from oak and chaparral vegetation around 59 ka to pine-juniper-cypress woodlands by 24 ka, then to a closed-cone juniper-ponderosa forest with abundant winter snow during the last glacial maximum (24–14 ka). During the last glacial-interglacial transition from 14 to 10 ka, the landscape returned to dominant oak-chaparral and coastal sagebrush with pulses of alder, and in the past 10,000 years, the region has been vegetated by the modern assemblage of oak-chaparral-herbaceous vegetation. Coltrain et al. (2004) used stable isotope analyses and found evidence of increased seasonal aridity during the last interglacial and previous glacial.

Thus, the Rancho La Brea sample provides a well-dated sequence of fossils that experienced selective pressure as climates changed over the past 35,000 years, comparable to the differences in climate seen by some of these same genera in their geographic spread across latitude to form modern clines. These rules include well-known ecological clines such as Bergmann's rule, where more cold-adapted species or subspecies tend to have larger body sizes to conserve body heat, and Allen's rule, where more cold-adapted species or subspecies tend to have shorter and more robust limbs and other appendages (such as ears) than those of warmer climates, again to conserve body heat. If geographic clines and chronoclines are truly comparable, we should expect the Rancho La Brea fossils to exhibit changes in size or shape as predicted by Bergmann's rule and Allen's rule in the modern latitudinal clines of these same genera.

The measurement protocols and landmarks were detailed in the papers cited above. Only unbroken adult specimens (as indicated by the complete fusion of the epiphyses of bones) were used, so no immature or juvenile specimens were capable of distorting the data set. Data were statistically analyzed and plotted in Microsoft Excel. We used the Shapiro-Wilk test to determine which samples were normally distributed. For parametric samples, we used ANOVA to see if there was a significant change in size between samples of different ages. For samples that were nonparametric, we used the Kruskal-Wallis test in order to find out whether there was a significant change in size or shape between samples. Prothero et al. (2012) also performed a time-series analysis of all these data, based on the techniques described by Hunt (2006, 2007). They found no instances of progressive change through time or size changes correlated with climate events. Instead, all of the data showed either stasis or random walks, as defined by the criteria of this time series method.

Results

Many of the fossil species found at Rancho La Brea have living descendants in the same genus (and sometimes species) that exhibit considerable modern clinal geographic variation. Examples include the following:

Wolves

Among modern subspecies of the gray wolf *Canis lupus*, populations from Alaska and Canada are 3–6 times as large as those from tropical latitudes (Hunter and Barrett, 2011). Linden (2011) studied the most common mammal in the tar pits, the dire wolf *Canis dirus*. He sampled all the abundantly preserved bones of the skeleton of *Canis dirus*, especially femora, humeri, tibiae, astragali, and metapodials. He found no significant differences in size or shape between the pits, with the exception of Pit 13 (16,000 years in age), which yielded aberrantly small specimens. This is the same conclusion obtained by O'Keefe (2008, 2010), who found that the Pit 13 dire wolf skull sample also seemed small and stunted compared to all other pit samples. The full reason behind this anomalous sample from 16 ka is not understood, but it is important to note that it is *not* the sample closest in age to the last glacial maximum (20 ka), when wolves might be expected to grow larger due to Bergmann's rule, or any other significant climatic event. All of the other pit samples except Pit 13 are statistically indistinguishable from one another. Thus, the dire wolf sample shows stasis in every size measurement of every skeletal element measured in all but one sample, and does not match the prediction of the climate hypothesis of larger, more robust specimens about 20 ka.

Bison

The northern bison subspecies, the wood bison (*Bison bison athabascae*), is much larger and with more robust limbs than the more southerly subspecies, the plains bison (*Bison bison bison*) (McDonald, 1981; Nowak, 1991). Raymond and Prothero (2011) studied *Bison antiquus* from Rancho La Brea. Only the MC3-4 and MT3-4 metapodials (cannon bones), calcanea, astragali, patellae, and humeri were sufficiently abundant in enough well dated pits for this kind of analysis. As detailed in Raymond and Prothero (2011), there were no statistically significant differences in size among any of the pit samples of bison from Rancho La Brea (fig. 14.1A), and shape (as measured by robustness of the long bones) was also static, based on ANOVA of the entire sample.

Equus

Among living horses (Groves, 1974; Nowak, 1991), there are some intraspecific size trends that suggest the influence of Bergmann's rule. For example, among the wild asses (Groves, 1974), the kiang (*E. kiang*) of the Tibetan Plateau weighs up to 400 kg, while the desert-dwelling African wild ass (*E. asinus*) weighs about 250 kg. In other *Equus* species, the trend is less obvious. The cold steppe–dwelling Przewalski's horse (*Equus ferus*

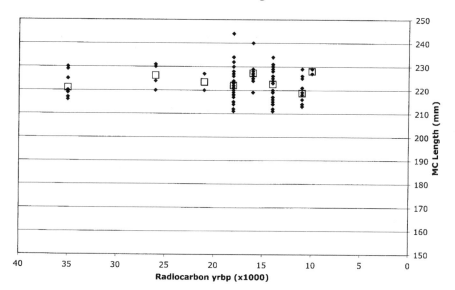

Figure 14.1 Representative graphs of bone dimensions of different La Brea taxa plotted against the age of the pit sample.
Small symbols indicate individual specimens, large open symbols are the means for each pit sample. A. MC3-4 (cannon bone) lengths of *Bison antiquus* (after Raymond and Prothero, 2011). B. MC3 (cannon bone) length of *Equus "occidentalis"* (after DeSantis et al., 2011). C. *Smilodon fatalis* MC3 length. D. *Smilodon fatalis* humerus robustness (C and D after Madan et al., 2011). E. Femur length of *Aquila chrysaetos* (after Molina and Prothero, 2011).

przewalskii) weighs about 200–300 kg, whereas the more tropical zebras tend to weigh 170–270 kg. DeSantis et al. (2011) analyzed the most common elements of the La Brea horse *Equus "occidentalis"* (the proper trivial name is still controversial): the cannon bones (MC3 and MT3), astragali, and patellae. As reported by DeSantis et al. (2011), there was no statistically significant difference among any of the pit samples as established by ANOVA (fig. 14.1B).

Camelids

Bergmann's rule is very apparent among living camelids (Nowak, 1991; Franklin, 1983). The steppe-dwelling Bactrian camel (*Camelus bactrianus*) weighs about 600–1040 kg, while the desert-dwelling dromedary camel (*C. dromedarius*) weighs only 450–680 kg. In the wild New World lamine camelids, the guanaco (*Lama guanicoe*), which inhabits the cold mountains

and steppes of Patagonia, weighs about 100–120 kg, whereas the more tropical mountain and grassland taxon, the vicuña (*Vicugna vicugna*) weighs
only 35–65 kg. Although specimens of *Camelops hesternus* are common at
Rancho La Brea, there were not enough long limb bones (humeri, radii, ul-
nae, femora, tibiae), nor even the digit 3-4 metapodials ("cannon bones"),

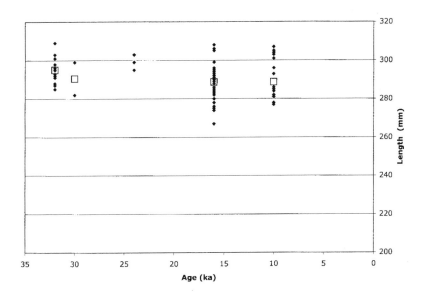

Smilodon MC3 length

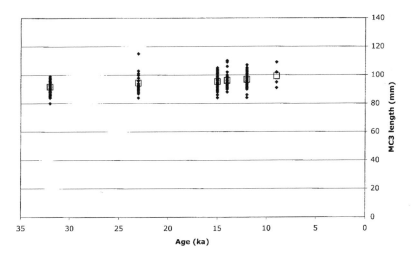

Figure 14.1 (*continued*)

Smilodon humerus robustness

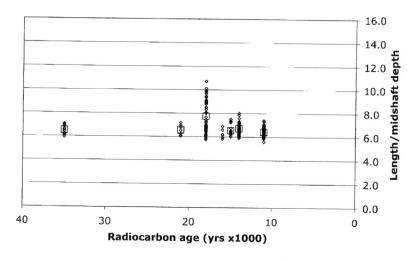

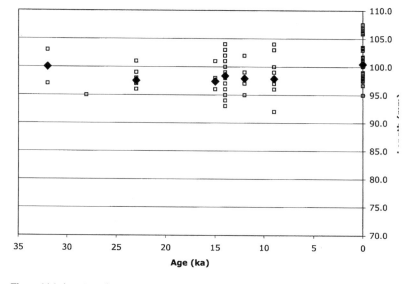

Figure 14.1 (*continued*)

from enough different pits to use these elements in our study. However, there are large samples from multiple pits of astragali, calcanea, cuboids, and patellae (Webb, 1965). When these samples were plotted by radiocarbon age, there were no statistically significant differences among samples in any of these bones (DeSantis et al., 2011).

Felids

Although not all felids demonstrate Bergmann's rule of larger body size in colder climates, the American puma (or cougar or mountain lion) (*Felis concolor*) does vary in body size by latitude, with the largest in the higher latitudes of North and South America, and the smallest in the tropics (Agustin Iriarte et al., 1990; Sunquist and Sunquist, 2002). Similarly, among the many subspecies of the tiger (*Panthera tigris*), the largest are the cold-adapted Siberian tigers (227 kg in weight), while the smallest are the tropical subspecies such as the Sumatran tiger (75–140 kg in weight) or the Indochinese tiger (110–140 kg in weight) (Sunquist and Sunquist, 2002). Madan et al. (2011) analyzed multiple dimensions of most of the common limb bones of both the sabertoothed cat, *Smilodon fatalis* (fig. 14.1C, D), and the Ice Age "lion" or "jaguar," *Panthera atrox*, including humeri, femora, patellae, tibiae, astragali, third metacarpals (MC3) and third metatarsals (MT3), all from La Brea. They found no statistically significant changes in either size or robustness of any of the variables from any of these elements among all the pits sampled.

Birds

Among birds examined in this study, the bald eagle (*Haliaeetus leucocephalus*) and the golden eagle (*Aquila chrysaetos*) show considerable clinal variation, with larger-bodied subspecies in the high latitudes in both Siberia and North America today (Brown, 1968; Johnsgard, 1990). Molina and Prothero (2011) studied the most common bird of Rancho La Brea, the golden eagle. They analyzed the large sample of right tarsometatarsi (TMT), the lower leg bone, which is the most commonly preserved skeletal element in most fossil birds. They measured five variables from almost 700 specimens from numerous pits, and found no statistically significant differences among the pit samples as determined by ANOVA (fig. 14.1E).

Caracara plancus, the extant southern caracara, is well known to have larger body sizes in the southern cold regions of South America than it does in the tropics (Brown, 1968; Johnsgard, 1990). Given these strong size and shape trends in modern populations, there is every reason to suspect that populations that experienced dramatic cooling or warming in the Pleistocene might show similar trends. Fragomeni and Prothero (2011) looked at the caracara, turkey, and the bald eagle fossils found at Rancho La Brea. They measured multiple dimensions of the TMTs, and found no evidence of size or shape changes in these limb bones through the entire time span.

Finally, the great horned owl (*Bubo virginianus*) shows some clinal variation as well, with specimens from the higher latitudes in North America larger than those from lower latitudes (McGillivray, 1989). But Madan et al. (2015) found no evidence of larger body sizes at the last glacial maximum, or any significant differences in size at any interval in the past 35,000 years; all of the La Brea great horned owls are in the same size range as their modern descendants.

Discussion

This lack of change in size and shape of the limb bones in Rancho La Brea bison, horses, camels, cats, wolves, and six species of birds over the past 35,000 years of environmental change contradicts the predictions of Bergmann's rule, which suggests that within a species or between species, the body mass tends to increase with latitude and colder climate. This also contradicts Allen's rule, which predicts shorter and more robust limbs in colder climates. As Pleistocene paleontologists have long known and documented, lack of directional phenotypic response through dramatic climate changes seems to be a prevalent occurrence among Pleistocene species despite the rapid changes of their environment and habitats (Barnosky, 1994, 2005; Lister, 2004).

Despite this knowledge, the pervasive stasis of nearly all Pleistocene mammals is a conundrum that has not been satisfactorily explained. Eldredge and Gould (1972) suggested that the stasis might be due to developmental constraints and canalization, although Gould (2002) rejected that notion. Bennett (1990, 1997) proposed that the 10–100 ky climatic changes during the glacial-interglacial cycle were too rapid and did not allow time for adaptation. However, adaptation on a much faster scale has been demonstrated: the Galapagos finches changed in a matter of years in response to a small-scale climatic change. The Rancho La Brea data sets described above have a time resolution of 1,000–10,000, and therefore should capture comparable change in slower-breeding large animal populations. The measurements showing stasis, then, are unlikely to be explained by sluggish response to a fast-moving evolutionary optimum.

A more plausible idea is that no matter how severely an environment changes, most large animals have the ability to adapt to a wide range of environments, and therefore are not responsive to local short-term changes in climate (Lieberman et al.1995; Lieberman and Dudgeon, 1996; Eldredge et al., 2005). This may be appropriate for the large mammals and birds at Rancho La Brea, but not for many cases where small mammals show stasis across climate change (e.g., Barnosky, 1994, 2005; Lister, 2004; Prothero and Heaton, 1996).

Clearly, the processes that form geographic clines in the modern world are not the same as those that may or may not cause change in fossil lineages through time. Equating geographic clines in space with chronoclines through time and climate change is inappropriate. The fossil record shows that most species are quite stable in morphology over time, and resistant to environmental changes causing small-scale clinal change through time. The prevalence of such stasis, as pointed out by the species sorting models and the hierarchical approach to speciation (summarized in Prothero, 2013), rules out the notion that simple gradients in environments, as seen in modern geographic clines, are that important to the process of speciation as revealed in the fossil record.

As Prothero (1999, 2012) pointed out, there is also a big difference between studies like this, which examine detailed changes in morphology of many lineages over short time intervals of severe climate change, and studies that only count the presence or absence of taxa from time intervals that may average 2–3 million years in length. Fine-scale studies show the pervasiveness of stasis in the face of climate change in fossil mammals and in many other taxa, as predicted by the punctuated equilibrium model (Gould and Eldredge, 1977; Jablonski, 2000, 2008). But coarse-scale studies that simply count presence or absence of taxa in long time-bin intervals do seem to respond to some sort of climatic signal (Janis, 1984, 1993; Vrba, 1985, 1993; Van Dam et al., 2006; Badgley et al., 2008; DeSantis et al., 2009; Woodburne et al., 2009; Blois and Hadly, 2009; Blois et al., 2010; Figueirido et al., 2011; Secord et al., 2012). Such patterns would appear to be species sorting in action (Jablonski, 2000, 2008).

Acknowledgments

We thank the staff of the Page Museum, especially John Harris, Aisling Farrell, and Chris Shaw for allowing access to their collections. We thank Ken Campbell, Jr., for access to the Page Museum bird collection, and for letting us use the specimen database of La Brea birds. We thank Edward

Linden and others who have helped us in these studies for amassing the huge data sets and performing the analyses on them. We thank John Harris, Robin O'Keefe, and Edward Linden for their comments and suggestions, and Bruce MacFadden, Warren Allmon, and two anonymous reviewers for critiquing the paper.

References

Agustin Iriarte, J., Franklin, W. J., Johnson, W. E., and Redford, K. H., 1990, Biogeographic variation of food habits and body size of the American puma. Oecologia 85(2), 185–190.

Badgley, C., Barry, J. C., Morgan, M. E., Nelson, S. V., Behrensmeyer, A. K., Cerling, T. E., and Pilbeam, D., 2008, Ecological changes in Miocene mammalian record show impact of prolonged climatic forcing. Proceedings of National Academy of Sciences USA, 105, 12145–12149.

Barnosky, A. D., 1994, Defining climate's role in ecosystem evolution: clues from late Quaternary mammals. Historical Biology, 18, 173–190.

Barnosky, A. D., 2005, Effects of Quaternary climatic change on speciation of mammals. Journal of Mammalian Evolution, 12 (1/2), 247–264.

Bennett, K. D., 1990, Milankovitch cycles and their effects on species in ecological and evolutionary time. Paleobiology, 16, 11–21.

Bennett, K. D., 1997, Evolution and Ecology: The Pace of Life. Cambridge University Press, Cambridge.

Blois, J. L., and Hadly, E. A. 2009. Mammalian response to Cenozoic climate change. Annual Reviews of Earth and Planetary Science, 37, 1–28.

Blois, J. L., McGuire, J. L., and Hadly, E. A. 2010. Small mammal diversity loss in response to late-Pleistocene climatic change. Nature, 465, 771–774.

Brown, L., 1968, Eagles, Hawks, and Falcons of the World. McGraw-Hill, New York.

Coltrain, J. B., Harris, J. M., Cerling, T. E., Ehleringer, J. R., Dearing, M., Ward, J., and Allen, J., 2004, Rancho La Brea stable isotope biogeochemistry and its implications for the palaeoecology of the late Pleistocene, coastal southern California. Palaeogeography, Palaeoclimatology, Palaeoecology, 205, 199–219.

DeSantis, L. R., Feranec, R. S., and MacFadden, B. J. 2009. Effects of global warming on ancient mammalian communities and their environments. PloS One, 4, 5750–5757.

DeSantis, S. N., Prothero, D. R., and Gage, G. L., 2011, Size and shape stasis in late Pleistocene horses and camels from Rancho La Brea during the last glacial-interglacial cycle. New Mexico Museum of Natural History Bulletin, 53, 505–510.

Eldredge, N., and Gould, S. J., 1972, Punctuated equilibria: an alternative to phyletic gradualism. In: Schopf, T. J. M. (ed.), Models in Paleobiology. Freeman, San Francisco, pp. 82–115.

Eldredge, N., Thompson, J. N., Brakefield, P. M., Gavrilets, S., Jablonski, D.,

Jackson, J. B. C., Lenski, R. E., Lieberman, B. S., McPeek, M. A., and Miller, W., III, 2005, The dynamics of evolutionary stasis. Paleobiology, 31, 133–145.

Endler, J. A. 1977. Geographic Variation, Speciation, and Clines. Princeton University Press, Princeton, NJ.

Figueirido, B., Janis, C. M., Perez-Claros, J. A., de Renzi, M., and Palmqvist, P. 2011. Cenozoic climate change influences mammalian evolutionary dynamics. Proceedings of National Academy of Sciences USA, 109, 722–727.

Fragomeni, A., and Prothero, D. R., 2011, Stasis in late Quaternary birds from the La Brea tar pits during the last glacial-interglacial cycle. New Mexico Museum of Natural History Bulletin, 53, 511–516.

Franklin, W. D., 1983, Contrasting socioecologies of South America's wild camelids: the vicuña and the guanaco. American Society of Mammalogists Special Publication, 7, 573–629.

Futuyma, D. 2009. Evolution (2nd ed.). Sinauer, Sunderland, MA.

Geary, D. H., 2009, The legacy of punctuated equilibrium. In: Allmon, W. D., Kelley, P. H., and Ross, R. M. (eds.), Stephen Jay Gould: Reflections on His View of Life. Oxford University Press, Oxford, pp. 127–147.

Gould, S. J., 2002, The Structure of Evolutionary Theory. Harvard University Press, Cambridge, Massachusetts.

Gould, S. J., and N. Eldredge. Punctuated equilibria: the tempo and mode of evolution reconsidered. Paleobiology 3:115–151.

Groves, C. P., 1974, Horses, Asses, and Zebras in the Wild. David and Charles Publishers, Newton Abbott, England.

Hallam, A., 2009, The problem of punctuational speciation and trends in the fossil record. In: Ruse, M., and Sepkoski, D. (eds.), The Paleobiological Revolution. University of Chicago Press, Chicago, pp. 423–432.

Heusser, L., 1998, Direct correlation of millennial-scale changes in western North American vegetation and climate with changes in the California Current system over the past 60 kyr. Paleoceanography, 13, 252–262.

Hunt, G. 2006, Fitting and comparing modes of phyletic evolution: random walks and beyond. Paleobiology, 32, 578–601.

Hunt, G., 2007, The relative importance of directional change, random walks, and stasis in the evolution of fossil lineages. Proceedings of the National Academy of Sciences, USA, 104, 18404–18408.

Hunter, L., and Barrett, P., 2011, A Field Guide to the Carnivores of the World. New Holland Publishers, Amsterdam.

Jablonski, D., 2000, Micro- and macroevolution: scale and hierarchy in evolutionary biology and paleobiology. Paleobiology, 26, 15–52.

Jablonski, D., 2008, Species selection: theory and data. Annual Review of Ecology, Evolution, and Systematics, 39, 501–524.

Jackson, J. B. C., and Cheetham, A. H., 1999, Tempo and mode of speciation in the sea. Trends in Ecology and Evolution, 14, 72–77.

Janis, C. M., 1984, The significance of fossil ungulate communities as indicators of vegetation structure and climate. In: Brenchley, P. J. (ed.), Fossils and Climate. John Wiley and Sons, New York, pp. 85–104.

Janis, C. M., 1993, Tertiary mammal evolution in the context of changing climates, vegetation, and tectonic events. Annual Reviews of Ecology and Systematics, 24, 467–500.

Janis, C. M., 1997, Ungulate teeth, diets and climate changes at the Eocene/Oligocene. Zool.-Annal. Complex Syst. 100, 203–220.

Johnsgard, P. A., 1990, Hawks, Eagles, and Falcons of North America. Smithsonian Institution Press, Washington, D.C.

Lazarus, D. B., 1983, Speciation in pelagic Protista and its study in the planktonic microfossil record: a review. Paleobiology, 9, 327–340.

Levinton, J. S. 2001. Genetics, Paleontology, and Macroevolution. Cambridge University Press, Cambridge.

Lieberman, B. S., and Dudgeon, S., 1996, An evaluation of stabilizing selection as a mechanism for stasis. Palaeogeography, Palaeoclimatology, Palaeoecology, 127, 229–238.

Lieberman, B. S., Brett, C. E., and Eldredge, N., 1995, A study of stasis and change in two species lineages from the Middle Devonian of New York State. Paleobiology, 21, 15–27.

Liebers, D., de Knijff, P., and Helbig, A. J. 2004. The herring gull complex is not a ring species. Proceedings of the Royal Society of London B, 271, 893–901.

Linden, E., 2111, Morphological response of dire wolves (*Canis dirus*) of Rancho la Brea over time due to climate changes. Journal of Vertebrate Paleontology, 31(supplement to #3), 144.

Lister, A. M., 2004, The impact of Quaternary ice ages on mammalian evolution. Philosophical Transactions of the Royal Society of London, B 359, 221–241.

Madan, M., Prothero, D. R., and Sutyagina, A., 2011, Did Rancho La Brea large felids (*Panthera atrox* and *Smilodon fatalis*) change in size or shape during the Late Pleistocene? New Mexico Museum of Natural History and Science Bulletin, 53, 554–563.

Madan, M., Prothero, D. R., and Syverson, V. J., 2015, Stasis in great horned owls from the La Brea tar pits during the last glacial-interglacial cycle. New Mexico Museum of Natural History and Science Bulletin, 67, 221–226.

Marcus, L. F., and Berger, R., 1984, The significance of radiocarbon dates for Rancho La Brea. In: Martin, P. S., and Klein, R. G. (eds.), Quaternary Extinctions: A Prehistoric Revolution. University of Chicago Press, Chicago, pp. 159–188.

Mayr, R., 1942, Systematics and Origin of Species. Columbia University Press, New York.

McDonald, J. N., 1981, North American Bison—Their Classification and Evolution. University of California Press, Berkeley.

McGillivray, W. B., 1969, Geographic variation in size and reverse size dimorphism of the Great Horned Owl in North America. Condor, 91, 777–786.

Molina, S., and Prothero, D. R., 2011, Evolutionary stasis in late Pleistocene golden eagles. New Mexico Museum of Natural History Bulletin, 53, 564–569.

Nowak, R. M., 1991, Walker's Mammals of the World (5th ed.). Johns Hopkins University Press, Baltimore.

O'Keefe, F. R., 2008, Population-level response of the dire wolf, *Canis dirus*, to climate change in the Upper Pleistocene. Journal of Vertebrate Paleontology, 28 (supplement to #3), 122A.

O'Keefe, F. R., 2010, Craniodental measures of dire wolf population health imply rapid extinction in the Los Angeles Basin. Journal of Vertebrate Paleontology, 30 (supplement to #3), 141A.

O'Keefe, F. R., Fet, E. V., and Harris, J. M., 2009, Compilation, calibration, and synthesis of faunal and floral radiocarbon dates, Rancho La Brea, California. Contributions in Science, Natural History Museum of Los Angeles County, 518, 1–16.

Princehouse, P., 2009, Punctuated equilibrium and speciation: what does it mean to be a Darwinian? In: Ruse, M., and Sepkoski, D. (eds.), The Paleobiological Revolution. University of Chicago Press, Chicago, pp. 149–175.

Prothero, D. R., 1999, Does climatic change drive mammalian evolution? GSA Today, 9(9), 1–5.

Prothero, D. R. 2012. Cenozoic mammals and climate change: the contrast between coarse-scale versus high-resolution studies explained by species sorting. Geosciences, 2, 24–41.

Prothero, D. R. 2013. Bringing Fossils to Life: An Introduction to Paleobiology (3rd ed.). Columbia University Press, New York.

Prothero, D. R., and Heaton, T. H., 1996, Faunal stability during the early Oligocene climatic crash. Palaeogeography, Palaeoclimatology, Palaeoecology, 127, 239–256.

Prothero, D. R., and Raymond, K. R., 2011, Stasis in late Pleistocene ground sloths (*Paramylodon harlani*) from Rancho La Brea, California. New Mexico Museum of Natural History Bulletin, 53, 624–628.

Prothero, D. R., Syverson, V. J., Raymond, K. R., Madan, M., Molina, S., Fragomeni, A., DeSantis, S., Sutyagina, A., and Gage, G. L., 2012, Shape and size stasis in late Pleistocene mammals and birds from Rancho La Brea during the last glacial-interglacial cycle. *Quaternary Science Reviews* 56, 1–10.

Raymond, K. R., and Prothero, D. R., 2011, Did climate change affect size in late Pleistocene bison? New Mexico Museum of Natural History Bulletin, 53, 636–640.

Ridley, M. 2004. Evolution (2nd ed.). Oxford University Press, Oxford.

Ruse, M., and Sepkoski, D. (eds.), 2009, The Paleobiological Revolution. University of Chicago Press, Chicago.

Secord, R., Bloch, J. I., Chester, S. G. B., Boyer, D. M., Wood, A. R., Wing, S. L., Kraus, M. J., McInerney, F. A., and Krigbaum, J. 2012. Evolution of the earliest horses driven by climate change in the Paleocene-Eocene Thermal Maximum. Science 335 (6071): 959–961.

Simpson, G. G., 1951. The species concept. Evolution 5, 285–298.

Stebbins, R. C., 1949, Speciation in salamanders of the plethodontid genus *Ensatina*. University of California Publications in Zoology 48, 377–526.

Strickberger, M. 2007. Evolution (3rd ed.). Jones and Bartlett, London.

Sunquist, M., and Sunquist, F., 2002, Wild Cats of the World. University Of Chicago Press, Chicago.

Syverson, V. J., and Prothero, D. R., 2010, Evolutionary patterns in late Quaternary California condors. PalArch Journal of Vertebrate Paleontology, 7(10), 1–18

Van Dam, J. A., Abdul Aziz, H., Alvarez Sierra, M. A., Hilgen, F. J., Ostende, L. V. H., Lourens, L. J., Mein, P., van der Meulen, A. J., and Pelaez-Campomanes, P., 2006, Long-period astronomical forcing of mammalian turnover. Nature, 443, 687–691.

Vrba, E. S., 1985, Environment and evolution: alternative causes of temporal distribution of evolutionary events. South African Journal of Science, 81, 229–236.

Vrba, E. S., 1993, Turnover-pulses, the Red Queen, and related topics. American Journal of Science, 293A, 418–452.

Wake, D. B., 1997, Incipient species formation in salamander of the *Ensatina* complex. Proceedings of the National Academy of Sciences USA 94, 7761–7767.

Ward, J. W., Harris, J. M., Cerling, T. E., Wiedenhoeft, A., Lott, M. J., Dearing, M., Coltrain, J. B., and Ehleringer, J. R., 2005, Carbon starvation in glacial trees recovered from the La Brea tar pits, southern California. Proceedings of the National Academy of Sciences USA, 102(3), 690–694.

Warter, J. K., 1976, Late Pleistocene plant communities—evidence from Rancho La Brea tar pits. Symposium Proceedings on the Plant Communities of Southern California. Native Plant Society Special Publication, 2, 32–39.

Webb, S. D., 1965, The osteology of *Camelops*. Bulletin of the Los Angeles County Museum, Science, 1, 1–54.

Woodburne, M. O., Gunnell, G. F., and Stucky, R. K. 2009. Climate directly influences Eocene mammal faunal dynamics in North America. Proceedings of National Academy of Sciences USA, 106, 13399–13403.

Contributors

Warren D. Allmon
Paleontological Research
  Institution, and Department
  of Earth and Atmospheric
  Sciences
Cornell University
Ithaca, NY 14850
USA

William I. Ausich
School of Earth Sciences
Ohio State University
Columbus, OH 43210
USA

William E. Bemis
Department of Ecology and
  Evolutionary Biology
Cornell University
Ithaca, NY 14853
USA

Ann F. Budd
Department of Earth and
  Environmental Sciences
University of Iowa
Iowa City, IA 52242
USA

Sylvana DeSantis
Department of Geology
Occidental College
Los Angeles, CA 90041
USA

Torbjørn Ergon
Centre for Ecological and
  Evolutionary Synthesis
Department of Biosciences
University of Oslo
0316, Oslo
Norway

Rodney M. Feldmann
Department of Geology
Kent State University
Kent, OH 44242
USA

Ashley Fragomeni
Department of Vertebrate
  Paleontology
Natural History Museum
Los Angeles, CA 90007
USA

Gina L. Gage
Division of Geological and
    Planetary Sciences
California Institute of
    Technology
Pasadena, CA 91125
USA

Steven J. Hageman
Department of Geology
Appalachian State University
Boone, NC 28608
USA

Melanie J. Hopkins
Division of Paleontology
American Museum of Natural
    History
New York, NY 10034
USA

Scott Lidgard
Integrative Research Center
Field Museum of Natural
    History
Chicago, IL, 60605
USA

Lee Hsiang Liow
Centre for Ecological and
    Evolutionary Synthesis
Department of Biosciences
University of Oslo
0316, Oslo
Norway

Meena Madan
School of Earth Sciences
University of Bristol
Bristol, BS8 1TQ
UK

William Miller III
Department of Geology
Humboldt State University
Arcata, CA 95521
USA

Sarah Molina
Department of Geology
Occidental College
Los Angeles, CA 90041
USA

John M. Pandolfi
Centre for Marine Science
    and School of Biological
    Sciences
Australian Research Council
    Centre of Excellence for
    Coral Reef Studies
University of Queensland
Brisbane, QLD 4072
Australia

Donald R. Prothero
Department of Vertebrate
    Paleontology
Natural History Museum
Los Angeles, CA 90007
USA

Kristina R. Raymond
Don Sundquist Center for
    Excellence in Paleontology
East Tennessee State
    University
Johnson City, TN 37614
USA

Scott D. Sampson
Vice President of Research &
    Collections and Chief
    Curator
Denver Museum of Science
    and Nature
Denver, CO 80205
USA

Carrie E. Schweitzer
Department of Geology
Kent State University at Stark
North Canton, OH 44720
USA

David Sepkoski
Max Planck Institute for the
    History of Science
14195 Berlin
Germany

Alycia L. Stigall
Department of Geological
    Sciences and Ohio
    Center for Ecology and
    Evolutionary Studies
Ohio University
Athens, OH 45701
USA

Anastasiya Sutyagina
Department of Geology
Occidental College
Los Angeles, CA 90041
USA

Valerie J. Syverson
Museum of Paleontology
University of Michigan
Ann Arbor, MI 48109
USA

Margaret M. Yacobucci
Department of Geology
Bowling Green State
    University
Bowling Green, OH 43403
USA

Index